GLOBAL RISK GOVERNANCE

International Risk Governance Council (IRGC) Book Series on Global Risk Governance

Series Editor

Christopher Bunting
Secretary General
IRGC, Geneva

Aim and Scope of this Series

The aim of this series is to provide a forum for work by academics and practitioners of risk assessment and risk management from varied disciplines and sectors on the identification, evaluation and governance of emerging global risks.

The work presented in the series draws on projects undertaken by the International Risk Governance Council and, particularly, projects undertaken conducted the leadership of members of the IRGC's Scientific and Technical Council.

Books published in the series will be of interest to policy and decision makers around the world in government, industry, and civil society, as well as to academics and students working in this field.

About IRGC

The IRGC is an independent organisation based in Geneva, Switzerland, whose purpose is to help the understanding and management of emerging global risks that have impacts on human health and safety, the environment, the economy and society at large. IRGC's work includes developing concepts of risk governance, anticipating major risk issues and providing risk governance policy recommendations for key decision makers.

IRGC's goal is to bring the principles of integrated risk governance of important emerging, systemic risks to the highest levels of decision making. It believes that by combining forces, governments, industry, academia and international and large non-governmental organisations can together develop and implement the best options for governing global risks as well as maximise public trust in the process. Coordinated and coherent policy making, regulation, research agendas and communication will be required.

The IRGC's priorities and project work are identified and led by our Board and Scientific & Technical Council. Both groups are composed of renowned business people, policy makers and academic leaders representing countries throughout the world; the membership of both bodies may be found on the adjacent pages.

Further information on IRGC, our current and past projects, and our sources of income, may be obtained from our website www.irgc.org.

Global Risk Governance

Concept and Practice Using the IRGC Framework

Edited by

Ortwin Renn

University of Stuttgart and DIALOGIK, gGmbH, Stuttgart, Germany

and

Katherine D. Walker

IRGC, Geneva, Switzerland

Springer

A C.I.P. Catalogue record for this book is available from the Library of Congress.

ISBN 978-1-4020-6798-3 (HB)
ISBN 978-1-4020-6799-0 (e-book)

Published by Springer,
P.O. Box 17, 3300 AA Dordrecht, The Netherlands.

www.springer.com

Printed on acid-free paper

IRGC Scientific & Technical Council Members

Table of Contents

PART 2. A FRAMEWORK FOR RISK GOVERNANCE: CRITICAL REVIEWS

Foreword by Peter Sutherland

A Business Perspective on IRGC's Risk Governance Framework

I first learned of the work of IRGC in early 2005 when I was made aware of a different kind of risk management conference. My interest grew when I learned that the conference was to be held in Beijing, China. Insurance Australia Group (IAG), where I work, has business interests in China, and is committed to expanding its presence in the Chinese market, one of the most exciting and fastest-growing markets in the world.

Having signed up for the conference, I was then invited by IRGC's General Secretary Chris Bunting to give an insurance perspective on new technologies. And now, some 18 months on, Chris has kindly asked me to offer a business perspective on IRGC's risk governance framework. I am of course delighted and privileged to do so. And so to the task at hand . . .

Why does risk governance matter to IAG? Risk management is core business for insurance companies. In order to be there when a claim is made by a customer, the company must understand and price risk appropriately. One of the pillars of IAG's purpose is appropriately pricing the risk associated with future events. This is crucial to the Group's long-term sustainability. In addition, the market makes a working assumption that risks will be properly managed by the company and that there will be no major surprises.

The broader community, too, expects that insurance companies will adapt to the changing nature of risk. A sustainable insurance business seeks to reduce risk in the community – through advocacy and engagement with government; financial and non-financial contributions to community initiatives that seek to address causal factors giving rise to insurable risk; and targeted involvement in customer education programs that support and reward sustainable practices or product choices.

As IAG grows in scale and complexity, so too does the complexity of the risk governance issues faced by the company. In this context, IRGC's risk governance methodology provides important insight and tools to help us manage a changing and more complex risk profile.

My work at IAG involves continually adapting the company's risk management framework to the changing internal and external environment. Perhaps the biggest learning for me from this experience is this: any system of risk management that ignores or underestimates the 'socio-cultural' dimension of risk cannot in my view call itself 'integrated'.

More broadly, we have all observed the evolution of risk management frameworks from the traditional or statistical models common to financial services to the current focus on 'Enterprise Risk Management', or ERM. This development represents a logical and natural response to growing complexity, uncertainty and ambiguity associated with 21st century corporate life.

Hence my particular interest in IRGC's contribution to this topic, *Risk Governance – Towards an Integrative Approach*. The first thing that struck me about the IRGC approach was the language used. The title 'risk governance' sets the framework apart from other risk frameworks I have seen by framing IRGC's approach broadly and inclusively. Second, the categorisation of 'risk-related knowledge' along the spectrum of 'simple', 'complex', 'uncertain' and 'ambiguous' is an important contribution. On reflection, I am surprised that such a logical idea had not, to my knowledge, been introduced previously. This characterisation should greatly assist risk management professionals in their consideration of new risks and patterns of risk faced by global businesses.

The more I delved into IRGC's second White Paper *Nanotechnology Risk Governance*, the more I sensed I was travelling in unknown territory. This was not just because of the topic itself (regarding which I confess to be a curious novice!) but rather because the White Paper reflected a risk management discourse operating at a plane beyond my normal experience in business. The concepts require we 'mere mortals' to stretch our thinking and engage in issues involving multiple frames of reference. I do not propose to comment on the actual nanotechnology risks and proposed mitigation strategies contained in IRGC's second White Paper (one of the test applications in this volume is based on IRGC's work in the field of nanotechnology risk governance) other than to observe that they look very sensible to the lay observer!

IRGC's White Paper on risk governance brings fresh, insightful and challenging perspectives to those engaged in transboundary risk management. Moreover, there would appear to be a real appetite on the part of IRGC to tackle some of the truly 'big' global inter-generational issues, using the dispassionate lens of risk governance. The real challenge for IRGC seems to me to be taking others such as governments, NGOs and multinationals on this complex journey. This won't be an impossible task – the risk issues tackled by IRGC will invariably demand coordinated action by national governments.

In closing, I would like to offer some musings on the future. The role of the 'Chief Risk Officer' is now seen as a necessity in large, multi-jurisdiction companies, particularly those engaged in financial services. The role brings together seemingly disparate risk disciplines to better equip the enterprise to see risk coming.

It is interesting to speculate whether this kind of role may gain traction beyond the business realm. Will we see 'Country Risk Officers' being employed by national governments and/or NGOs to bring more unified and consistent approaches to dealing with risks, particularly transboundary risks? I think this is a distinct possibility. Moreover, IRGC's framework could be applied to manage risk at political and transboundary levels. At the very least, it is suggestive of the need for more systematic risk management at the higher levels of government. In the event of, for example, a pandemic outbreak globally, one wonders whether traditional/corporate risk frameworks would cope with the scale of risk issues that would proliferate.

A further observation concerns the evolution of transboundary risk governance structures. The UN Security Council, for example, was established in an environment where the dominant global risk was the threat of war and nuclear weapons. Since that time, transboundary risks have multiplied and arguably represent larger risks than the risk of war. Will we see new global structures addressing this new risk complexity? Will we see the emergence of a UN 'Risk Management Council' that seeks to bring a common risk governance framework to the work of all groups engaged in dealing with global risk issues?

There are, of course, no clear answers to these kinds of questions. However, the IRGC framework represents a new and important contribution to the broad body of work on risk management and sheds light on the kind of risk framework necessary to address 21st century global risk issues. It challenges the sustainability and scalability of generally accepted approaches adopted by the commercial sector. I look forward to road-testing some of IRGC's concepts within my company!

Peter Sutherland
Head of Group Risk & Compliance
Insurance Australia Group
Sydney, Australia

Foreword by Jan Mattingly

Fresh Thinking for Risk Management Practitioners

If you're like me, you are always on the lookout for new knowledge in the risk management field, ideas that will help to inspire and propel you forward in your day to day work in managing risk. I first came across the work of IRGC as a result of a recommendation by noted risk management iconoclast Felix Kloman, Former Publisher, Risk Management Reports. In 2005, Felix had identified the work of the group as internationally noteworthy. I made a mental note to find and read the White Paper, 'Risk Governance – Towards an Integrative Approach'.

A few weeks later, I recall reading the document with the usual rapid scan. It was one document of the usual weekly onslaught that I wanted to review as new and possibly useful risk information. Several key concepts and figures caught my eye and prompted me to read and re-read it in greater detail.

I had not expected what I found: original and fresh thinking, and concepts which carefully and coherently expressed the range of complexity and nuances involved in risk assessment in a new and succinct way. In developing this thought piece, IRGC provided leadership to the risk community in the approach used to research, engage and collaborate with others.

The paper spoke to the influence of risk perception in characterising risk assessment in a way that was crisper and more cogent that other publications I (and perhaps you) have read.

There were several other concepts expressed in the work of IRGC relating to risk governance which are important to the risk profession and to those who wish to de-mystify the management of risk in a way that adds value to organisational performance and indeed, our society.

As a Canadian participant on the work of the ISO 31000 Standard for Managing Risk in Organisation with a corporate risk management and non-science background, my committee colleagues and I have been interested and motivated by the work of IRGC when looking to create original and fresh input to the new Standard.

For example, some of the inspiration for the risk assessment guidance in the new standard has its root in the work of IRGC and contributors like Peter Graham. In particular, the paper shed light on the complex yet simple notion of scaleable risk assessment. How many organisations use one type of risk assessment method for all types of risks? Our working group used IRGC findings as input for lively discussion and debate on the topic, surely a climate for innovation and creativity.

In my daily work as a practitioner working with organisations to modify and strengthen their risk management activities enterprise-wide, I have found good value in the work of IRGC in the key concepts of risk governance and scaleable risk assessment: these concepts resonate with senior executives and business leaders alike, moving us all a little closer to de-mystifying the art and science of enterprise-wide risk management.

For the community of risk management professionals worldwide, the work of IRGC should be a permanent part of your risk library if only for the breadth of perspectives and worldwide collaboration that it represents on a spectrum of key topics for our profession, such as the Risk Governance Framework.

Surely the best way to inspire inspiration in others is to be inspiring and for this reason I wish to recognise the work of IRGC and express my appreciation to its members for inspiring me. I suggest that you have made a noteworthy contribution to the community at large through careful, clear and cogent thinking on time-worn topics of our profession. I trust that readers will be similarly inspired by the pages that follow.

Jan Mattingly, BA, CRM, RF, CIP, ABCP
Risk Management Practitioner, Canada and
ISO 31000 Working Group Member

Foreword by David E. Slavin

A Better Platform for Global Risk Debates

Today there is a growing public fascination with risk. Politicians, protesters, and pundits use the term and its supporting data, in loose and diverging ways. Often these data are used creatively by people who wish to take opposing positions about a technology, its uses or social and environmental impact. This battle of ideology can make it very hard for sections of the general public to gauge the merits of a proposed technology.

And technology based industries are required to develop new technologies on behalf of investors for a profit. That is part of our society's cycle for training and rewarding highly skilled employees, growing investment and pension funds, and of course generating tax revenues. Funding can be either internal (shareholders/ society) or external venture capital (investors/society) and, in some sectors, it is mostly government funding (taxpayer/society) or variations thereof. In all of these cases a return on investment is always required. So in order to optimise these investments as far as possible – commercial marketing strategies can be deployed as they are very effective. Here, pure or clean scientific arguments about a technology are clouded in commercial or political fog at best, deliberate smokescreen at worst. This does nothing for public trust of industry and further complicates the picture.

Where a technology has, or more importantly is perceived to have, potential safety hazards, there is generally a system of regulatory approval prior to public exposure. Although this differs between and amongst industries and countries, the essential elements are a series of ongoing presentations by industry to an expert government regulator with a varied degree of public consultation and appeal. This so called risk assessment process typically concerns only public safety but increasingly, arguments concerning social utility and cost from part of the approval process. Here decision making is hugely affected by the trust triangle between industry sectors, expert government regulators and society and complicated by social amplification factors, etc. Traditional bodies such as the professions, politicians and businesses have become less trusted by society. This distrust is often fuelled by actions of 'trusted' NGOs.

When policy decisions occur in the public sphere, the three domains (the ideological, the commercial and public safety) discussed in the previous paragraphs all come together. Risk communication initiatives can do much to help, but in many cases groups within society remain unconvinced or even strongly opposed to the very intent behind an activity or technology. No amount of risk management process will square the circle of even a legally mandated technology without social license. The fate of the nuclear industry in the UK provides the best example – nanotechnology may not be far behind!

I think this is where a lack of an agreed and trusted, understandable and explicit framework for risk discussion and decision causes us the greatest problems today. Those problems are manifest for industry by gross imbalances of risk mitigation efforts between different industrial sectors. Often this is achieved by the misuse of the Precautionary Principle to seduce society to aspire to a zero risk situation. This approach imposes duties and burdens from which no commercial enterprise can prosper. Rather living by the old adage 'better safe than sorry', we may live in a 'safe and sorry' society.

So what might be the answer for us to most properly use the innovations and opportunities that present themselves to us?

I think that we need to address the 'social-license' element of our decision making. This is why I am attracted to the IRGC approach as it is a unique platform for global debate and addresses societal concerns head-on. Could IRGC become that independent, trusted third party we sorely need to set risk tolerability frameworks to place decisions in context? It could reflect different views and practices, broker the different interests, and provide balanced risk governance strategies. With the IRGC's innovative inclusion of pre-assessment and an explicit concern assessment in addition to more usual risk assessment, it may very well have developed a tool kit to achieve success.

Industry, as part of our society, needs this sort of approach and guidance in order to make the huge investments in future technology a success for all stakeholders in society.

David E. Slavin
Head of Business Innovation Unit
Pfizer Global research and Development
Sandwich, England, UK

Introduction

Japanese government planners set out in the 1960s to build a barrage on the Nagara River, one of the last major freeflowing rivers in Japan. Conceived during a period of rapid growth in the Japanese economy, the barrage was part of a national effort to ensure adequate water supplies for future economic development as well as to reduce flooding risks to downstream communities. A string of lawsuits brought by groups concerned about the impact of the dam on ecological and fisheries resources resulted in costly delays: the dam was not completed for more than 25 years.

The 1990s witnessed the start of a kind of biotech gold rush toward the use of genetic modification (GM) as tool to develop more productive crops through the introduction of herbicide, insect and disease resistance to feed a growing world. Opponents of the rapid deployment of GM crops have raised concerns about the safety of the technology and about its socio-economic, cultural, and ethical implications. The debate over this issue divided the world – for example, the US allowed the development of GM crops to move forward and now accounts for over half the GM crops grown worldwide whereas the European Union only recently lifted a de facto moratorium imposed in 1998 and now authorises products on a case by case basis. Worldwide, the development and use of GM crops is still barely covered by a patchwork of regulations and guidelines, ranging from strict prohibition to none at all, and creating its own sets of disparities and risks.

What went wrong?

These two examples illustrate just some of the potential breakdowns in *risk governance*, the complex process by which risks are identified, assessed, communicated, and managed. The Japanese authorities, by focusing on one set of risks, failed to consider the broader set of risks created by the Nagara River Estuary barrage. In the GM crop case, differences in regulatory approaches, disparities in the influence of various stakeholders, and the role of the media are pointed to as some the problems – or as the successes, depending on one's point of view. Whereas the authorities in Japan could begin to address the governance deficit in their own country by instituting a process for involving the views of different stakeholders in large scale technological projects like the Nagara River Estuary Barrage, solutions to the risk governance challenges posed by GM crops will require a more global perspective.

Global risks, global opportunities, global risk governance challenges. The International Risk Governance Council (IRGC) was established because of widespread concern within the public sector, the corporate world, academia, the media, and society at large that the increasing complexity and interdependence of the world we live in and the risks we face would make the development and implementation of adequate risk governance strategies ever more difficult.

How does one organisation begin to tackle this enormous issue? IRGC began by asking what could be learned from existing risk governance approaches around the world. What has worked well? How, where and when do problems arise? A team of social and natural scientists, engineers, and lawyers undertook a thorough examination of the fundamental principles and structures that guide the way emerging risks have been identified, assessed, managed, and communicated. From this process, IRGC developed and proposed a framework for risk governance that was then subjected to rigorous peer review prior to its publication in September 2005.[1] IRGC next invited formal comments from several experts and from the public, examined the framework carefully in the context of a series of diverse case studies, and documented carefully what lessons this input offered. The result is this volume, *Global Risk Governance: Concept and Practice Using the IRGC Framework*, the first in a series to be published in association with Springer, Dordrecht, the Netherlands. The volume has four parts:

Part 1: The IRGC White Paper on Risk Governance

The first chapter presents the risk governance framework as described in IRGC's 2005 White Paper, *Risk Governance – Towards an Integrative Approach.* The framework was the culmination of a major effort involving numerous individuals, both members of IRGC's Scientific and Technical Council and other leading authorities from around the world. Its purpose is to support IRGC's investigation of risk issues, the governance processes and structures pertaining to them, and the development of policy recommendations for addressing important deficits in risk governance.

What are the innovative features of the framework and how does it differ from those that were analysed in the examination described above?

- *A better definition of risk governance.* A significant part of IRGC's early work was a study of the principles of good *governance* – how the many different groups in society, from governments to individuals – collectively make decisions. These principles underpin IRGC's view that *risk governance* includes the actors, rules, conventions, processes, and mechanisms concerned with how relevant risk information is collected, analysed and communicated

[1] IRGC White Paper No. 1, *Risk Governance – Towards an Integrative Approach*, IRGC, Geneva, 2005.

and management decisions are taken. Risk governance thus extends beyond the three conventionally recognised elements of *risk analysis* (risk assessment, risk management and risk communication)[2] and thus includes matters of institutional design and role, organisational capacity, stakeholder involvement, collaborative decision making and political accountability on the part of public bodies and corporate responsibility on the part of private enterprises. It also includes the requirement on the part of government, commercial and civil society actors for the development and use of scientific knowledge within the risk governance process.

- *A simple, but comprehensive framework.* The framework's process for dealing with risk comprises five phases: *pre-assessment*; *risk appraisal*, *risk characterisation/evaluation*; *risk management*; and *risk communication*. We also distinguish between a management sphere (containing decision making and implementation) and an assessment sphere (containing risk appraisal). The pre-assessment, characterisation/evaluation and communication phases are in both spheres because, although we strongly endorse the separation of risk appraisal and management, these three other phases need the combined efforts of the people responsible for both. We position risk communication at the centre of the framework to reflect its crucial role throughout – rather than at a particular point of – the entire process. The IRGC framework is, therefore, deliberately open, interlinked and iterative.
- *A truly interdisciplinary approach.* The framework urges risk governance institutions, in their *appraisal* of risks, to consider input from a broader base of scientific knowledge. Not only knowledge about the physical impacts of technologies, natural events or human activities that is the typical basis for *risk assessment* but also knowledge about the concerns that people associate with these sources of risk. This *concern assessment* is a social science activity aimed at providing a comprehensive diagnosis of concerns, expectations and worries that individuals, groups or different cultures may link to the hazards which, in turn, are a key input to assessing a risk's acceptability and to designing appropriate risk management strategies.
- *An idea of inclusive governance.* Inclusive governance is seen as a necessary, although not sufficient, prerequisite for tackling risks and, consequently, requires the productive and meaningful involvement of all stakeholders, in particular, civil society. It is based on the assumption that all stakeholders have something positive to contribute to the process of risk governance.

---

[2] National Research Council, 1996, *Understanding Risk: Informing Decisions in a Democratic Society*, National Academy Press, Washington DC; Codex Alimentarius Commission, 2005, *Procedural Manual*, Fifteenth edition, Joint FAO/WHO Food Standards Programme, Rome, World Health Organisation/Food and Agriculture Organisation of the United Nations, 2005; Regulation (EC) No 178/2002 (OJ 2002, L31/1) as amended by Regulation (EC) No 1642/2003 (OJ 2003, L 245/4).

Part 2: Critical Comments

When IRGC published the White Paper on Risk Governance in 2005, it envisioned the document with its underlying framework as a work-in-progress, a focus for comment and debate. IRGC clearly recognised that no framework can emerge fully formed to deal with the broad array of risk problems facing society today. The organisation expected the need for refinement and revision and therefore welcomed the process of thoughtful and critical review.

Since that time, the IRGC risk governance framework has received considerable attention from the risk assessment and risk management communities. It has been presented at numerous conferences and symposia in Europe, North America and Asia. The original White Paper has been reprinted twice and many hundreds of people have accessed and downloaded the document from our website. The feedback obtained from many of these people has been invaluable to IRGC's understanding of the framework's strengths and weaknesses.

The critical reviews discussed in this section of the book encompass both formal and informal comments received by IRGC following its presentations. The first four chapters present the formal written comments commissioned from four individuals with extensive experience in risk assessment and/or management:

- Ragnar Lofstedt, Professor of Risk Management and Director of the King's Centre for Risk Management International Policy Institute;
- Eugene A. Rosa, Professor of Sociology, Edward R. Meyer Professor of Natural Resource and Environmental Policy, Washington State University;
- Robin Cantor, Managing Director, Navigant Consulting, Inc. and Past President of the Society for Risk Analysis;
- Warner North, President and Principal Scientist, Northworks, Inc., Adjunct Professor, Stanford University.

The final chapter provides a synopsis of the numerous informal comments IRGC has received from respected individuals and institutions in the field of risk governance. The commentators represent a diverse mix of stakeholder groups: the international academic community, international and national regulatory institutions, industry, risk research and environmental NGOs.

This section of the book does not provide the authors' or IRGC's response to these comments but, rather, provides an opportunity to acknowledge them and to honour those individuals who have taken time and effort to compose thoughtful and constructive statements about the IRGC framework.

Part 3: Case Studies

A small but diverse set of case studies were commissioned to assess how well the IRGC risk governance framework both supports the comprehensive understanding of a risk and facilitates the development of policy options. While most of the studies were retrospective, each provided several important insights into how risk governance might have been improved or could still be improved:

- *Listeria in raw milk soft cheese* (Andrew Knight et al., Food Safety Policy Center, Michigan University, USA). This case study illustrates how important framing of a risk issue is to both how the risk is managed and how successfully the chosen risk management strategy is implemented. While one framing – 'illness prevention' – informed the decision to ban the use of unpasteurised milk in making soft cheese in the US, the 'consumer sovereignty' framing lies behind a minority whose behaviour knowingly disregards the law.
- *Genetically modified (GM) crops* (Joyce Tait, University of Edinburgh, Scotland). Tait raises several important issues for emerging technologies, among them the role that framing has in determining the regulatory path and even ultimate commercial success of a new technology. She emphasises the importance of defining processes for the responsible involvement of stakeholders and the introduction of evidence to the governance process.
- *Nagara River Estuary Barrage conflict* (Norio Okada et al., Disaster Prevention Research Institute, University of Kyoto, Japan). This case study provides relevant lessons for governments today about the dangers of failing to consider other stakeholder concerns (environment, fisheries, etc.) in the planning stages of large-scale projects. It illustrates the need for stakeholder feedback systems during project development that allow for changes to occur.
- *Nature-based tourism* (Caroline Kuenzi, IRGC, Switzerland and Jeff McNeely, World Conservation Union). This case study presents a complex problem for risk governance in which the risks and benefits of nature-based tourism and responsibilities for managing them span a diverse group of individuals, private enterprises, government agencies, non-governmental agencies, countries and intergovernmental organisations. Multi-faceted strategies for risk management will be required.
- *Acrylamide in food* (Sabine Bonneck, Cologne, Germany). Bonneck traces the crisis that erupted across Europe when acrylamide was discovered in food products. It provides important insights for improvements in risk communication and the involvement of stakeholders in the risk management process.
- *Energy security for the Baltic Region* (Warner North, Northworks Inc. and Stanford University, USA). In his preliminary 'pre-assessment' of the complex problem of energy security in the Baltic region, North lays out the difficult risk tradeoffs that must be considered and the challenges of balancing the interests of the different governments and political and civil society actors.
- *Nanotechnology* (Mihail Roco, National Science Foundation, USA; Ortwin Renn, University of Stuttgart and DIALOGIK gGmbH, Germany and Alexander Jäger, Interdisciplinary Research Unit on Risk Governance and Sustainable Technology Development (ZIRN), Stuttgart, Germany). Unlike the other case studies, the chapter on nanotechnology is not retrospective, but also offers insights using the IRGC framework for risk governance of the still emerging, newer generations of nanotechnology and their applications.

IRGC recognises that such retrospective analysis is only one tool for evaluating the IRGC framework and that it has its own limitations. Ultimately, a true test of any model is how well it performs when used proactively and IRGC will continue to use its framework to support its work in understanding emerging risks and the development of risk governance recommendations for policy makers.

Part 4: Lessons Learned

In the final chapter, IRGC has carefully considered the many constructive, critical comments received (Part 2), the experiences from the case studies (Part 3) and has laid out the lessons learned. There are many. IRGC's goal in this chapter is not to respond in detail to every comment but to acknowledge the common themes that have emerged regarding both broad conceptual issues as well as the practical aspects of each phase of the framework. The authors hope that these 'lessons learned' will both guide IRGC's ongoing refinement of the framework and assist others who may be encountering the framework for the first time.

Although work remains to be done, IRGC has accomplished the task of creating a broad conceptual framework that incorporates key principles for sound risk governance. It provides a structure, within or around which particular risks may be investigated, discussed by stakeholders, communicated, and managed. By laying a clear rationale for taking into account not only scientific evidence, economic considerations, but also risk perceptions, social concerns and societal values, the IRGC framework attempts to provide a more comprehensive and integrated view of risk governance than other approaches have in the past.

The IRGC risk governance framework is not a manual; ultimately experts with specialised training will be needed to carry out the tasks necessary for the governance of particular risks. However, IRGC does hope that this more comprehensive framework will assist decision makers in asking the right questions, questions that will help them avoid the pitfalls of the past and to develop more effective risk governance strategies for the increasingly complex risks – and opportunities – we face in the world.

Ortwin Renn
University of Stuttgart and DIALOGIK gGmbH, and
Member of the IRGC Scientific and Technical Council

and

Katherine Walker
IRGC, Geneva, Switzerland

Acknowledgements

This book is a product of a major collaborative effort. Part 1, which presents the original IRGC risk governance framework, owes a substantical debt to Peter Graham who did the research and writing of the annexes. Although not reproduced here, the annexes laid important groundwork for the development of the framework. Substantial input was provided by five background papers that were commissioned to inform discussions at a project workshop held in Ismaning, Germany, in fall 2004. In particular, the paper by Jean-Pierre Contzen on organisational capacity has been largely adopted for the section on organisational capacity building. Caroline Kuenzi edited the whole manuscript carefully and added several paragraphs and, with Chris Bunting, provided text for the case examples. Howard Kunreuther provided valuable material to the section on interpretative ambiguity and interdependencies. The members of the IRGC risk governance project gave helpful advice and constructive feedback in all stages of completing the manuscript. Those members are: Lutz Cleemann, Jean-Pierre Contzen, Harry Kuiper, Peter Graham, Wolfgang Kröger, Joyce Tait and Jonathan Wiener. The editors are also indebted to the members of the IRGC Scientific and Technical Council for their feedback, and particularly to Manuel Heitor who acted as review coordinator. They are also grateful to the five anonymous reviewers of the manuscript who provided constructive criticism and suggestions for improvement. Additional reviews and input to the White Paper were also received from Eugene Rosa, Chris Bunting, Paul Stern, Granger Morgan, Marion Dreyer, Juergen Hampel, Alexander Jäger, Pia-Johanna Schweizer and the participants at the above-mentioned workshop.

We are very grateful to each of the reviewers of the White Paper who gave their time both to speak at the Society for Risk Analysis conference in 2005 and to prepare formal written versions of their thoughtful and constructive comments that appear in Part 2. Each of the case study authors, whose chapters appear in Part 3 of the book, similarly deserve special thanks for the care with which they analysed their subjects within the context of the IRGC framework. Their insights to some of the practical aspects of applying the framework were extremely helpful.

We wish to thank Timothy Walker for taking on the heroic task of reviewing the entire book, pushing us to think carefully about risk governance, and providing many other insightful comments.

Last, but not least, all of the authors owe a special thanks to Christopher Bunting at IRGC, without whose constant motivation and support, this book would not have been possible.

Part 1. A Framework for Risk Governance

Chapter 1
White Paper on Risk Governance: Toward an Integrative Framework[1]

Ortwin Renn
University of Stuttgart, Stuttgart, Germany and DIALOGIK gGmbH, Stuttgart, Germany

Purpose and Objectives of This White Paper

This document aims to guide the work of the International Risk Governance Council and its various bodies in devising comprehensive and transparent approaches to 'govern' a variety of globally relevant risks. Globally relevant risks include *trans-boundary* risks, i.e. those that originate in one country and affect other countries (such as air pollution), *international* risks, i.e. those that originate in many countries simultaneously and lead to global impacts (such as carbon dioxide emissions for climate change) and ubiquitous risks, i.e. those that occur in each country in similar forms and may necessitate a co-ordinated international response (such as car accidents or airline safety). To this end the document and the framework it describes provide a common *analytic structure for investigating and supporting the treatment of risk issues* by the relevant actors in society. In doing so, the focus is not restricted to how governmental or supranational authorities deal with risk but equal importance is given to the roles of the corporate sector, science, other stakeholders as well as civil society – and their interplay. The analytic structure will, it is hoped, facilitate terminological and conceptual clarity, consistency and transparency in the daily operations of IRGC and assure the feasibility of comparative approaches in the governance of risks across a broad range of hazardous events and activities. In particular, this document is meant to assist members of IRGC in their tasks to provide scientifically sound, economically feasible, legally and ethically justifiable and politically acceptable advice to IRGC's targeted audiences. It is also to support IRGC in its effort to combine the best available expertise in the respective field with practical guidance for both risk managers and stakeholders.

[1] This chapter is the main body of a complete work with the same title published by IRGC in 2005. The IRGC document contains in addition three brief case studies and a series of appendices detailing other risk governance schemes. It can be downloaded from our website: http://www.irgc.org/spip/IMG/pdf/IRGC_WP_No_1_Risk_Governance_(reprinted_version).pdf

O. Renn and K. Walker (eds.), Global Risk Governance: Concept and Practice Using the IRGC Framework, 3–73.

The overall objective of this document is to establish a comprehensive and consistent yet flexible prototype analytic framework and unified set of guidance for improved risk governance. This framework integrates the following components:

- harmonised terminology with respect to key terms and concepts;
- a robust and coherent concept of framing and characterising the essential physical as well as social elements of coping with risks, including both the classic components (i.e. risk assessment, risk management and risk communication) as well as the contextual aspects such as a wider framework of risk appraisal, governance structure, risk perception, regulatory style and organisational capacity;
- a categorisation and enhancement of approaches to risk assessment and risk management including suggestions for basic safety principles and integrated appraisal and management strategies based on scientific analysis, precautionary considerations and vulnerability assessment;
- inclusion of risk-benefit evaluation and risk-risk tradeoffs;
- a conceptual framework for integrating civil society (stakeholders from the corporate sector, NGOs, associations, science communities as well as representatives of the public) in risk governance;
- principles of 'good' risk governance;
- requirements for improving risk governance capacity including the new perspective of integrated disaster risk management (IDRM).

This document draws on an initial compilation and critical review of work already available in this area (e.g. existing risk taxonomies and, in particular, guidance documentation on risk and risk governance) as well as on an acknowledgement of commonalities and differences between these approaches. A selection of these guidelines, manuals, standards, government reports etc. is summarised in annexes to the original document which may be downloaded from IRGC's website (see Footnote 1). A glossary defining the terms and concepts used throughout this White Paper may be found at the end of the chapter.

Target Audience of This White Paper

The primary audience of this document is IRGC itself which will use it as an analytic blueprint for its further work and will implement its recommendations within future IRGC projects. After a period of intense testing within several IRGC projects and empirical analysis of its use in different risk contexts and cultural environments, it is anticipated that a further revised version of this White Paper can offer assistance and guidance to senior risk managers and decision makers as well as risk practitioners outside of IRGC in their daily efforts to identify, assess, manage and monitor risk.

An important niche could, for instance, derive from the active transfer and dissemination of this body of knowledge to key actors in politics and society in developing countries and those in transition. Many of these countries are only now starting to formally think about issues of risk governance and IRGC's providing them with relevant information and knowledge could provide valuable insights and, possibly, help them to avoid some of the pitfalls inherent in dealing with risk. A flexible yet harmonised framework might also be of benefit to both the government and industry sectors in OECD countries, since its main thrust is to provide logically coherent and sensible guidance for conducting risk appraisals (including risk assessment as well as concern assessment), for steering risk management and for improving risk governance structures in a variety of risk areas and socio-political cultures. Therefore, this document particularly addresses all those actors who will benefit from more direct cooperation with other stakeholders and from integrated risk governance procedures. Indeed, specifying the role of these actors within an integrated framework of risk governance is one of the main goals of IRGC, and this goal has also inspired the framework presented in this document. While it is clear that each risk field or 'case' under consideration is different in that it requires further specifications and adjustments, it is nonetheless hoped that the framework presented herein can serve as a 'default option' from which one can and should deviate if necessary.

Scope of the Proposed Framework

This document covers a wide range of both risks and governance structures. **Risk** is understood in this document as an uncertain consequence of an event or an activity with respect to something that humans value (definition originally in: Kates et al. 1985: 21). Risks always refer to a combination of two components: the likelihood or chance of potential consequences and the severity of consequences of human activities, natural events or a combination of both. Such consequences can be positive or negative, depending on the values that people associate with them. IRGC is not covering all risk areas but confines its efforts to (predominantly negatively evaluated) risks that lead to physical consequences in terms of human life, health, and the natural and built environment. It also addresses impacts on financial assets, economic investments, social institutions, cultural heritage or psychological well-being as long as these impacts are associated with the physical consequences.[2] In addition to the strength and likelihood of these consequences, the framework emphasises the distribution of risks over time, space and populations. In particular, the timescale of appearance of adverse effects is very important and links risk governance to sustainable development (delayed effects).

In this document we distinguish risks from **hazards**. Hazards describe the potential for harm or other consequences of interest. These potentials may never even materialise if, for example, people are not exposed to the hazards or if the targets are made resilient against the hazardous effect (such as immunisation). In conceptual terms, hazards characterise the *inherent properties of the risk agent and related*

[2] Although IRGC focuses on physical risks and their secondary implications, the framework may also be extended to allow for the investigation of financial, social or political risks as primary risk consequences.

processes, whereas risks describe the *potential effects that these hazards are likely to cause on specific targets such as buildings, ecosystems or human organisms and their related probabilities*.

Table 1 provides a systematic overview of the sources of risks or hazards that potentially fall within the scope of IRGC's work programme. The purpose of this overview is to lay out the variety of sources of risks rather than to claim that the categories proposed are exhaustive or mutually exclusive (see review of classification in Morgan et al. 2000). Furthermore, IRGC places most attention on risk areas of global relevance (i.e. transboundary, international and ubiquitous risks) which additionally include large-scale effects (including low-probability, high-consequence outcomes), require multiple stakeholder involvement, lack a superior decision-making authority and involve the potential to cause wide-ranging concerns and outrage.

The IRGC has as one of its primary responsibilities the provision of expertise and practical advice in dealing with a novel type of risk, which the OECD has labelled '**systemic risks**' (OECD 2003). This term denotes the embeddedness of any risk to human health and the environment in a larger context of social, financial and economic consequences and increased interdependencies both across risks and between their various backgrounds. Systemic risks are at the crossroads between natural events (partially altered and amplified by human action such as the emission of greenhouse gases), economic, social and technological developments and policy-driven actions, both at the domestic and the international level. These new interrelated and interdependent risk fields also require a new form of handling risk, in which data from different risk sources are either geographically or functionally integrated into one analytical perspective. Handling systemic risks requires a holistic approach to hazard identification, risk assessment, concern assessment, tolerability/acceptability judgements and risk management. Investigating systemic risks goes beyond the usual agent-consequence analysis and focuses on interdependencies and spill-overs between risk clusters.

Risk in a Broader Context

The focus on risk should be seen as a segment of a larger and wider perspective on how humans transform the natural into a cultural environment with the aims of improving living conditions and serving human wants and needs (Turner et al. 1990). These transformations are performed with a purpose in mind (normally a benefit to those who initiate them). When implementing these changes, intended (or tolerated) and unintended consequences may occur that meet or violate other dimensions of what humans value. Risks are not taken for their own sake; rather more they are, actively or passively, incurred because of their being an integral factor in the very activity that is geared towards achieving the particular human need or purpose. In this context, it is the major task of risk assessment to identify and explore, preferably in quantitative terms, the types, intensities and likelihood of the (normally undesired)

Table 1 Risks taxonomy according to hazardous agents.

- **Physical Agents**
  - Ionising radiation
  - Non-ionising radiation
  - Noise (industrial, leisure, etc.)
  - Kinetic energy (explosion, collapse, etc.)
  - Temperature (fire, overheating, overcooling)
- **Chemical Agents**
  - Toxic substances (thresholds)
  - Genotoxic/carcinogenic substances
  - Environmental pollutants
  - Compound mixtures
- **Biological Agents**
  - Fungi and algae
  - Bacteria
  - Viruses
  - Genetically modified organisms
  - Other pathogens
- **Natural Forces**
  - Wind
  - Earthquakes
  - Volcanic activities
  - Drought
  - Flood
  - Tsunamis
  - (Wild) fire
  - Avalanche
- **Social-Communicative Hazards**
  - Terrorism and sabotage
  - Human violence (criminal acts)
  - Humiliation, mobbing, stigmatising
  - Experimentation with humans (such as innovative medical applications)
  - Mass hysteria
  - Psychosomatic syndromes
- **Complex Hazards (Combinations)**
  - Food (chemical and biological)
  - Consumer products (chemical, physical, etc.)
  - Technologies (physical, chemical, etc.)
  - Large constructions such as buildings, dams, highways, bridges
  - Critical infrastructures (physical, economic, social-organisational and communicative)

consequences related to a risk. In addition, these consequences are associated with special concerns that individuals, social groups or different cultures may attribute to these risks. They also need to be assessed for making a prudent judgement about the tolerability or acceptability of risks. Once that judgement is made, it is the task of risk management to prevent, reduce or alter these consequences by choosing appro-

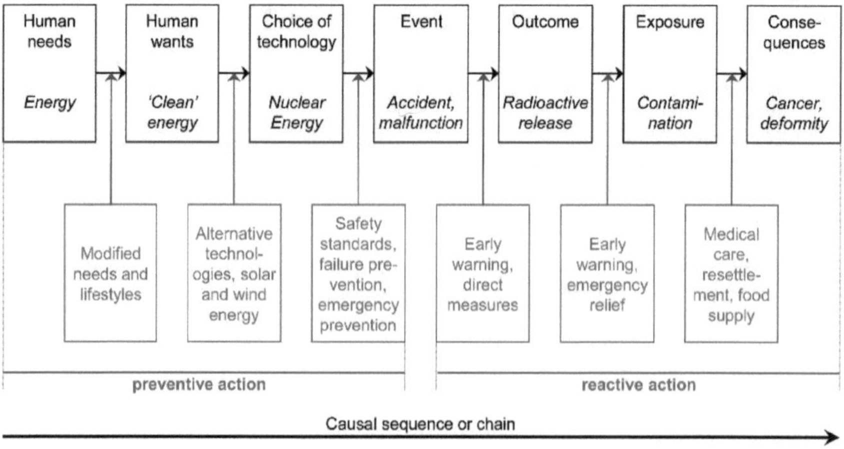

Fig. 1 Seven steps of a risk chain: The example of nuclear energy (from Hohenemser et al. 1983).

priate actions. As obvious as this distinction between risk and concern assessment (as a tool of gaining knowledge about risks) and risk management (as a tool for handling risks) appears at first glance, the *distinction becomes blurred* in the actual risk governance process.

This blurring is due to the fact that assessment starts with the respective risk agent or source and tries to both identify potential damage scenarios and their probabilities and to model its potential consequences over time and space, whereas risk management oversees a much larger terrain of potential interventions (Stern and Fineberg 1996; Jasanoff 1986: 79f; 2004). Risk management may alter human wants or needs (so that the agent is not even created or continued). It can suggest substitutes or alternatives for the same need. It can relocate or isolate activities so that exposure is prevented, or it can make risk targets less vulnerable to potential harm. Figure 1 illustrates this larger perspective for technological risks and lists the possible intervention points for risk management (taken from Hohenemser et al. 1983). Risk assessment and management are therefore not symmetrical to each other: management encompasses a much larger domain and may even occur before assessments are performed. It is often based on considerations that are not affected by or part of the assessment results. In more general terms, risk management refers to the creation and evaluation of options for initiating or changing human activities or (natural and artificial) structures with the objective being to increase the net benefit to human society and prevent harm to humans and what they value. The identification of these options and their evaluation is guided by systematic and experiential knowledge gained and prepared for this purpose by experts and stakeholders. A major proportion of that relevant knowledge comprises the results of risk assessments. However, risk managers also need to act in situations of 'non-knowledge' or insufficient knowledge about potential outcomes of human actions or activities.

The most complex questions emerge, however, when one looks at how society and its various actors actually handle risk. In addition to knowledge gained through risk assessments and/or option generation and evaluation through risk management, the decision-making structure of a society is itself highly complicated and often fragmented. Apart from the structure itself – the people and organisations that share responsibility for assessing and managing risk – one must also consider the need for sufficient organisational capacity to create the necessary knowledge and implement the required actions, the political and cultural norms, rules and values within a particular societal context and the subjective perceptions of individuals and groups. These factors leave their marks on the way risks are treated in different domains and socio-political cultures. To place risk within a context of – sometimes closely interwoven – decision-making structures such as those prevalent in governments and related authorities, in the corporate sector and industry, in the scientific community and in other stakeholder groups is of central concern to IRGC.

In the last decade the term '**governance**' has experienced tremendous popularity in the literature on international relations, comparative political science, policy studies, sociology of environment and technology as well as risk research.[3] On a national scale, *governance describes structures and processes for collective decision-making involving governmental and non-governmental actors* (Nye and Donahue 2000). Governing choices in modern societies is seen as an interplay between governmental institutions, economic forces and civil society actors (such as NGOs). At the global level, *governance embodies a horizontally organised structure of functional self-regulation encompassing state and non-state actors bringing about collectively binding decisions without superior authority* (cf. Rosenau 1992; Wolf 2002). In this perspective non-state actors play an increasingly relevant role and become more important, since they have decisive advantages of information and resources compared to single states.

It is useful to differentiate between *horizontal and vertical governance* (Benz and Eberlein 1999; Lyall and Tait 2004). The horizontal level includes the relevant actors in decision-making processes within a defined geographical or functional segment (such as all relevant actors within a community, region, nation or continent); the vertical level describes the links between these segments (such as the institutional relationships between the local, regional and state levels).

'*Risk governance*' involves the 'translation' of the substance and core principles of governance to the context of risk and risk-related decision-making. In IRGC's understanding, risk governance includes the totality of actors, rules, conventions, processes, and mechanisms concerned with how relevant risk information is collected, analysed and communicated and management decisions are taken. Encompassing the combined risk-relevant decisions and actions of both governmental and private actors, risk governance is of particular importance in, but not restricted to, situations where there is no single authority to take a binding risk management decision but where, instead, the nature of the risk requires the collaboration of, and co-ordination

[3] According to Rhodes (1996) there are six separate uses of the term governance: as minimal state, as corporate governance, as new public management, as good governance, as social-cybernetic systems and as self-organised networks.

between, a range of different stakeholders. Risk governance however not only includes a multifaceted, multi-actor risk process but also calls for the consideration of contextual factors such as institutional arrangements (e.g. the regulatory and legal framework that determines the relationship, roles and responsibilities of the actors and co-ordination mechanisms such as markets, incentives or self-imposed norms) and political culture, including different perceptions of risk.

When looking at risk governance structures it is not possible to include all the variables that may influence the decision-making process; there are too many. Therefore it is necessary to limit one's efforts to those factors and actors that, by theoretical reasoning and/or empirical analysis, are demonstrably of particular importance with respect to the outcome of risk governance. IRGC has highlighted the following aspects of risk governance which extend beyond risk assessment and risk management:

- the structure and function of various actor groups in initiating, influencing, criticising and/or implementing risk policies and decisions;
- risk perceptions of individuals and groups;
- individual, social and cultural concerns associated with the consequences of risk;
- the regulatory and decision-making style (political culture);
- the requirements with respect to organisational and institutional capabilities for assessing, monitoring and managing risks (including emergency management).

In addition to these analytical categories, this document also addresses best practice and normative aspects of what is needed to improve governance structures and processes (EU 2001a). With respect to best practice it is interesting to note that often risk creators, in particular when directly affected by the risk they generate, engage in risk reduction and avoidance out of self-interest or on a voluntary basis (e.g. industry 'gentleman's agreements', self-restriction, industry standards). Other stakeholders' efforts in risk governance therefore have to be coordinated with what is tacitly in place already. The emphasis here is on cooperative models of public-private partnerships forming a governance system that aims at effective, efficient and fair risk management solutions.[4]

Before Assessment Starts

Risks are *mental 'constructions'* (OECD 2003: 67). They are not real phenomena but originate in the human mind. Actors, however, creatively arrange and reassemble signals that they get from the 'real world' providing structure and guidance to an ongoing process of reality enactment.[5] So risks represent what people observe in

[4] Excluded from this document are such topics as crisis intervention, crisis communication, emergency planning and management and post-accidental relief. They will be covered in a separate document at a later stage.

[5] I am indebted to Gene Rosa for giving me guidance on keeping a healthy balance between a relativist and realist version of risk. For further reading refer to Rosa (1998). It should be noted

reality and what they experience. The link between risk as a mental concept and reality is forged through the experience of actual harm (the consequence of risk) in the sense that human lives are lost, health impacts can be observed, the environment is damaged or buildings collapse. The invention of risk as a mental construct is contingent on the belief that human action can prevent harm in advance. Humans have the ability to design different futures, i.e. construct scenarios that serve as tools for the human mind to anticipate consequences in advance and change, within constraints of nature and culture, the course of actions accordingly.

The status of risk as a mental construct has major implications on how risk is looked at. Unlike trees or houses, one cannot scan the environment, identify the objects of interest, and count them. Risks are created and selected by human actors. What counts as a risk to someone may be an act of God to someone else or even an opportunity for a third party. Although societies have over time gained experience and collective knowledge of the potential impacts of events and activities, one cannot anticipate all potential scenarios and be worried about all the many potential consequences of a proposed activity or an expected event. By the same token, it is impossible to include all possible options for intervention. Therefore societies have been *selective* in what they have chosen to be worth considering and what to ignore (Thompson et al. 1990; Douglas 1990; Beck 1994: 9ff.). Specialised organisations have been established to monitor the environment for hints of future problems and to provide early warning of some potential future harm. This selection process is not arbitrary. It is guided by cultural values (such as the shared belief that each individual life is worth protecting), by institutional and financial resources (such as the decision of national governments to spend money or not to spend money on early warning systems against highly improbable but high-consequence events) and by systematic reasoning (such as using probability theory for distinguishing between more likely and less likely events or methods to estimate damage potential or distribution of hazards in time and space).

Based on these preliminary thoughts, a systematic review of risk-related actions needs to start with an analysis of what major societal actors such as e.g. governments, companies, the scientific community and the general public select as risks

---

that this White Paper takes no stand on the *controversial issue of constructivism versus realism of evidence and values* (this topic is extensively reviewed in Mayo and Hollander 1991, specific positions in Bradbury 1989; Douglas 1990; Shrader-Frechette 1991b; 1995; Wynne 1992; Laudan 1996; Jasanoff 2004). Whether the evidence collected represents human ideas about reality or depicts representations of reality is of no importance for the distinction between evidence and values that is suggested throughout the document. Handling risks will inevitably be directed by evidence claims (what are the causes and what are the effects?) and normative claims (what is good, acceptable and tolerable?). It is true that providing evidence is always contingent on existing normative axioms and social conventions. Likewise, normative positions are always enlightened by assumptions about reality (Ravetz 1999). The fact that evidence is never value-free and that values are never void of assumptions about evidence does not compromise the need for a functional distinction between the two. For handling risks one is forced to distinguish between what is likely to be expected when selecting option x rather than option y, on one hand, and what is more desirable or tolerable: the consequences of option x or option y, on the other hand. It is hence highly advisable to maintain the classic distinction between evidence and values and also to affirm that justifying claims for evidence versus values involves different routes of legitimisation and validation.

and what types of problems they label as risk problems (rather than opportunities or innovation potentials, etc.). In technical terms this is called '**framing**'. Framing in this context encompasses the selection and interpretation of phenomena as relevant risk topics (Tversky and Kahneman 1981; van der Sluijs et al. 2003; Goodwin and Wright 2004). The process of framing is already part of the governance structure since official agencies (for example food standard agencies), risk and opportunity producers (such as the food industry), those affected by risks and opportunities (such as consumer organisations) and interested bystanders (such as the media or an intellectual elite) are all involved and often in conflict with each other when framing the issue. What counts as risk may vary among these actor groups. Consumers may feel that all artificial food additives pose a risk, whereas industry may be concerned about pathogens that develop their negative potential due to the lack of consumer knowledge about food storage and preparation. Environmental groups may be concerned with the risks of industrial food versus organic food. Whether a consensus evolves about what requires consideration as a relevant risk depends on the legitimacy of the selection rule. The acceptance of selection rules rests on two conditions: first, all actors need to agree with the *underlying goal* (often legally prescribed, such as prevention of health detriments, or guarantee of an undisturbed environmental quality, for example purity laws for drinking water); secondly, they need to agree with the *implications derived from the present state of knowledge* (whether and to what degree the identified hazard impacts the desired goal). Even within this preliminary analysis, dissent can result from conflicting values as well as conflicting evidence, and, in particular, from the inadequate blending of the two. Values and evidence can be viewed as the two sides of a coin: the values govern the selection of the goal whereas the evidence governs the selection of cause-effect claims. Both need to be properly investigated when analysing risk governance but it is of particular importance to understand the values shaping the interests, perceptions and concerns of the different stakeholders as well as to identify methods for capturing how these concerns are likely to influence, or impact on, the debate about a particular risk. The actual measurements of these impacts should then be done in the most professional manner, including the characterisation of uncertainties (Keeney 1992; Pidgeon and Gregory 2004; Gregory 2004).

A second part of the pre-assessment phase concerns the institutional means of *early warning and monitoring*. Even if there is a common agreement of what should be framed as (a) risk issue(s), there may be problems in monitoring the environment for signals of risks. This is often due to a lack of institutional efforts to collect and interpret signs of risk and deficiencies in communication between those looking for early signs and those acting upon them. The recent tsunami catastrophe in Asia provides a more than telling example of the discrepancy between the capability to have early warning systems and the decision to install or use them. It is therefore important to look at early warning and monitoring activities when investigating risk governance.

In many risk governance processes, information about risks are pre-screened and then allocated to different assessment and management routes. In particular, industrial risk managers search for the most efficient strategy to deal with risks. This includes prioritisation policies, protocols for dealing with similar causes of risks, and optimal models combining risk reduction and insurance. Public risk regulators often use pre-screening activities to allocate risks to different agencies or to pre-defined procedures. Sometimes risks may seem to be less severe and it may be adequate to cut short risk or concern assessment. In a pending crisis situation, risk management actions may need to be taken before any assessment is even carried out. A full analysis should therefore include provisions for *risk screening* and the selection of different routes for risk assessment, concern assessment and risk management. This aspect has been called '*risk assessment policy*' in the Codex Alimentarius. It is meant to guide the assessment process in terms of assessment and management protocols, methods of investigation, statistical procedures and other scientific conventions used in assessing risks or selecting risk reduction options. A screening process may also be employed when characterising risks according to complexity, uncertainty and ambiguity as we will explain later.

Another major component of pre-assessment is the *selection of conventions and procedural rules* needed for a comprehensive scientific appraisal of the risk, i.e. for assessing the risk and the concerns related to it (see below). Any such assessment is based on prior informed yet subjective judgements or conventions articulated by the scientific community or a joint body of risk assessors and managers. Those judgements refer to (Pinkau and Renn 1998; van der Sluijs et al. 2004: 54ff.):

- the social definition of what is to be regarded as adverse, (for example by defining the 'No Adverse Effect Level' in food (NOAEL));
- the selection rule determining which potentially negative effects should be considered in the risk governance process knowing that an infinite number of potential negative outcomes can be theoretically connected with almost any substance, activity, or event;
- the aggregation rule specifying how to combine various effects within a one-dimensional scale, for example early fatalities, late fatalities, cancer, chronic diseases and so on;
- the selection of the testing and detection methods which are presently used in risk assessment, for example the use of genomics for calculating risk from transgenic plants;
- the selection of valid and reliable methods for measuring perceptions and concerns;
- the determination of models to extrapolate high dose effects to low dose situations, for example linear, quadro-linear, exponential or other functions or assumptions about thresholds or non-thresholds in dose-response relationships;
- the extrapolation of the results of animal data to humans;
- assumptions about exposure or definition of target groups;
- the handling of distributional effects which may cover inter-individual, intergroup, regional, social, time-related and inter-generational aspects.

These judgements reflect the consensus among the experts or are common products of risk assessment and management (for example by licensing special testing methods). Their incorporation in guiding scientific analyses is unavoidable and this does not discredit the validity of the results. Yet it is essential that risk managers and interested parties are informed about these conventions and understand their rationale. On one hand knowledge about these conventions can lead to a more cautious apprehension of what the assessments mean and imply, on the other hand they can convey a better understanding of the constraints and conditions under which the results of the various assessments hold true.

In summary, Table 2 provides a brief overview of the four components of pre-assessment. The table also lists some indicators that may be useful as heuristic tools when investigating different risk governance processes. The choice of indicators is not exhaustive and will vary depending on risk source and risk target. Listing the indicators serves the purpose of illustrating the type of information needed to perform the task described in each step. The title 'pre-assessment' does not mean that these steps are always taken before assessments are performed. Rather they are logically located in the forefront of assessment and management. They should also not be seen as sequential steps but as elements that are closely interlinked. As a matter of fact, and depending on the situation, early warning might precede problem framing.[6]

Risk Assessment

The purpose of risk assessment is the generation of knowledge linking specific risk agents with uncertain but possible consequences (Lave 1987; Graham and Rhomberg 1996). The final product of risk assessment is an estimation of the risk in terms of a probability distribution of the modelled consequences (drawing on either discrete events or continuous loss functions). The different stages of risk assessment vary from risk source to risk source. Many efforts have been made to produce a harmonised set of terms and conceptual phase-model that would cover a wide range of risks and risk domains (cf. Codex Alimentarius 2001; National Research Council 1982, 1983; Stern and Fineberg 1996; EU 2000, 2003).[7] The most recent example is the risk guidance book by the International Programme on Chemical Safety (IPCS) and WHO (IPCS and WHO 2004). Although there are clear differences in structuring the assessment process depending on risk source and organisational culture, there is an agreement on basically three core components of risk assessment:

- an identification and, if possible, estimation of hazard;
- an assessment of exposure and/or vulnerability;
- an estimation of risk, combining the likelihood and the severity of the targeted consequences based on the identified hazardous characteristics and the exposure/vulnerability assessment.

[6] It should also be noted that early warning may of course also benefit from 'non-systematic' findings and incidental/accidental reporting.

[7] An extended review of a large variety of risk taxonomies is summarised in Annex A to the original White Paper, downloadable from our website:
http://www.irgc.org/spip/IMG/pdf/IRGC_WP_No_1_Risk_Governance_(reprinted_version).pdf

Table 2 Components of pre-assessment in handling risks.

Pre-assessment Components	Definition	Indicators
1 Problem framing	Different perspectives of how to conceptualise the issue	• dissent or consent on goals of selection rule • dissent or consent on relevance of evidence • choice of frame (risk, opportunity, fate)
2 Early warning	Systematic search for new hazards	• unusual events or phenomena • systematic comparison between modelled and observed phenomena • novel activities or events
3 Screening (risk assessment and concern assessment policy)	Establishing a procedure for screening hazards and risks and determining assessment and management route	• screening in place? • criteria for screening: – hazard potential – persistence – ubiquity, etc. • criteria for selecting risk assessment procedures for: – known risks – emergencies, etc. • criteria for identifying and measuring social concerns
4 Scientific conventions for risk assessment and concern assessment	Determining the assumptions and parameters of scientific modelling and evaluating methods and procedures for assessing risks and concerns	• definition of no adverse effect levels (NOAEL) • validity of methods and techniques for risk assessments • methodological rules for assessing concerns

As we have seen before, it is crucial to distinguish between hazards and risks. Correspondingly, **identification** (i.e. establishing cause-effect link) and **estimation** (determining the strength of the cause-effect link) need to be performed for hazards and risks separately. The estimation of risk depends on an exposure and/or vulnerability assessment. **Exposure** refers to the contact of the hazardous agent with the target (individuals, ecosystems, buildings, etc.). **Vulnerability** describes the various degrees of the target to experience harm or damage as a result of the exposure (for example: immune system of target population, vulnerable groups, structural deficiencies in buildings, etc.). In many cases it is common practice to combine hazard and risk estimates in scenarios that allow modellers to change parameters and include different sets of context constraints.

The basis of risk assessment is the systematic use of analytical – largely probability-based – methods which have been constantly improved over the past years. Probabilistic risk assessments for large technological systems, for instance, include tools such as fault and event trees, scenario techniques, distribution models based on Geographic Information Systems (GIS), transportation modelling and empirically driven human-machine interface simulations (IAEA 1995; Stricoff 1995). With respect to human health, improved methods of modelling inter-individual variation (Hattis 2004), dose-response relationships (Olin et al. 1995) and exposure assessments (USEPA 1997) have been developed and successfully applied. The processing of data is often guided by inferential statistics and organised in line with decision analytic procedures. These tools have been developed to generate knowledge about cause-effect relationships, estimate the strength of these relationships, characterise remaining uncertainties and ambiguities and describe, in quantitative or qualitative form, other risk or hazard related properties that are important for risk management (IAEA 1995; IEC 1993). In short, risk assessments specify what is at stake, calculate the probabilities for (un)wanted consequences, and aggregate both components into a single dimension (Kolluru 1995: 2.3f). In general, there are five methods for calculating probabilities:

- collection of statistical data relating to the performance of a risk source in the past (actuarial extrapolation);
- collection of statistical data relating to components of a hazardous agent or technology. This method requires a synthesis of probability judgements from component failure to system performance (probabilistic risk assessments, PRA);
- epidemiological or experimental studies which are aimed at finding statistically significant correlations between an exposure of a hazardous agent and an adverse effect in a defined population sample (probabilistic modelling);
- experts', or decision makers' best estimates of probabilities, in particular for events where insufficient statistical data is available (normally employing Bayesian statistical tools);
- scenario techniques by which different plausible pathways from release of a harmful agent to the final loss are modelled on the basis of worst and best cases or estimated likelihood for each consequence at each knot).

All these methods are based either on the past performance of the same or a similar risk source or an experimental intervention. The possibility that the circumstances of the risk situation vary over time in an unforeseeable way and that people will thus make decisions in relation to changing hazards – sometimes they may even change in an unsystematic, unpredictable manner – leads to unresolved or remaining uncertainty (second order uncertainty). One of the main challenges of risk assessment is the systematic characterisation of these remaining uncertainties. They can partly be modelled by using inferential statistics (confidence interval) or other simulation methods (such as Monte Carlo), but often they can only be described in qualitative terms. Risk analysts consequently distinguish between *aleatory and epistemic un-*

certainty: epistemic uncertainty can be reduced by more scientific research[8] while aleatory uncertainty will remain fuzzy regardless of how much research is invested in the subject (Shome et al. 1998). Remaining uncertainties pose major problems in the later stages of risk characterisation and evaluation as well as risk management since they are difficult to integrate in formal risk-benefit analyses or in setting standards.

There is no doubt that risk assessment methods have matured to become sophisticated and powerful tools in coping with the potential harm of human actions or natural events (Morgan 1990). Its worldwide application, however, in dealing and managing risks is far from reflecting this degree of power and professionalism. At the same time, there are new challenges in the risk field that need to be addressed by the risk assessment communities. These challenges refer to (cf. Brown and Goble 1990; Hattis and Kennedy 1990; Greeno and Wilson 1995; Renn 1997):

- widening the scope of effects for using risk assessment, including chronic diseases (rather than focusing only on fatal diseases such as cancer or heart attack); risks to ecosystem stability (rather than focusing on a single species); and the secondary and tertiary risk impacts that are associated with the primary physical risks;
- addressing risk at a more aggregated and integrated level, such as studying synergistic effects of several toxins or constructing a risk profile over a geographic area that encompasses several risk causing facilities;
- studying the variations among different populations, races, and individuals and getting a more adequate picture of the ranges of sensibilities with respect to environmental pollutants, lifestyle factors, stress levels, and impacts of noise;
- integrating risk assessment in a comprehensive technology assessment or option appraisal so that the practical value of its information can be phased into the decision-making process at the needed time and that its inherent limitations can be compensated through additional methods of data collection and interpretation; and
- developing more forgiving technologies that tolerate a large range of human error and provide sufficient time for initiating counteractions.

Table 3 lists the three generic components of risk assessment and provides an explanation for the terms as well as a summary list of indicators that can be used in the different risk contexts for performing the respective task. As with Table 2, the choice of indicators is not exhaustive and serves the purpose of illustrating the type of information needed to perform the task described in each step. The three components are normally performed sequentially but, depending on circumstances, the order may be changed. Often, exposure assessments are done before hazards are

---

[8] There are many tools available to model epistemic uncertainty. The Dutch guidance document on uncertainty assessment and communication lists the following tools: sensitivity analysis, error propagation methods, Monte Carlo Analysis, NUSAP (numeral, unit, spread, assessment, pedigree), expert elicitation, scenario analysis, PRIMA (pluralistic framework of integrated uncertainty management and risk analysis) and checklists for model quality assistance (van der Sluijs et al. 2004).

Table 3 Generic components of risk assessment.

Assessment Components	Definition	Indicators
1 Hazard identification and estimation	Recognising potential for adverse effects and assessing the strength of cause-effect relationships	• properties such as flammability, etc. – persistence – irreversibility – ubiquity – delayed effects – potency for harm • dose-response relationships
2 Exposure/vulnerability assessment	Modelling diffusion, exposure and effects on risk targets	• exposure pathways • normalised behaviour of target • vulnerability of target
3 Risk estimation	• *Quantitative*: probability distribution of adverse effects • *Qualitative*: combination of hazard, exposure, and qualitative factors (scenario construction)	• expected risk value(s) (individual, collective) • xx% confidence interval • risk description • risk modelling as function of variations in context variables and parameters

estimated. If, for example, exposure can be prevented, it may not be necessary to perform any sophisticated hazard estimate.

Generic Challenges for Risk Assessment

Risk assessment is confronted with three major challenges that can be best described using the terms 'complexity', 'uncertainty' and 'ambiguity'. These three challenges are not related to the intrinsic characteristics of hazards or risks themselves but to the *state and quality of knowledge available* about both hazards and risks. Since risks are mental constructs, the quality of their explanatory power depends on the accuracy and validity of their (real) predictions. Unlike some other scientific constructs, validating the results of risk assessments is particularly difficult because, in theory, one would need to wait indefinitely to prove that the probabilities assigned to a specific outcome were correctly assessed. If the number of predicted events is frequent and the causal chain obvious (as is the case with car accidents), validation is relatively simple and straightforward. If, however, the assessment focuses on risks where cause-effect relationships are difficult to discern, where effects are rare and/or difficult to interpret, and where variations in both causes and effects are obscuring the results, the validation of the assessment results becomes a major problem. In such instances, assessment procedures are needed to characterise the existing knowledge with respect to complexity, remaining uncertainties and ambiguities (WBGU 2000, 195ff.; Klinke and Renn 2002).

- **Complexity** refers to the difficulty of identifying and quantifying causal links between a multitude of potential causal agents and specific observed effects. The nature of this difficulty may be traced back to interactive effects among these agents (synergism and antagonisms), long delay periods between cause and effect, inter-individual variation, intervening variables, and others. Risk assessors have to make judgements about the level of complexity that they are able to process and about how to treat intervening variables (such as lifestyle, other environmental factors, psychosomatic impacts, etc.). Complexity is particularly pertinent in the phase of estimation with respect to hazards as well as risks. Examples of highly complex risk include sophisticated chemical facilities, synergistic effects of potentially toxic substances, failure risk of large interconnected infrastructures and risks of critical loads to sensitive ecosystems.
- **Uncertainty** is different from complexity but often results from an incomplete or inadequate reduction of complexity in modelling cause-effect chains. Whether the world is inherently uncertain is a philosophical question that we will not pursue here. It is essential to acknowledge in the context of risk assessment that human knowledge is always incomplete and selective and thus contingent on uncertain assumptions, assertions and predictions (Functowicz and Ravetz 1992; Laudan 1996; Bruijn and ten Heuvelhof 1999). It is obvious that the modelled probability distributions within a numerical relational system can only represent an approximation of the empirical relational system with which to understand and predict uncertain events (Cooke 1991). It therefore seems prudent to include other, additional, aspects of uncertainty (Morgan and Henrion 1990; van Asselt 2000: 93–138; van der Sluijs et al. 2003). Although there is no consensus in the literature on the best means of disaggregating uncertainties, the following categories appear to be an appropriate means of distinguishing the key components of uncertainty:
  - *target variability* (based on different vulnerability of targets);
  - *systematic and random error in modelling* (based on extrapolations from animals to humans or from large doses to small doses, statistical inferential applications, etc.);
  - *indeterminacy or genuine stochastic effects* (variation of effects due to random events, in special cases congruent with statistical handling of random errors);
  - *system boundaries* (uncertainties stemming from restricted models and the need for focusing on a limited amount of variables and parameters);
  - *ignorance or non-knowledge* (uncertainties derived from lack or absence of knowledge).

  The first two components of uncertainty qualify as epistemic uncertainty and therefore can be reduced by improving the existing knowledge and by advancing the present modelling tools. The last three components are genuine uncertainty components of aleatory nature and thus can be characterised to some extent using scientific approaches but cannot be further resolved. If uncertainty, in particular the aleatory components, plays a large role then the estimation of risk becomes fuzzy. The validity of the end results is questionable and, for risk management purposes, additional information is needed such as a subjective confidence level in the risk estimates, potential alternative pathways of cause-effect relationships, ranges of reasonable estimates, loss scenarios and others. Examples for high uncertainty, particularly aleatory uncertainty, include many natural disasters such as earthquakes, possible health effects of mass pollutants below the threshold of statistical significance, acts of violence such as terrorism and sabotage and long-term effects of introducing genetically modified species into the natural environment.

- (Interpretative and normative) ambiguity is the last term in this context. Whereas uncertainty refers to a lack of clarity over the scientific or technical basis for decision-making (interpretative and normative) ambiguity is a result of divergent or contested perspectives on the justification, severity or wider 'meanings' associated with a given threat (Stirling 2003). The term 'ambiguity' may be misleading because it has different connotations in everyday English language.[9] In relation to risk governance it is understood as 'giving rise to several meaningful and legitimate interpretations of accepted risk assessments results'. It can be divided into **interpretative ambiguity** (different interpretations of an identical assessment result: e.g. as an adverse or non-adverse effect) and *normative ambiguity* (different concepts of what can be regarded as tolerable referring e.g. to ethics, quality of life parameters, distribution of risks and benefits, etc.). A condition of ambiguity emerges where the problem lies in agreeing on the appropriate values, priorities, assumptions, or boundaries to be applied to the definition of possible outcomes. What does it mean, for example, if neuronal activities in the human brain are intensified when subjects are exposed to electromagnetic radiation? Can this be interpreted as an adverse effect or is it just a bodily response without any health implication? Many scientific disputes in the fields of risk assessment and management do not refer to differences in methodology, measurements or dose-response functions, but to the question of what all of this means for human health and environmental protection. High complexity and uncertainty favour the emergence of ambiguity, but there are also quite a few simple and highly probable risks that can cause controversy and thus ambiguity. Examples for high interpretative ambiguity include low dose radiation (ionising and non-ionising), low concentrations of genotoxic substances, food supplements and hormone treatment of cattle. Normative ambiguities can be associated, for example, with pass-

[9] With respect to risk and decision-making the term ambiguity has been used with various meanings. Some analysts refer to ambiguity as the conflicting goals of participants in the process (Skinner 1999), others use the term ambiguity when they refer to the inability to estimate probabilities of an event occurring (Gosh and Ray 1997: Ho et al. 2002; Stirling 2003). In the context of the present framework ambiguity denotes the variability in interpretation and normative implications with respect to accepted evidence.

ive smoking, nuclear power, pre-natal genetic screening and genetically modified food.

Risk Perception

Since risk is a mental construct there are a wide variety of construction principles for conceptualising risk. Different disciplines within the natural and social sciences have formed their own concepts of risk; stakeholder groups, driven by interest and experience, have developed their specific perspective on risk; and, last but not least, representatives of civil society as well as the general public are responding to risks according to their own risk constructs and images. These images are called '**perceptions**' in the psychological and social sciences and they have been intensely researched in relation to risk – as have their underlying factors (Covello 1983; Slovic 1987; Boholm 1998; Rohrmann and Renn 2000). Risk perceptions belong to the contextual aspects that risk managers need to consider when deciding whether or not a risk should be taken as well as when designing risk reduction measures.

First of all it is highly important to know that human behaviour is primarily driven by perception and not by facts or by what is understood as facts by risk analysts and scientists. Most cognitive psychologists believe that perceptions are formed by common sense reasoning, personal experience, social communication and cultural traditions (Brehmer 1987; Drottz-Sjöberg 1991; Pidgeon et al. 1992; Pidgeon 1998). In relation to risk it has been shown that humans link certain expectations, ideas, hopes, fears and emotions with activities or events that have uncertain consequences. People do, however, not use completely irrational strategies to assess information, but, most of the time, follow relatively consistent patterns of creating images of risks and evaluating them. These patterns are related to certain evolutionary bases of coping with dangerous situations. Faced with an eminent threat, humans react with four basic strategies: *flight*, *fight*, *play dead* and, if appropriate, *experimentation* (on the basis of trial and error).

In the course of cultural evolution the basic patterns of perception were increasingly enriched with cultural patterns. These cultural patterns can be described by so-called *qualitative evaluation characteristics* (Slovic 1992). They describe properties of risks or risky situations going beyond the two classical factors of risk assessment based on which risk is usually judged, i.e. level of probability and degree of possible harm. Here, psychologists differentiate between two classes of qualitative perception patterns: on the one hand *risk-related patterns*, which are based on the properties of the source of risk; on the other hand *situation-related patterns*, based on the idiosyncrasies of the risky situation (Fischhoff et al. 1978; Slovic 1987, 1992).

One example of a risk-related pattern is the perceived 'dread' of the consequences of a possible harmful event. If, for example, a person is riding in a car and thinking about possible accidents, s/he will always be under the impression s/he would, with high probability, get away unscathed in a car accident ('fender-bender mentality'). However, if the same person is sitting in an airplane s/he will be under the impression that if something happens here there is no getting away. This feeling of apprehensiveness does not subside even when this person knows the odds and is convinced that statistically many more people die in car accidents than in airplane crashes.

Situation-related patterns of perception include aspects such as voluntariness and the ability to exercise self-control. If a person is of the opinion that s/he can control the risk, then s/he will perceive it as less serious. This mode of thinking frequently takes effect where eating habits are concerned. People believe they can easily do without sweets, alcohol or other food considered unhealthy, if only they wanted to. However, mostly harmless chemical food additives are perceived as a threat to one's health. With respect to collective risks, people show special concern for risks that they believe are not adequately controlled by public authorities (as in the case of GMOs).

Considered together these qualitative evaluation characteristics can be subdivided into a limited number of consistent risk perception classes. In literature they are also called **semantic risk patterns**. The following patterns were examined particularly thoroughly (Renn 2004a):

- risks posing an immediate threat such as nuclear energy or large dams;
- risks dealt with as a twist of fate such as natural disasters;
- risks presenting a challenge to one's own strength such as sports activities;
- risk as a gamble such as lotteries, stock exchanges, insurances;
- risks as an early indication of insidious danger such as food additives, ionising radiation, viruses.

These patterns have functions similar to drawers in a filing cabinet. When faced with a new risk or when obtaining new information about a risk, most people try to file this new information into one of the existing drawers.[10] In addition to the cognitive processing of risk characteristics and risk situations, studies have shown that people tend to *stigmatise risk sources* that are associated with specific dreadful associations (Kunreuther and Heal 2003). A salient example of stigma is the reaction to products that are deemed to be carcinogenic, although there is often limited, if any, scientific evidence to support this position. The mere suspicion that a substance could cause cancer is often sufficient for generating fear and asking for strict regulatory actions. Stigmatisation leads to a cycle of public outrage and regulatory responses feeding into the process that has been described as social amplification of risk (Kasperson et al. 1988, 2003). Stimulated by media reporting, the public's perception of the risk is often amplified in ways that are difficult to explain if one were focusing on the standard elements of any technical risk assessment – probability and direct losses.

The problems associated with risk perception are compounded because of the difficulty individuals have in interpreting low probabilities when making their decisions (Kunreuther et al. 2001). In fact, there is evidence that people may not even

[10] The 'drawers' cannot be treated in detail here since this would exceed the scope of this document (more information in Streffer et al. 2003: 269ff.).

want data on the likelihood of an event occurring. If people do not think probabilistically, how do they make their choices? Psychological research has revealed the following patterns of drawing inferences about probabilities and risks (Tversky and Kahneman 1974; Ross 1977; Kahneman and Tversky 1979; Renn 2004a):

- The easier and faster a risk is recognised, the more conscious individuals are of it and the greater is the chance of its probability being overestimated. If, for example, an individual has known someone who died after being struck by lightning, that individual will perceive the risk of being struck by lightning as being particularly large (*availability bias*).
- The more a risk provokes associations with known events, the more likely its probability will be overestimated. This is why, for example, the use of the term 'incinerating' in waste disposal facilities readily evokes an association with harmful chemicals, especially dioxins and furans, even if there is no way that they could be released into the environment by the facilities concerned (*anchoring effect*).
- The more constant and similar the losses from risk sources, the more likely the impact of average losses will be underestimated. While road traffic accidents are not deemed acceptable, they are more or less passively accepted. If the average annual number of road deaths in a given country were to occur at one point in time instead of being spread out over the year, then a considerably greater level of rejection could be expected. Thus, people are not indifferent as regards the *distribution of risks over time*: they prefer even loss distribution over individual disasters (Kahneman and Tversky 1979).
- The greater the uncertainty of loss expectation, the more likely the average loss assessment will be in the region of the median of all known loss expectations. In this way, loss expectations in objectively low risks are often overestimated while objectively high risks are often underestimated (*assessment bias*).

While important for actually evaluating and managing a risk, overestimation or underestimation of loss expectations is not, however, the most important aspect of risk perception. Instead the context-dependent nature of risk assessment is the deciding factor. This context includes the qualitative risk evaluation characteristics, the semantic images and the stigma effects. More recently, psychologists have also discovered that affect and emotions play an important role in people's decision processes (Slovic et al. 2002; Loewenstein et al. 2001). These factors are particularly relevant when individuals face a decision that involves a difficult trade-off between attributes or where there is interpretative ambiguity as to what constitutes a 'right' answer. In these cases, people often appear to resolve problems by focusing on those cues that send the strongest affective signals (Hsee and Kunreuther 2000).

The most important policy question is how to treat risk perceptions in a policy arena that includes responses of different actors and the general public (Slovic et al. 1982; Fischhoff 1985, 1995). There are two suggestions, from opposite ends of a spectrum. The first position states that the scientific concepts of risk are the only ones that can claim inter-subjective validity and applicability and, therefore, requires risk managers to obtain an assurance that (erroneous) risk perceptions are corrected via risk communication and education (Cross 1998; Coglianese 1999). The second position states that there is no overarching universally applicable quality criterion available in order to evaluate the appropriateness or validity of risk concepts. As a result, scientific concepts (often called *narratives* in this school of thought) should compete with concepts of stakeholders and public groups (Liberatore and Funtowicz 2003). If collective decisions on risk are necessary, the concept that is used to make these decisions should be negotiated among all relevant concept holders. None of these groups, including the science communities, is allowed to claim any privileged position in this negotiation.

IRGC has strong reservations with respect to both positions. IRGC advocates an approach by which the elements of what matters to the different groups when they conceptualise risk should be regarded as equally legitimate factors for inclusion within risk governance (see also Gigerenzer and Selten 2001). This implies, for example, that if people are willing to accept higher risks when they are in control of them, then this preference cannot be de-legitimised by professional economists who favour cost-effectiveness studies that treat all risks equally. In identifying aspects of concern and worry all groups in society have the same right to raise them and to bring them to the negotiation table. However, the question of the degree to which these concerns are met or violated by risk-bearing activities or events should be primarily answered by those who have the knowledge, skills and/or the experience to measure or estimate the strength of relationships between cause (or dose) and effect. It seems wrong to give equal standing to those who intuitively estimate risks and those who assess risks on the basis of systematic observation, empirical data collection and rigorous modelling, just as it seems wrong to dismiss non-factual perceptions purely because they appear irrational to those with expert knowledge. This said, IRGC wishes to emphasise that the proposed quality distinction between intuition and systematic knowledge does not predetermine a position in the philosophical debate on realism versus constructivism; the argument here is strictly focused on the structure and content of knowledge claims, not on claims about reality representation.[11]

This position has major impacts on risk policy making and communication. Policy making needs to, inter alia, organise systematic feedback from society and, equally, to include risk perceptions as an important input to deciding on whether something should be done about a certain risk and, if so, what (Jaeger et al. 2001). How this can be accomplished is explained in the next section on risk appraisal. Risk communication is also affected, in two ways: first, it is bound to elicit, and enable the exchange of, concerns and conceptual aspects of risk among and between all relevant actors, and, second, risk managers are well advised to ensure that the best available knowledge is widely distributed to those who raise these concerns.

[11] For more comprehensive arguments on this debate see footnote 4.

Risk Appraisal

The term *risk appraisal* has sometimes been used in the risk governance literature to include all knowledge elements necessary for risk characterisation and evaluation as well as risk management (Stirling 1998, 2003). For society to make prudent choices about risks, it is not enough to consider only the results of (scientific) risk assessment. In order to understand the concerns of the various stakeholders and public groups, information about both risk perceptions and the further implications of the direct consequences of a risk – including its social mobilisation potential (i.e. how likely is it that the activity will give rise to social opposition or protest?) – is needed and should be collected by risk management agents. In addition, other aspects of the risk causing activity that seem to be relevant for characterising and evaluating the risk and selecting risk reduction options should be pulled together and fed into the analysis. Based on such a wide range of information, risk managers can make more informed judgements and design the appropriate risk management options (Clark 2001).

Risk appraisal thus includes the scientific assessment of the risks to human health and the environment and an assessment of related concerns as well as social and economic implications. The appraisal process is and should be clearly dominated by scientific analyses – but, in contrast to the traditional risk governance model, the scientific process includes both the natural/technical as well as the social sciences, including economics. We envision risk appraisal as having two process stages: first, natural and technical scientists use their skills to produce the best estimate of the physical harm that a risk source may induce (as described in the chapter on risk assessment); secondly, social scientists and economists identify and analyse the issues that individuals or society as a whole link with a certain risk. For this purpose the repertoire of the social sciences such as survey methods, focus groups, econometric analysis, macro-economic modelling, or structured hearings with stakeholders may be used.

Based on the results of risk assessment and the identification of individual and social concerns this second process stage also investigates and calculates the *social and economic implications of risks*. Of particular interest in this context are financial and legal implications, i.e. economic losses and liabilities, as well as social responses such as political mobilisation. These secondary implications have been addressed by the concept of *social amplification of risk* (Kasperson et al. 2001, 2003). This concept is based on the hypothesis that events pertaining to hazards interact with psychological, social, institutional, and cultural processes in ways that can heighten or attenuate individual and social perceptions of risk and shape risk behaviour. Behavioural patterns, in turn, generate secondary social or economic consequences that extend far beyond direct harm to human health or the environment, including significant indirect impacts such as liability, insurance costs, loss of confidence in institutions, or alienation from community affairs (Burns et al. 1993). Such amplified secondary effects can then trigger demands for additional institutional responses and protective actions, or, conversely (in the case of risk attenuation), place impediments in the path of needed protective actions. Secondary impacts, whether amplified or not, are of major concern to those who are obliged to take over the costs or cope with the consequences of being accountable.

Risk appraisal intends to produce the best possible scientific estimate of the physical, economic and social consequences of a risk source. It should not be confused with direct stakeholder involvement which will be covered later. Involvement by stakeholders and the population is only desirable at this stage if knowledge from these sources is needed to improve the quality of the assessments.

In a recent draft document published by the UK Treasury Department (2004) the authors recommend a risk appraisal procedure that includes the results of risk assessment, the direct input from data on public perception and the assessment of social concerns. The document offers a tool for evaluating public concerns using six factors related to the hazard(s) leading to a risk, the risk's effects and its management:[12]

- perception of familiarity and experience with the hazard;
- understanding the nature of the hazard and its potential impacts;
- repercussions of the risk's effects on equity (inter-generational, intra-generational, social);
- perception of fear and dread in relation to a risk's effect;
- perception of personal or institutional control over the management of a risk;
- degree of trust in risk management organisations.

A similar list of appraisal indicators was suggested by a group of Dutch researchers and the Dutch Environmental Protection Agency (van der Sluijs et al. 2003, 2004). In the late 1990s, the German Council for Global Environmental Change (WBGU) has also addressed the issue of risk appraisal and developed a set of eight criteria to characterise risks beyond the established assessment criteria (WBGU 2000). These are:

- *Extent of damage*: Adverse effects in natural units, e.g. death, injury, production loss, etc.
- *Probability of occurrence*: Estimate of relative frequency, which can be discrete or continuous.
- *Incertitude*: How do we take account of uncertainty in knowledge, in modelling of complex systems or in predictability in assessing a risk?
- *Ubiquity*: Geographical dispersion of damage.
- *Persistence*: How long will the damage last?
- *Reversibility*: Can the damage be reversed?
- *Delay effects*: Latency between initial event and actual damage.
- *Potential for mobilisation*: The broad social impact. Will the risk generate social conflict or outrage, etc.?

After the WBGU proposal had been reviewed and discussed by many experts and risk managers, it was suggested to unfold the compact 'mobilisation index' and divide it into four major elements:

[12] Since the document has only been released in late 2004, reports about practical experiences regarding its implementation are not yet available.

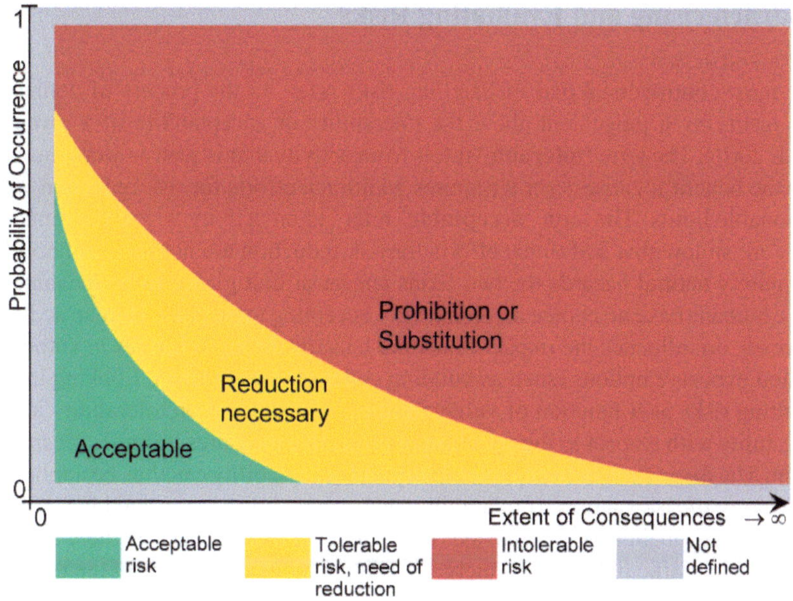

Fig. 2 Acceptable, tolerable and intolerable risks (traffic light model).

- *Inequity and injustice* associated with the distribution of risks and benefits over time, space and social status.
- *Psychological stress and discomfort* associated with the risk or the risk source (as measured by psychometric scales).
- *Potential for social conflict and mobilisation* (degree of political or public pressure on risk regulatory agencies).
- *Spill-over effects* that are likely to be expected when highly symbolic losses have repercussions on other fields such as financial markets or loss of credibility in management institutions.

These four sub-criteria reflect many factors that have been proven to influence risk perception as stated above.[13]

When dealing with complex, uncertain and/or ambiguous risks it is essential to complement data on physical consequences with data on secondary impacts, including social responses to risk, and insights into risk perception. The suggestions listed above can provide some orientation for the criteria to be considered. Depending on the risk under investigation, additional criteria can be included or proposed criteria neglected.

[13] A similar decomposition has been proposed by the UK government (Environment Agency 1998; Pollard et al. 2000).

Characterising and Evaluating Risks

The most controversial part of handling risks refers to the process of delineating and justifying a judgement about the tolerability or acceptability of a given risk (HSE 2001). The term '**tolerable**' refers to an activity that is seen as worth pursuing (for the benefit it carries) yet it requires additional efforts for risk reduction within reasonable limits. The term '**acceptable**' refers to an activity where the remaining risks are so low that additional efforts for risk reduction are not seen as necessary. For purely natural hazards the two terms appear at first glance to be meaningless, since humans have no choice in tolerating or accepting these risks. Human activities, however, do influence the impact of natural hazards through changes in vulnerability and exposure options (such as building codes or zoning laws). Looking into the resulting risks as a function of vulnerabilities, a judgement on tolerability and acceptability with respect to the selection of protective measures becomes meaningful again. The distinction between tolerability and acceptability can thus be applied to a large array of risk sources. If tolerability and acceptability are located in a risk diagram (with probabilities on the y-axis and extent of consequences on the x-axis), the well-known traffic light model emerges[14] (Figure 2). In this variant of the model the red zone signifies intolerable risk, the yellow one indicates tolerable risk in need of further management actions (in accordance with the 'as low as reasonably possible' – ALARP – principle) and the green zone shows acceptable or even negligible risk.

To draw the line between 'intolerable' and 'tolerable' as well as 'tolerable' and 'acceptable' is one of the most difficult tasks of risk governance. The UK Health and Safety Executive has developed a procedure for chemical risks based on risk-risk comparisons (Löfstedt 1997). Some Swiss cantons such as Basle County experimented with Round Tables as a means to reach consensus on drawing the two lines, whereby participants in the Round Table represented industry, administrators, county officials, environmentalists, and neighbourhood groups (RISKO 2000). Irrespective of the selected means to support this task, the judgement on acceptability or tolerability is contingent on making use of a variety of different knowledge sources. One needs to include the risk estimates derived from the risk assessment stage, and additional assessment data from the concern assessment within the appraisal stage.

Existing taxonomies of risk differ considerably in where they position the decision-making with regard to what is acceptable and what is tolerable within the overall risk process. Some assign it to the risk assessment part, others to the risk management part and others place it at the level of policy and option assessment, reaching far beyond the narrow risk acceptance criteria. For the generic approach to risk handling that this document pursues, the question of appropriate placement should be handled in a flexible manner.

[14] The traffic light model in this context is an illustrative means of mapping risks according to their tolerability or acceptability. The same metaphor has also been used to map the degree of controversy or normative ambiguity, for example in the area of siting mobile base stations (Kemp 1998; Kemp and Greulich 2004). The criticism that has been raised against using the traffic light model for addressing opposition to base stations is not relevant to the application of this model in the context of risk characterisation and evaluation.

Why? As with the framing part, judgements on acceptability rely on two major inputs: *values and evidence*. What society is supposed to tolerate or accept can never be derived from looking at the evidence alone. Likewise, evidence is essential if we are to know whether a value has been violated or not (or to what degree). With respect to values and evidence we can distinguish three cases: (i) ambiguity on evidence but not on values (interpretative ambiguity) (ii) ambiguity on values but not on evidence (normative ambiguity) and (iii) ambiguities on values and evidence.

Case 1: Interpretative ambiguity. In those cases where there is unanimous agreement about the underlying values and even the threshold of what is regarded as tolerable or acceptable, evidence in the form of risk estimates may be sufficient to locate the risk within the traffic light diagram. A judgement can then best be made by those who have most expertise in risk and concern assessments, in which case it makes sense to place this task within the domain of risk appraisal. The judgement will thus be based on best scientific modelling of epistemic uncertainties and the best qualitative characterisation of aleatory uncertainties. Characterisation also includes an analysis of the concerns associated with different outcomes and the likely secondary implications. It will be helpful for risk managers to receive best expert advice on potentially effective risk reduction measures and other management options that may lead to satisfactory results. It is, however, not the task of the risk appraisal team to make a selection of options, let alone decide on which option should be implemented.

Leaving the resolution of interpretative ambiguity to the risk and concern assessors places a major challenge to the science-based assessment process. It may be extremely difficult for experts to find an agreement on interpreting ambiguous results. It is not uncommon for the public to hear expert 1 say that there is 'nothing to worry about regarding a particular risk' while at the same time learning from expert 2 that 'this risk should be on your radar screen'. One way to capture these discrepancies in risk interpretations is to construct an *exceedance probability (EP) curve* (Grossi and Kunreuther 2005). An EP curve specifies the probabilities that certain level of losses will be exceeded. The losses can be measured in terms of dollars of damage, fatalities, illness or some other unit of analysis.

To illustrate with a specific example, suppose one was interested in constructing an EP curve for dollar losses to homes in Seattle from an earthquake. Using probabilistic risk assessment, one combines the set of events that could produce a given dollar loss and then determines the resulting probabilities of exceeding losses of different magnitudes. Based on these estimates, one can construct the mean EP depicted in Figure 3. By its nature, the EP curve inherently incorporates uncertainty associated with the probability of an event occurring and the magnitude of dollar losses. This uncertainty is reflected in the 5% and 95% confidence interval curves in the figure.

The EP curve also serves as an important tool for evaluating risk management options, thus assisting managers to optimise risk reduction. It puts pressure on experts to state the assumptions on which they are basing their estimates of the likelihood of certain events occurring and the resulting consequences. In fact, EP curves, such as those depicted in Figure 3, supplemented by a discussion of the nature of these

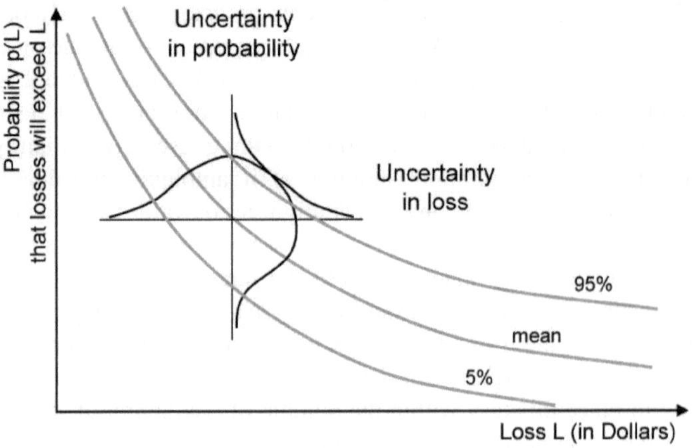

Fig. 3 Example of loss exceedance probability curves.

assumptions, should enable the assessors to both characterise interpretative ambiguities and to provide a framework for risk managers to test the efficiency of risk reduction options.

Case 2: Normative ambiguity. If the underlying values of what could be interpreted as tolerable or acceptable are disputed, while the evidence of what is at stake is clearly given and non-controversial, the judgement needs to be based on a discourse about values and their implications. Such a discourse falls clearly in the domain of risk management. A good example may be the normative implications of risks related to smoking. Science is very familiar with these risks and there is little uncertainty and interpretative ambiguity about dose-effect relationships. Yet there is considerable debate whether smoking is tolerable or not. Being a voluntary activity some countries leave it to the decision of each consumer while others initiate major activities to reduce and even ban smoking. Another example is wearing helmets on bicycles. The statistical data on this subject is rather straightforward; there are no major uncertainties or interpretative ambiguities. Yet many countries do dot want to impinge on the freedom of each cyclist to personally decide whether or not to wear a helmet, while other countries pursue a more paternalistic policy.

Case 3: Interpretative and normative ambiguity. A third case arises where both the evidence and the values are disputed. This would imply that assessors should engage in an activity to find some common ground for characterising and qualifying the evidence and risk managers need to establish agreement about the appropriate values and their application. A good example for this third case may be the interpretative and normative implications of global climate change. An international expert group such as the Intergovernmental Panel on Climate Change (IPCC) has gone through considerable effort to articulate a common characterisation of climatic risks and their uncertainties. Given the remaining uncertainties and the complexities of the causal relationships between greenhouse gases and climate change, it is then a question of values whether governments place their priorities on prevention or on mitigation (Keeney and McDaniels 2001).

Since the third of the above cases includes both of the other two, the process of judging the tolerability and acceptability of a risk can be structured into two distinct components: risk characterisation and risk evaluation. The first step, '**risk characterisation**', determines the evidence-based component for making the necessary judgement on the tolerability and/or acceptability of a risk; the step '**risk evaluation**' determines the value-based component for making this judgement. Risk characterisation includes tasks such as point estimates of risks, descriptions of remaining uncertainties (as undertaken for instance in climate change models or risk studies on endocrine disruptors) and potential outcome scenarios including the social and economic implications, suggestions for safety factors to include inter-target variation, assurance of compatibility with legal prescriptions, risk-risk comparisons, risk-risk trade-offs, identification of discrepancies between risk assessment and risk perceptions as well as of potential equity violations, and suggestions for reasonable standards to meet legal requirements (Stern and Fineberg 1996). The evidence collected and summarised here goes beyond the classic natural science reservoir of knowledge and includes economic and social science expertise. This is also the reason why in the process of risk characterisation an interdisciplinary team of scientists is needed to draw a complete picture of what is known and what is and may remain unknown. In the course of risk characterisation, scientists are asked to design a multi-criteria profile of the risk in question, make a judgement about the seriousness of the risk and suggest potential options to deal with the risk.

The second step, risk evaluation, broadens the picture to include pre-risk aspects such as choice of technology, social need for the specific risk agent (substitution possible?), risk-benefit balances, political priorities, potential for conflict resolution and social mobilisation potential. The main objective here is to arrive at a judgement on tolerability and acceptability based on balancing pros and cons, testing potential impacts on quality of life, discussing different development options for the economy and society and weighing the competing arguments and evidence claims in a balanced manner. It should be noted that this elaborate procedure is only necessary if tolerability and/or acceptability is disputed and if society faces major dissents and conflicts among important stakeholders. If so, the direct involvement of stakeholders and the public will be a prerequisite for successful risk governance.

The separation of evidence and values underlying the distinction between characterisation and evaluation is, of course, functional and not necessarily organisational. Since risk characterisation and evaluation are closely linked and each depends on the other, it may even be wise to perform these two steps simultaneously in a joint effort by both assessors and risk managers. As some analysts have pointed out (Löfstedt and Vogel 2001; Vogel 2003): the US regulatory system tends to favour an organisational combination of characterisation and evaluation, while European risk managers tend to maintain the organisational separation (particularly in the food area). IRGC takes no stance in this question: there are good reasons for both models, yet IRGC does insist on a functional distinction.

The distinction between the three challenges of risk assessment, i.e. complexity, uncertainty and ambiguity, can also assist assessors and managers in assigning, or dividing, the judgement task. If a given risk is characterised by high complexity, low remaining uncertainties and hardly any ambiguities (except for interpretative differences over an established scientific risk assessment result), it is wise to let the assessment team dominate the process of making tolerability/acceptability judgements. If, in contrast, the risk is characterised by major unresolved uncertainties and if the results lead to highly diverse interpretations of what they mean for society, it is advisable to let risk managers take the lead.

Table 4 summarises these two steps which, as we have indicated, can be closely interrelated and may be merged if the circumstances require it. The list of indicators again represents only a small selection of potential dimensions and is displayed here for illustrative purposes.

Risk Management

Risk management starts with a review of all relevant information, in particular that from the combined risk appraisal, consisting of both a risk assessment and concern assessment whereby the latter is based on risk perception studies, economic impact assessments and the scientific characterisation of social responses to the risk source. This information, together with the judgements made in the phase of risk characterisation and evaluation, form the input material on which risk management options are being assessed, evaluated and selected. At the outset, risk management is presented with three potential outcomes:

- *Intolerable situation*: this means that either the risk source (such as a technology or a chemical) needs to be abandoned or replaced or, in cases where that is not possible (for example natural hazards), vulnerabilities need to be reduced and exposure restricted.
- *Tolerable situation*: this means that the risks need to be reduced or handled in some other way within the limits of reasonable resource investments (ALARP, including best practice). This can be done by private actors (such as corporate risk managers) or public actors (such as regulatory agencies) or both (public-private partnerships).
- *Acceptable situation*: this means that the risks are so small – perhaps even regarded as negligible – that any risk reduction effort is unnecessary. However, risk sharing via insurance and/or further risk reduction on a voluntary basis present options for action which can be worthwhile pursuing even in the case of an acceptable risk.

With regard to these outcomes risk managers may either face a situation of unanimity, i.e. all relevant actors agree with how a given risk situation should be qualified, or a situation of conflict in which major actors challenge the classification under-

Table 4 Tolerability/acceptability judgement.

Assessment Components	**Definition**	**Indicators**
1 Risk characterisation	Collecting and summarising all relevant evidence necessary for making an informed choice on tolerability or acceptability of the risk in question and suggesting potential options for dealing with the risk from a scientific perspective	
	(a) risk profile	• risk estimates • confidence intervals • uncertainty measures • hazard characteristics • range of 'legitimate' interpretations • risk perceptions • social and economic implications
	(b) judging the seriousness of risk	• compatibility with legal requirements • risk-risk trade-offs • effects on equity • public acceptance
	(c) conclusions and risk reduction options	suggestions for: • tolerable risk levels • acceptable risk levels • options for handling risks
2 Risk evaluation	Applying societal values and norms to the judgement on tolerability and acceptability and, consequently, determining the need for risk reduction measures	• choice of technology • potential for substitution • risk-benefit comparison • political priorities • compensation potential • conflict management • potential for social mobilisation

taken by others. The degree of controversy is one of the drivers for selecting the appropriate instruments for risk prevention or risk reduction.

For a systematic analysis of the risk management process it is advisable to focus on tolerable risks and those where tolerability is disputed, for the other cases are fairly easy to deal with. In the case of intolerable risks – and often in the case of tolerable but highly disputed risks – risk managers should opt for prevention strategies as a means to replace the hazardous activity with another activity leading to identical and similar benefits. One should first make sure, however, that the replacement does not introduce more risks or more uncertainties than the agent that it replaces (Graham and Wiener, 1995; Wiener 1998). In the case of acceptable risks it should be left to private actors to initiate additional risk reduction or to seek insurance for covering potential but acceptable losses (although this does not eliminate the need for all concerned to have sufficient information and resources to do so). If risks are classified as tolerable, or if there is dispute as to whether they are tolerable or acceptable, risk management needs to design and implement actions that make these risks acceptable over time. Should this not be feasible then risk management, aided by communication, needs at least to credibly convey the message that major effort is undertaken to bring these risks closer to being acceptable. This task can be described in terms of classic decision theory, i.e. in the following steps (Morgan 1990; Keeney 1992; Hammond et al. 1999):

(a) *Identification and generation of risk management options*: Generic risk management options include risk avoidance, risk reduction, risk transfer and – also an option to take into account – self retention. Whereas to avoid a risk means either selecting a path which does not touch on the risk (e.g. by abandoning the development of a specific technology) or taking action in order to fully eliminate a certain risk, risk transfer deals with ways of passing the risk on to a third party. Self retention as a management option essentially means taking an informed decision to do nothing about the risk and to take full responsibility both for the decision and any consequences occurring thereafter. Risk management by means of risk reduction can be accomplished by many different means. Among them are:

- technical standards and limits that prescribe the permissible threshold of concentrations, emissions, take-up or other measures of exposure;
- performance standards for technological and chemical processes such as minimum temperatures in waste incinerators;
- technical prescriptions referring to the blockage of exposure (e.g. via protective clothing) or the improvement of resilience (e.g. immunisation or earthquake tolerant construction);
- governmental economic incentives including taxation, duties, subsidies and certification schemes;
- third party incentives, i.e. private monetary or in kind incentives;
- compensation schemes (monetary or in kind);
- insurance and liability;
- co-operative and informative options ranging from voluntary agreements to labelling and education programmes.

All these options can be used individually or in combination to accomplish even more effective risk reduction. Options for risk reduction can be initiated by private and public actors or both together.

(b) *Assessment of risk management options with respect to predefined criteria*: Each of the options will have desired and unintended consequences which relate to the risks that they are supposed to reduce. In most instances, an assessment should be done according to the following criteria:

- *Effectiveness*: Does the option achieve the desired effect?

- *Efficiency*: Does the option achieve the desired effect with the least resource consumption?
- *Minimisation of external side effects*: Does the option infringe on other valuable goods, benefits or services such as competitiveness, public health, environmental quality, social cohesion, etc.? Does it impair the efficiency and acceptance of the governance system itself?
- *Sustainability*: Does the option contribute to the overall goal of sustainability? Does it assist in sustaining vital ecological functions, economic prosperity and social cohesion?
- *Fairness*: Does the option burden the subjects of regulation in a fair and equitable manner?
- *Political and legal implementability*: Is the option compatible with legal requirements and political programmes?
- *Ethical acceptability*: Is the option morally acceptable?
- *Public acceptance*: Will the option be accepted by those individuals who are affected by it? Are there cultural preferences or symbolic connotations that have a strong influence on how the risks are perceived?

Measuring management options against these criteria may create conflicting messages and results. Many measures that prove to be effective may turn out to be inefficient or unfair to those who will be burdened. Other measures may be sustainable but not accepted by the public or important stakeholders. These problems are aggravated when dealing with global risks. What appears to be efficient in one country may not work at all in another country. Risk managers are therefore well advised to make use of the many excellent guidance documents on how to handle risk trade-offs and how to employ decision analytic tools for dealing with conflicting evidence and values (cf. Viscusi 1994; Wiener 1998; van der Sluijs et al. 2003; Goodwin and Wright 2004).

(c) *Evaluation of risk management options*: Similar to risk evaluation, this step integrates the evidence on how the options perform with regard to the evaluation criteria with a value judgement about the relative weight each criterion should be assigned. Ideally, the evidence should come from experts and the relative weights from politically legitimate decision makers. In practical risk management, the evaluation of options is done in close cooperation between experts and decision makers. As pointed out later, this is the step in which direct stakeholder involvement and public participation is particularly important and is therefore best assured by making use of a variety methods (Rowe and Frewer 2000; OECD 2002).

(d) *Selection of risk management options*: Once the different options are evaluated, a decision has to be made as to which options are selected and which rejected. This decision is obvious if one or more options turn out to be dominant (relatively better on all criteria). Otherwise, trade-offs have to be made that need legitimisation (Graham and Wiener 1995). A legitimate decision can be made on the basis of formal balancing tools (such as cost-benefit or multi-criteria-decision analysis), by the respective decision makers (given his decision is informed by a holistic view of the problem) or in conjunction with participatory procedures.

(e) *Implementation of risk management options*: It is the task of risk management to oversee and control the implementation process. In many instances implementation is delegated, as when governments take decisions but leave their implementation to other public or private bodies or to the general public. However, the risk management team has at any rate the implicit mandate to supervise the implementation process or at least monitor its outcome.

(f) *Monitoring of option performance*: The last step refers to the systematic observation of the effects of the options once they are implemented. The monitoring system should be designed to assess intended as well as unintended consequences. Often a formal policy assessment study is issued in order to explore the consequences of a given set of risk management measures on different dimensions of what humans value. In addition to generating feedback for the effectiveness of the options taken to reduce the risks, the monitoring phase should also provide new information on early warning signals for both new risks and old risks viewed from a new perspective. It is advisable to have the institutions performing the risk and concern assessments participate in monitoring and supervision so that their analytic skills and experience can be utilised in evaluating the performance of the selected management options.

These steps follow a logical sequence but can be arranged in different orders depending on both situation and circumstance. It might be helpful to visualise the steps not as a linear progression but as a circle forming an iterative process in which reassessment phases are intertwined with new options emerging, new crisis situations arising or new demands being placed on risk managers. Similarly, sometimes the assessment of different options causes the need for new options to be created in order to achieve the desired results. In other cases, the monitoring of existing rules impacts on the decision to add new criteria to the portfolio. Rarely do issues for risk appraisal and management thus follow the sequence used for the description of the process in this paper. Option generation, information processing, and options selection should indeed be seen as a dynamic process with many iterative loops.

Table 5 summarises the steps of risk management in accordance with the basic model used by decision theory. The list of indicators represents the most frequently used heuristic rules for selecting input and for measuring performance.

Risk Management Strategies

Based on the distinction between complexity, uncertainty, and ambiguity it is possible to design generic strategies of risk management to be applied to classes of risks, thus simplifying the risk management process as outlined above. One can distinguish four such classes:

Table 5 Generic components of risk management.

Management Components	**Definition**	**Indicators**
1 Option generation	Identification of potential risk handling options, in particular risk reduction, i.e. prevention, adaptation and mitigation, as well as risk avoidance, transfer and retention	• standards • performance rules • restrictions on exposure or vulnerability • economic incentives • compensation • insurance and liability • voluntary agreements • labels • information/education
2 Option assessment	Investigations of impacts of each option (economic, technical, social, political, cultural)	• effectiveness • efficiency • minimisation of side effects • sustainability • fairness • legal and political implementability • ethical acceptability • public acceptance
3 Option evaluation and selection	Evaluation of options (multi-criteria analysis)	• assignment of trade-offs • incorporation of stakeholders and the public
4 Option implementation	Realisation of the most preferred option	• accountability • consistency • effectiveness
5 Monitoring and feedback	• Observation of effects of implementation (link to early warning) • Ex-post evaluation	• intended impacts • non-intended impacts • policy impacts

- **Simple risk problems**: This class of risk problems requires hardly any deviation from traditional decision-making. Data is provided by statistical analysis, goals are determined by law or statutory requirements and the role of risk management is to ensure that all risk reduction measures are implemented and enforced. Traditional risk-risk comparisons (or risk-risk trade-offs), risk-benefit analysis and cost-effectiveness studies are the instruments of choice for finding the most appropriate risk reduction measures. Additionally, risk managers can rely on best practice and, in cases of low impact, on trial and error. It should be noted, however, that simple risks should not be equated with small or negligible risks. The major issues here are that the potential negative consequences are obvious, the values that are applied are non-controversial and the remaining uncertainties low. Examples are car accidents, known food and health risks, regularly reoccurring natural disasters or safety devices for high buildings.

- **Complex risk problems**: For this risk class major input for risk management is provided by the scientific characterisation of the risk. Complex risk problems are often associated with major scientific dissent about complex dose-effect relationships or the alleged effectiveness of measures to decrease vulnerabilities (for complexity refers to both the risk agent and its causal connections and the risk absorbing system and its vulnerabilities). The objective for resolving complexity is to receive a complete and balanced set of risk and concern assessment results that fall within the legitimate range of plural truth claims.

  In a situation where there is no complete data the major challenge is to define the factual base for making risk management or risk regulatory decisions. So the main emphasis is on improving the reliability and validity of the results that are produced in the risk appraisal phase. Risk and concern assessors as well as managers need to make sure that all relevant knowledge claims are selected, processed and evaluated. They may not get a single answer but they might be able to get a better overview on the issues of scientific controversy. If these efforts lead to an acknowledgement of wide margins of uncertainty, the management tools of the uncertainty strategy should be applied. If input variables to decision-making can be properly defined and affirmed, risk characterisation and evaluation can be done on the basis of risk-benefit balancing and normative standard setting (*risk-based/risk-informed regulation*). Traditional methods such as risk-risk-comparison, cost-effectiveness and cost-benefit analysis are also well-suited to facilitate the overall judgement for placing the risk in the traffic-light model (acceptable, tolerable or intolerable). These instruments, if properly used, provide effective, efficient and fair solutions with respect to finding the best trade-off between opportunities and risks. The choice of instruments includes all the classic options outlined in the section on risk management.

  It is, however, prudent to *distinguish management strategies for handling the risk agent* (such as a chemical or a technology) *from those needed for the risk absorbing system* (such as a building, an organism or an ecosystem). Addressing complex structures of risk agents requires methods for improving causal modelling and data quality control. With respect to risk absorbing systems the emphasis is on the improvement of *robustness*[15] in responding to whatever the target is going to be exposed to. Measures to improve robustness include inserting conservatisms or safety factors as an assurance against individual variation (normally a factor of 10-100 for occupational risk exposure and 100–1000 for public risk exposure), introducing redundant and diverse safety

[15] The terms robustness and resilience have different meanings in different contexts. In most of the *natural hazard literature*, robustness is one of the main components of resilience. In much of the *cybernetic literature*, robustness refers to the insensitivity of numerical results to small changes, while resilience characterises the insensitivity of the entire system against surprises. Our suggestion for distinguishing the two comes close to the cybernetic use of the terms.

devices to improve structures against multiple stress situations, reducing the susceptibility of the target organism (example: iodine tablets for radiation protection), establishing building codes and zoning laws to protect against natural hazards as well as improving the organisational capability to initiate, enforce, monitor and revise management actions (high reliability, learning organisations).

- **Risk problems due to high unresolved uncertainty**: If there is a high degree of remaining uncertainties, risk management needs to incorporate hazard criteria (which are comparatively easy to determine), including aspects such as reversibility, persistence, and ubiquity, and select management options empowering society to deal even with worst case scenarios (such as containment of hazardous activities, close monitoring of risk-bearing activities, securing reversibility of decisions in case risks turn out to be higher than expected). According to IRGC, the management of risks characterised by multiple and high uncertainties should be guided by the *precautionary approach*. Since high unresolved uncertainty implies that the (true) dimensions of the risks are not (yet) known, one should pursue a cautious strategy that allows learning by restricted errors. The main management philosophy for this risk class is to allow small steps in implementation (containment approach) that enable risk managers to stop or even reverse the process as new knowledge is produced or the negative side effects become visible. The primary thrust of precaution is to avoid irreversibility (Klinke and Renn 2002).[16]

  With respect to risk absorbing systems, the main objective is to make these systems resilient so they can withstand or even tolerate surprises. In contrast to robustness, where potential threats are known in advance and the absorbing system needs to be prepared to face these threats, *resilience* is a protective strategy against unknown or highly uncertain hazards. Instruments for resilience include the strengthening of the immune system, diversification of the means for approaching identical or similar ends, reduction of the overall catastrophic potential or vulnerability even in the absence of a concrete threat, design of systems with flexible response options and the improvement of conditions for emergency management and system adaptation. Robustness and resilience are closely linked but they are not identical and require partially different types of actions and instruments.

- **Risk problems due to interpretative and normative ambiguity**: If risk information is interpreted differently by different stakeholders in society – i.e. there are different viewpoints about the relevance, meaning and implications of factual explanations and predictions for deciding about the tolerability of a risk as well as management actions – and if the values and priorities of what should be protected or reduced are subject to intense controversy, risk management needs to address the causes for these conflicting views (von Winterfeldt and

[16] The link between precaution and irreversibility was also mentioned in the aforementioned latest report on risk management by the UK Treasury Department (2004).

Edwards 1984).

Genetically modified organisms for agricultural purposes may serve as an example to illustrate the intricacies related to ambiguity. Surveys on the subject demonstrate that people associate high risks with the application of gene technology for social and moral reasons (Hampel and Renn 2000). Whether the benefits to the economy balance the costs to society in terms of increased health risks, was not mentioned as a major concern of the polled public. Instead, people disagreed about the social need for genetically modified food in western economies where abundance of conventional food is prevalent. They were worried about the loss of personal capacity to act when selecting and preparing food, about the long-term impacts of industrialised agriculture and the moral implications of tampering with nature (Sjöberg 1999). These concerns cannot be addressed by either scientific risk assessments or by determining the right balance between over- and under-protection. The risk issues in this debate focus on the differences between visions of the future, basic values and convictions, and the degree of confidence in the human ability to control and direct its own technological destiny. These wider concerns require the inclusion within the risk management process of those who express or represent them.

Risk managers should thus initiate a broader societal discourse to enable participative decision making. These discursive measures are aimed at finding appropriate conflict resolution mechanisms capable of reducing the ambiguity to a manageable number of options that can be further assessed and evaluated. The main effort of risk management is hence the organisation of a suitable discourse combined with the assurance that all stakeholders and public groups can question and critique the framing of the issue as well as each element of the entire risk chain.

Table 6 provides a summary of these four risk strategies and lists the instruments and tools that are most appropriate for the respective strategy. Again it should be emphasised that the list of strategies and instruments is not exhaustive and can be amended if the case requires it.

Managing Interdependencies

In an interdependent world, the risks faced by any individual, company, region or country depend not only on its own choices but also on those of others. Nor do these entities always face one risk at a time: they may need to find strategies to deal with a series of interrelated risks that are often ill-defined or outside of their control. In the context of terrorism, the risks faced by any given airline, for example, are affected by lax security at other carriers or airports. There are myriad settings that demonstrate similar interdependencies, including many problems in computer and network security, corporate governance, investment in research, and vaccination. Because interdependence does not require proximity, the antecedents to catastrophes can be

Table 6 Risk characteristics and their implications for risk management.

Knowledge Characterisation	Management Strategy	Appropriate Instruments	Stakeholder Participation
1 'Simple' risk problems	*Routine-based*: (tolerability/acceptability judgement)	→ Applying 'traditional' decision- making • Risk-benefit analysis • Risk-risk trade-offs	Instrumental discourse
	(risk reduction)	• Trial and error • Technical standards • Economic incentives • Education, labelling, information • Voluntary agreements	
2 Complexity-induced risk problems	*Risk-informed*: (risk agent and causal chain)	→ Characterising the available evidence • Expert consensus seeking tools: – Delphi or consensus conferencing – Meta analysis – Scenario construction, etc. • Results fed into routine operation	Epistemological discourse
	Robustness-focused: (risk absorbing system)	→ Improving buffer capacity of risk target through: • Additional safety factors • Redundancy and diversity in designing safety devices • Improving coping capacity • Establishing high reliability organisations	
3 Uncertainty-induced risk problems	*Precaution-based*: (risk agent)	→ Using hazard characteristics such as persistence, ubiquity etc. as proxies for risk estimates. Tools include: • Containment • ALARA (as low as reasonably achievable) and ALARP (as low as reasonably possible) • BACT (best available control technology), etc.	Reflective discourse
	Resilience-focused: (risk absorbing system)	→ Improving capability to cope with surprises • Diversity of means to accomplish desired benefits • Avoiding high vulnerability • Allowing for flexible responses • Preparedness for adaptation	
4 Ambiguity-induced risk problems	*Discourse-based*:	→ Application of conflict resolution methods for reaching consensus or tolerance for risk evaluation results and management option selection • Integration of stakeholder involvement in reaching closure • Emphasis on communication and social discourse	Participative discourse

quite distinct and distant from the actual disaster, as was the case of the September 11, 2001 attacks when security failures at Boston's Logan Airport led to crashes at the World Trade Center (WTC), the Pentagon, and in rural Pennsylvania. The same was true in the case of the August 2003 power failures in the Northeastern US and Canada, where the initiating event occurred in Ohio, but the worst consequences were felt hundreds of miles away. Similarly, a disease in one region can readily spread to other regions with which it has contact, as was the case with the rapid spread of SARS from China to its trading partners.

The more interdependencies there are within a particular setting (be this a set of organisational units, companies, a geographical area or a number of countries, etc.) and the more that this setting's entities – or participants – decide not to invest in risk reduction while being able to contaminate other entities, the less incentive each potentially affected participant will have to invest in protection. At the same time, however, each participant would have been better off had all the other participants invested in risk-reducing measures. In other words, weak links may lead to suboptimal behaviour by everyone.[17]

For situations in which participants are reluctant to adopt protective measures to reduce the chances of catastrophic losses due to the possibility of contamination from weak links in the system, a solution might be found in a public-private partnership. This is particularly true if the risks to be dealt with are associated with competing interpretations (ambiguities) as to what type of co-operation is required between different epistemic communities as well as risk management agencies in order to deal with various knowledge and competing value claims. Public-private partnerships also provide an interesting alternative in cases in which perceptions differ strongly and external effects are to be expected.

One way to structure such a partnership is to have government standards and regulations coupled with third party inspections and insurance to enforce these measures. Such a management-based regulatory strategy will not only encourage the addressees of the regulation, often the corporate sector, to reduce their risks from e.g. accidents and disasters. Indeed, it equally shifts the locus of decision-making from the government regulatory authority to private companies which are as a result required to do their own planning as to how they will meet a set of standards or regulations (Coglianese and Lazer 2003). This, in turn, can enable companies to choose those means and measures which are most fit for purpose within their specific environment and, eventually, may lead to a superior allocation of resources compared to more top-down forms of regulation. The combination of third party inspections in conjunction with private insurance is consequently a powerful combination of public oversight and market mechanisms that can convince many companies of the advantages of implementing the necessary measures to make their plants safer and encourage the remaining ones to comply with the regulation to avoid being caught and prosecuted.

Highly interdependent risks that can lead to stochastic contamination of third parties pose a specific challenge for global risk management (i.e. the management

[17] A more formal game theoretic treatment of this problem has been published in Kunreuther and Heal (2003).

of transboundary, international and ubiquitous risks). Due to the often particularly decentralised nature of decision-making in this area, a well balanced mix of consensual (e.g. international agreements and standards, gentleman's agreements), coercive (e.g. government regulation) and incentive-based (e.g. emission certificates) strategies is necessary to deal with such risk problems. Again these strategies can be best developed in close – international and transnational – cooperation between the public and the private sector. To generate the background knowledge for such cooperation and to facilitate its realisation is one of the prime goals of IRGC.

Stakeholder Involvement and Participation

Our emphasis on governance rather than governments or administrations is meant to underline the importance that IRGC places on the inclusion of stakeholders and public groups within the risk handling process and, consequently, on the establishment of adequate public-private partnerships and participatory processes. In the context of this framework we define **stakeholders** as socially organised groups that are or will be affected by the outcome of the event or the activity from which the risk originates and/or by the risk management options taken to counter the risk. Involving stakeholders is not enough, however. Other groups, including the media, cultural elites and opinion leaders, the non-organised *affected* public and the non-organised *observing* public, all have a role to play in risk governance.

Each decision-making process has two major aspects: what and whom to include on the one hand and what and how to select (closure) on the other hand (Hajer and Wagenaar 2003; Stirling 2004). *Inclusion and selection* are therefore the two essential parts of any decision or policy making activity. Classic decision analysis has been offering formal methods for generating options and evaluating these options against a set of predefined criteria. With the advent of new participatory methods, the two issues of inclusion and selection have become more complex and sophisticated than purported in these conventional methods.

The present framework advocates the notion of inclusive governance, in particular with respect to global and systemic risks. First and foremost this means that the four major actors in risk decision making, i.e. *political*, *business*, *scientific* and *civil society players*, should jointly engage in the process of framing the problem, generating options, evaluating options, and coming to a joint conclusion. This has also been the main recommendation of the EU White Paper on European Governance (EU 2001a). This document endorses transparency and accountability through formal consultation with multiple actors as a means for the European Union to address the various frames of governance issues and to identify culture-sensitive responses to common challenges and problems. Similarly to the actors determining the governance of a political union, it is obvious that the actors participating in risk-related decision-making are guided by particular interests which derive not only from the fact that some of them are risk producers – whereas others are exposed to it – but, equally, from their individual institutional rationale and perspective. Such vested interests require specific consideration and measures so that they are made transparent and, if possible, can be reconciled. Inclusive governance, as it relates to the inclusion part of decision-making, requires that (Trustnet 1999; Webler 1999; Wynne 2002):

- there has been a major attempt to involve representatives of all four actor groups (if appropriate);
- there has been a major attempt to empower all actors to participate actively and constructively in the discourse;
- there has been a major attempt to co-design the framing of the (risk) problem or the issue in a dialogue with these different groups;
- there has been a major attempt to generate a common understanding of the magnitude of the risk (based on expertise of all participants) as well as the potential risk management options and to include a plurality of options that represent the different interests and values of all parties involved;
- there has been a major effort to conduct a forum for decision-making that provides equal and fair opportunities for all parties to voice their opinion and to express their preferences; and
- there has been a clear connection between the participatory bodies of decision-making and the political implementation level.

If these conditions are met, evidence shows that actors, along with developing faith in their own competence, use the opportunity and start to place trust in each other and have confidence in the process of risk management (Kasperson et al. 1999; Viklund 2002; Beierle and Cayford 2002: 30f.). This is particularly true for the local level where the participants are familiar with each other and have more immediate access to the issue (Petts 1997). Reaching consensus and building trust on highly complex and controversial subjects such as global change is, however, much more difficult. Being inclusive and open to social groups does not guarantee, therefore, constructive cooperation by those who are invited to participate. Some actors may reject the framing of the issue and choose to withdraw. Others may benefit from the collapse of an inclusive governance process. It is essential to monitor these processes and make sure that particular interests do not dominate the deliberations and that rules can be established and jointly approved to prevent destructive strategising.

Inclusive governance needs to address the second part of the decision-making process as well, i.e. reaching closure on a set of options that are selected for further consideration, while others are rejected. Closure does not mean to have the final word on a development, a risk reduction plan or a regulation. Rather, it represents the product of a deliberation, i.e. the agreement that the participants reached. The problem is that the more actors, viewpoints, interests and values are included and thus represented in an arena, the more difficult it is to reach either a consensus or some other kind of joint agreement. A second set of criteria is thus needed, to evaluate the process by which closure of debates (be they final or temporary) is brought forth as well as the quality of the decision or recommendation that is generated through the closure procedure.

The first aspect, the quality of the closure process itself, can be subdivided into the following dimensions (Webler 1995; Wisdon and Willis 2004):

- Have all arguments been properly treated? Have all truth claims been fairly and accurately tested against commonly agreed standards of validation?
- Has all the relevant evidence, in accordance with the actual state-of-the-art knowledge, been collected and processed?
- Was systematic, experiential and practical knowledge and expertise adequately included and processed?
- Were all interests and values considered and was there a major effort to come up with fair and balanced solutions?
- Were all normative judgements made explicit and thoroughly explained? Were normative statements derived from accepted ethical principles or legally prescribed norms?
- Were all efforts undertaken to preserve plurality of lifestyles and individual freedom and to restrict the realm of collectively binding decisions to those areas in which binding rules and norms are essential and necessary to produce the wanted outcome?

Turning to the issues of outcome, additional criteria need to be addressed. They have been discussed in the political science and governance literature for a long time (Dryzek 1994; Rhodes 1997). They are usually stated as comprising effectiveness, efficiency, accountability, legitimacy, fairness, transparency, acceptance by the public and ethical acceptability. They largely coincide with those that have been postulated earlier for the assessments of risk management options.

When contemplating the requirements for inclusion, closure process and outcome quality, IRGC is convinced that:

- more inclusive procedures enrich the generation of options and perspectives, and are therefore more responsive to the complexity, uncertainty and ambiguity of the risk phenomena which are being assessed;
- more rational closure processes provide fairer and socially and culturally more adaptive and balanced judgements;
- the combination of voluntary and regulatory actions in form of public-private partnerships can be improved through early and constructive involvement procedures; and
- the outcomes derived from these procedures are of higher quality in terms of effectiveness, efficiency, legitimacy, fairness, transparency, public acceptance and ethical acceptability than the outcomes of conventional decision-making procedures.

The potential benefits resulting from stakeholder and public involvement depend, however, on the quality of the participation process. It is not sufficient to gather all interested parties around a table and merely hope for the catharsis effect to emerge spontaneously. In particular, it is essential to treat the time and effort of the participating actors as scarce resources that need to be handled with care and respect (Chess et al. 1998). The participation process should be designed so that the various actors are encouraged to contribute to the process in those areas in which they feel they are competent and can offer something to improve the quality of the final product.

In this respect the four risk classes discussed earlier, i.e. simple, complex, high uncertainty and high ambiguity risk problems, support generic suggestions for participation (Renn 2004b):

- *Simple risk problems*: For making judgements about simple risk problems a sophisticated approach to involve all potentially affected parties is not necessary. Most actors would not even seek to participate since the expected results are more or less obvious. In terms of cooperative strategies, an '**instrumental discourse**' among agency staff, directly affected groups (such as product or activity providers and immediately exposed individuals) as well as enforcement personnel is advisable. One should be aware, however, that often risks that appear simple turn out to be more complex, uncertain or ambiguous as originally assessed. It is therefore essential to revisit these risks regularly and monitor the outcomes carefully.
- *Complex risk problems*: The proper handling of complexity in risk appraisal and risk management requires transparency over the subjective judgements and the inclusion of knowledge elements that have shaped the parameters on both sides of the cost-benefit equation. Resolving complexity necessitates a discursive procedure during the appraisal phase with a direct link to the tolerability and acceptability judgement and risk management. Input for handling complexity could be provided by an '**epistemological discourse**' aimed at finding the best estimates for characterising the risks under consideration. This discourse should be inspired by different science camps and the participation of experts and knowledge carriers. They may come from academia, government, industry or civil society but their legitimacy to participate is their claim to bring new or additional knowledge to the negotiating table. The goal is to resolve cognitive conflicts. Exercises such as Delphi, Group Delphi and consensus workshops would be most advisable to serve the goals of an epistemological discourse (Webler et al. 1991; Gregory et al. 2001).
- *Risk problems due to high unresolved uncertainty*: Characterising risks, evaluating risks and designing options for risk reduction pose special challenges in situations of high uncertainty about the risk estimates. How can one judge the severity of a situation when the potential damage and its probability are unknown or highly uncertain? In this dilemma, risk managers are well advised to include the main stakeholders in the evaluation process and ask them to find a consensus on the extra margin of safety in which they would be willing to invest in exchange for avoiding potentially catastrophic consequences. This type of deliberation called '**reflective discourse**' relies on a collective reflection about balancing the possibilities for over- and under-protection. If too much protection is sought, innovations may be prevented or stalled; if we go for too little protection, society may experience unpleasant surprises. The classic question of 'how safe is safe enough' is replaced by the question of 'how much uncertainty and ignorance are the main actors willing to accept in exchange for

some given benefit'. It is recommended that policy makers, representatives of major stakeholder groups, and scientists take part in this type of discourse. The reflective discourse can take different forms: round tables, open space forums, negotiated rule-making exercises, mediation or mixed advisory committees including scientists and stakeholders (Amy 1983; Perrit 1986; Rowe and Frewer 2000).

- *Risk problems due to high ambiguity*: If major ambiguities are associated with a risk problem, it is not enough to demonstrate that risk regulators are open to public concerns and address the issues that many people wish them to take care of. In these cases the process of risk evaluation needs to be open to public input and new forms of deliberation. This starts with revisiting the question of proper framing. Is the issue really a risk problem or is it in fact an issue of lifestyle and future vision? The aim is to find consensus on the dimensions of ambiguity that need to be addressed in comparing risks and benefits and balancing the pros and cons. High ambiguities require the most inclusive strategy for participation since not only directly affected groups but also those indirectly affected have something to contribute to this debate. Resolving ambiguities in risk debates requires a '**participative discourse**', a platform where competing arguments, beliefs and values are openly discussed. The opportunity for resolving these conflicting expectations lies in the process of identifying common values, defining options that allow people to live their own vision of a 'good life' without compromising the vision of others, to find equitable and just distribution rules when it comes to common resources and to activate institutional means for reaching common welfare so all can reap the collective benefits instead of a few (coping with the classic commoners' dilemma).[18] Available sets of deliberative processes include citizen panels, citizen juries, consensus conferences, ombudspersons, citizen advisory commissions, and similar participatory instruments (Dienel 1989; Fiorino 1990; Durant and Joss 1995; Armour 1995; Applegate 1998).

Categorising risks according to the quality and nature of available information on risk may, of course, be contested among the stakeholders. Who decides whether a risk issue can be categorised as simple, complex, uncertain or ambiguous? It is possible that no consensus may be reached as to where to locate a specific risk. In those cases, a detailed (worst-case) analysis of possibilities of monitoring and surveillance may constitute the only achievable compromise (reversible removal of risk sources etc., timely detection of adverse effects, strength of surveillance systems). The best means, however, to deal with this conflict is to provide for stakeholder involvement when allocating the different risks into these four categories. This task can be located in the phase of screening as the third component of pre-assessment. Allocating risks to the four categories needs to be done before the assessment procedures start. Over the course of further analysis of risks and concerns the categorisation may change since new data and information is being collected that may necessitate a re-ordering of the risk. Yet the risk governance system that is proposed in this document builds

[18] For a more detailed analysis of participatory methods for reaching consensus refer to Barber (1984), Webler (1999) or Jaeger et al. (2001).

upon the need to classify risks at the beginning and allocate them to different routes of appraisal, characterisation, evaluation and management. It seems prudent to have a screening board perform this challenging task. This board should consist of members of the risk and concern assessment team, of risk managers and key stakeholders (such as industry, NGOs and representatives of related regulatory or governmental agencies). The type of discourse required for this task is called **design discourse**. It is aimed at selecting the appropriate risk and concern assessment policy, defining priorities in handling risks, organising the appropriate involvement procedures and specifying the conditions under which the further steps of the risk handling process will be conducted.

Figure 4 provides an overview of the different requirements for participation and stakeholder involvement for the four classes of risk problems and the design discourse. As is the case with all classifications, this scheme shows an extremely simplified picture of the involvement process and it has been criticised for being too rigid in its linking of risk characteristics (complexity, uncertainty, and ambiguity) and specific forms of discourse and dialogue (van Asselt 2005). In addition to the generic distinctions shown in the graph below, it may for instance be wise to distinguish between participatory processes based on risk agent or risk absorbing issues. To conclude these caveats, the purpose of this scheme is to provide general orientation and explain a generic distinction between ideal cases rather than to offer a strict recipe for participation.

Risk Communication

Given the arguments about risk perception and stakeholder involvement, IRGC believes strongly that effective communication has to be at the core of any successful activity to assess and manage risks. The field of risk communication initially developed as a means of investigating how best expert assessments could be communicated to the public so that the tension between public perceptions and expert judgement could be bridged. In the course of time this original objective of educating the public about risks has been modified and even reversed as the professional risk community realised that most members of the public refused to become 'educated' by the experts but rather insisted that alternative positions and risk management practices should be selected by the professional community in their attempt to reduce and manage the risks of modern technology (Plough and Krimsky 1987).

In a recent review of risk communication, William Leiss identified *three phases in the evolution of risk communication practices* (Leiss 1996: 85ff.). The first phase of risk communication emphasised the necessity to convey probabilistic thinking to the general public and to educate the laypersons to acknowledge and accept the risk management practices of the respective institutions. The most prominent instrument of risk communication in phase I was the application of risk comparisons. If anyone was willing to accept x fatalities as a result of voluntary activities, she or he should be obliged to accept another voluntary activity with less than x fatalities. However,

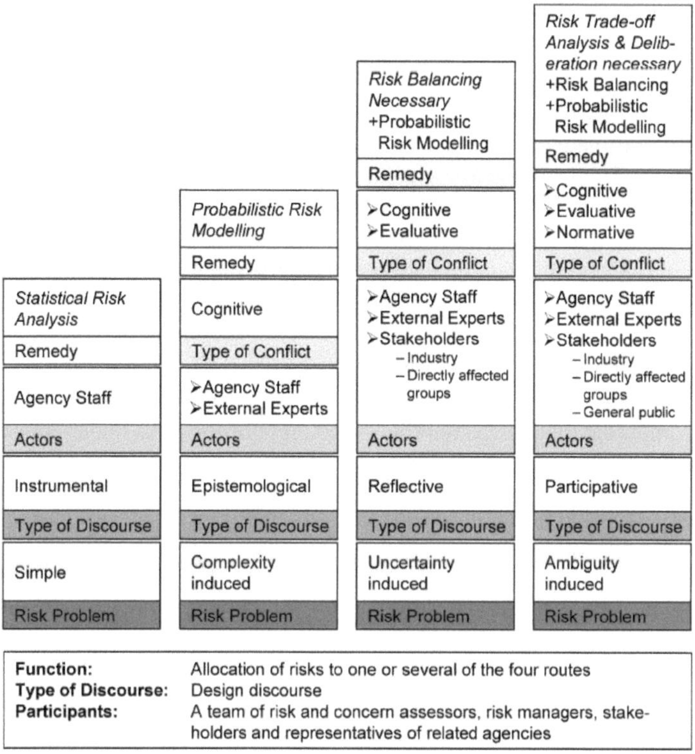

Fig. 4 The risk management escalator and stakeholder involvement (from simple via complex and uncertain to ambiguous phenomena).

this logic failed to convince audiences: people were unwilling to abstract from the context of risk-taking and the corresponding social conditions and they also rejected the reliance on expected values as the only benchmarks for evaluating risks. When this attempt at communication failed, phase II was initiated. This emphasised persuasion and focused on public relations efforts to convince people that some of their behaviour was unacceptable (such as smoking and drinking) since it exposed them to high risk levels, whereas public worries and concerns about many technological and environmental risks (such as nuclear installations, liquid gas tanks, or food additives) were regarded as overcautious due to the absence of any significant risk level. This communication process resulted in some behavioural changes at the personal level: many people started to quit some unhealthy habits. However, it did not convince a majority of these people that the current risk management practices for most of the technological facilities and environmental risks were indeed the politically appropriate response to risk. The one-way communication process of conveying a message to the public in carefully crafted, persuasive language produced little effect. Most respondents were appalled by this approach or simply did not believe the message, regardless how well it was packaged, so phase III evolved. This current phase of risk communication stresses a two-way communication process in which not only are members of the public expected to engage in a social learning process, but so are the risk managers as well. The objective of this communication effort is to build mutual trust by responding to the concerns of the public and relevant stakeholders. The ultimate goal of risk communication is to assist stakeholders in understanding the rationale of risk assessment results and risk management decisions, and to help them arrive at a balanced judgement that reflects the factual evidence about the matter at hand in relation to their own interests and values (OECD 2002). Good practices in risk communication help stakeholders to make informed choices about matters of concern to them and to create mutual trust (Hance et al. 1988; Lundgren 1994).

Risk communication is needed throughout the whole risk handling chain, from the framing of the issue to the monitoring of risk management impacts. The precise form of communication needs to reflect the nature of the risks under consideration, their context and whether they arouse, or could arouse, societal concern. Communication has to be a means to both ensure that:

- those who are central to risk framing, risk appraisal or risk management understand what is happening, how they are to be involved, and, where appropriate, what their responsibilities are; and,
- others outside the immediate risk appraisal or risk management process are informed and engaged.

The first task of risk communication, i.e. facilitating an exchange of information among risk professionals, has often been underestimated in the literature. A close communication link between risk/concern assessors and risk managers, particularly in the phases of pre-assessment and tolerability/acceptability judgement, is crucial for improving overall governance. Similarly, co-operation among natural and social scientists, close teamwork between legal and technical staff and continuous communication between policy makers and scientists are all important prerequisites for enhancing risk management performance. This is particularly important for the initial screening phase where the allocation of risks is performed.

The second task, that of communicating risk appropriately to the outside world, is also a very challenging endeavour. Many representatives of stakeholder groups and, particularly, members of the affected and non-affected public are often unfamiliar with the approaches used to assess and manage risks and/or pursue a specific agenda, trying to achieve extensive consideration of their own viewpoints. They face difficulties when asked to differentiate between the potentially dangerous properties of a substance (hazards) and the risk estimates that depend on both the properties of the substance, the exposure to humans, and the scenario of its uses (Morgan et al. 2002). Also complicating communication is the fact that some risks are acute, with severe effects that are easy to recognise, whereas others exert adverse effects only weakly but over a long period of time. Yet other risks' effects only start to show after an initial delay. Finally, it is no easy task to convey possible synergies of exposures to industrial substances with other factors that relate to lifestyle (e.g. nutrition, smoking, use of alcohol).

Effective communication, or the non-existence thereof, has a major bearing on how well people are prepared to face and cope with risk. Limited knowledge of, and involvement in, the risk management process can lead to inappropriate behaviour in emergency or risk-bearing situations (for example, when facing a pending flood or handling contaminated food or water). There is also the risk of failed communication: consumers or product users may misread or misunderstand risk warnings or labels so that they may, through ignorance, expose themselves to a larger risk than necessary. This is particularly prevalent in countries with high rates of illiteracy and unfamiliarity with risk-related terms. Providing understandable information to help people cope with risks and disasters is, however, only one function of risk communication. Most risk communication analysts list four major functions (Morgan et al. 1992; OECD 2002):

- *Education and enlightenment*: inform the audience about risks and the handling of these risks, including risk and concern assessment and management.
- *Risk training and inducement of behavioural changes*: help people cope with risks and potential disasters.
- *Creation of confidence in institutions responsible for the assessment and management of risk*: give people the assurance that the existing risk governance structures are capable of handling risks in an effective, efficient, fair and acceptable manner (such credibility is crucial in situations in which there is a lack of personal experience and people depend on neutral and disinterested information). It should be kept in mind, however, that trust cannot be produced or generated, but only be accumulated by performance, and that it can be undermined by the lack of respect for an individual within such an institution.
- *Involvement in risk-related decisions and conflict resolution*: give stakeholders and representatives of the public the opportunity to participate in the risk appraisal and management efforts and/or be included in the resolution of conflicts about risks and appropriate risk management options.

For all four functions, risk communication needs to address the following topics:

- explain the concept of probability and stochastic effects;
- explain the difference between risk and hazard;
- deal with stigmatised risk agents or highly dreadful consequences (such as nuclear waste or cancer);
- cope with long-term effects;
- provide an understanding of synergistic effects with other lifestyle factors;
- address the problem of remaining uncertainties and ambiguities;
- cope with the diversity of stakeholders and parties in the risk appraisal and management phase;
- cope with inter-cultural differences within pluralist societies and between different nations and cultures.

Although risk communication implies a stronger role for risk professionals to provide information to the public rather than vice versa, it should be regarded as a mutual learning process in line with the requirements that Leiss postulated for phase III. Concerns, perceptions and experiential knowledge of the targeted audience(s) should thus guide risk professionals in their selection of topics and subjects: it is not the task of the communicators to decide what people need to know but to respond to the questions of what people want to know ('right to know' concept, see Baram 1984). Risk communication requires professional performance both by risk and communication experts. Scientists, communication specialists and regulators are encouraged to take a much more prominent role in risk communication, because effective risk communication can make a strong contribution to the success of a comprehensive and responsible risk management programme. IRGC will invest much of its resources and efforts in contributing to the improvement of current risk communication practices and in itself acting as an effective risk communicator.

Wider Governance Issues: Organisational Capacity

The above framework covering the areas of risk framing (i.e. pre-assessment), appraisal (including risk assessment as well as the assessment of risk-related concerns and the non-physical secondary implications of risk), characterisation/evaluation, management and communication concludes this document's analysis of the classic components of handling risks. Looking at organisational capacity opens a new set of wider risk governance issues which relate to the interplay between the governing actors and their capability to fulfil their role in the risk governance process.

In discussing the different components of risk appraisal and management, it was implicitly assumed that society has developed the institutional and organisational capability to perform all the tasks prescribed in each component – preferably in a matter-of-fact, objective manner. This is, of course, an ideal picture that masks the realities of the *political* context in which risk governance takes place. In particular, the framing of risk is exposed to many institutional and political forces who may wish to jump on the bandwagon of public dissent or media hype in order to push their own interests (Shubik 1991). Given the potential of risk perceptions to mobilise public outrage and, thus, to make it impossible for decision-makers not to listen, some actors in society may have an interest in orchestrating 'risk events', whereas others might have a major motivation for concealing risks or downplaying their impacts. Most political systems have responded to this manoeuvring by establishing independent risk assessment and sometimes management agencies, expecting that these are less likely to be influenced by public pressures. Although IRGC is well aware of the political context in which risk governance takes place, it cannot provide guidance on how to govern risk debates. What it can and intends to do, however, is give advice on how to base risk governance on the best available knowledge and practice. Such advice addresses, first, the process phases of risk appraisal, management and communication as stated above and, second, the strengthening of institutions and agencies so that they are empowered and resourced to perform their tasks in the most effective, efficient and fair manner. As the European Commission's White Paper on European Governance pointed out, the key ingredients of 'good' governance in this sense are openness, participation, accountability, effectiveness and coherence (EU 2001a: 10). These requirements are important for all countries but, in particular, for many transitional and most developing countries. Since the IRGC's scope includes offering assistance to these countries, its work includes criteria for how to analyse and improve organisational capacity and 'good' governance practices.

For the analysis of institutional capacity it is useful to distinguish between assets, skills and capabilities (cf. Paquet 2001). Assets form the social capital for risk governance in the form of knowledge bases and structural conditions for effective management. *Skills* refer to the quality of institutional and human performance in exploring, anticipating and dealing with existing and emerging risks. *Capabilities* describe the institutional framework necessary to translate assets and skills into successful policies. These three components constitute the backbone of institutional capacity for risk governance.

The assets include:

- *Rules, Norms, Regulations*: these establish rights and obligations. In the risk area, the existence of norms, standards, best practices, legal instruments, etc., has always been a major and often contentious issue, hence the importance of such assets. This is true not only with regard to their prescribing of how to deal with risk but also for the absence, or the lack of observance, of rules (e.g. with regard to the end use of new technologies) which itself constitutes an increasing factor of risk.
- *Resources*: these are not limited to financial resources but comprise of an appropriate physical infrastructure for managing risk as well as the availability of adequate information, including the means for information gathering and processing.
- *Competencies and Knowledge*: this involves providing the necessary education and training and establishing and maintaining a pool of experience and expertise. Education should not only be directed at specialists but should reach out to the general public, building a culture of awareness and prevention.
- *Organisational Integration*: the capacity to access and retrieve, in a combination tailored to individual cases, each of these first three types of assets. Organisational integration is a key element, without which otherwise worthy assets will struggle to achieve much.

Using an analogy from mathematics, the three first assets are additive while organisational integration is a multiplying factor. A non-existent organisational capability for integration would nullify the efficacy of the other factors.

Skills are related to the capacity of organisations and institutions to deal with evolving, sometimes chaotic, external conditions. Such conditions should not be considered as an eventuality that cannot be dealt with, but should, instead, be viewed as input parameters to the risk process that require adequate treatment. Skills should enable political, economic and civic actors to use effectively, and enhance the impact of, the available assets. They relate to:

- *Flexibility*, i.e. new ways to make sense of a dynamic situation – adapting to change, which in many cases means fighting against established practices and institutional inertia. An example to illustrate this point can be found in the current concern that city planning frequently still follows 19th century practices while the increase in magnitude and frequency of extreme climatic events associated with climate change should dictate a new approach.
- *Vision*, i.e. bringing new practices into a context that would not naturally generate them – anticipating change. This implies devoting more attention to advanced methodological approaches such as foresight and scenario planning, and a preparedness to think 'outside the box'.
- *Directivity*, i.e. reframing the whole perception of the way of life – driving change that impacts on the outside world rather than limiting oneself to preventing or mitigating the effects of external forces. Several environmental policies (e.g. ban on CFCs) and security policies (e.g. ban on Weapons of Mass Destruction) adopted at the international level reflect this approach.

Using the same mathematical analogy, the three factors constituting the skills are in an additive relationship with each other. Within that relationship they can exhibit different intensities as a function of the nature of external forces.

Capabilities, finally, constitute the framework in which assets enriched by skills can be exploited for developing and implementing successful risk governance policies. Capabilities can be conceptualised as a structure with several successive layers (Wolf 2005):

- *Relations* link users and sources of knowledge as well as those carrying the authority and those bearing the risk, notably civil society. As previously stated, the participation of civil society in risk governance is essential. Relations should thus be based on inclusive decision-making in order to alleviate, at the outset, any circumstances that generate dispute and conflict and consequently aggravate risk.
- *Networks* constitute, in terms of structures, a close co-operative structure that goes beyond relations. Halfway between self-organisation and hierarchy, networks determine close links between and among groups of principally equal actors.
- *Regimes* establish the rules of the game, the framework in which the actors should act. Both relations and networks are essential for forming and sustaining regimes.

Drawing on the mathematical analogy again, the factors constituting the capabilities are additive, each having a separate but complementary function in the overall build up of capabilities.

While each of these assets, skills and capabilities would lend itself to a more detailed discussion IRGC wishes to underline the major importance of risk education and training. In a world where 'human capital' – and in particular brainpower combined with inspiration, courage and a strong ability towards implementation – has largely become the life-blood of society's progress and prosperity, it is quite evident that one of the major keys to the successful handling of risk is in people's heads as well. Given the often systemic and global (transboundary, international and ubiquitous) nature of today's major risks, special 'competencies and knowledge' are required. Specialised in-depth expert knowledge in a restricted area or sector may no longer suffice to understand and counteract risks which spread the boundaries of academic disciplines and business sectors, have several layers of effects and are determined by a multitude of often interlinked factors. However, compartmentalised specialisation is what many educational systems still foster. This approach should, in fact, be replaced by one which emphasises risk appraisal and management in education at all levels and which considers risk under a broad and multi-disciplinary perspective. There is a particular need for this in the engineering, architecture and design disciplines where a primarily technical focus should be extended to health, safety and environmental risk. Such a new approach, fostering in fact a 'bird's eye perspective' with regard to risk, should be anchored in national science and education policies and should grow to become part of our scientific and technological culture.

All three factors – assets, skills and capabilities – are important variables when assessing and investigating risk governance structures in different countries or risk domains; they can also serve as guiding principles for identifying and researching deficiencies and providing assistance to improve capacity. It may even be possible, based on the above mathematical analogies, to construct an overall performance indicator that could help countries to evaluate their risk governance capacities and to use these elements as pathfinders for establishing new institutional frameworks to achieve improved structures for coping with risk.

The Role of Political Culture

When considering the wider environment of risk handling in modern societies, many classes of influential factors come into play. Only a few can be mentioned here. For example, the distinction between horizontal and vertical governance as introduced in the first section of this document can be helpful in describing and analysing cases of risk handling in different countries and contexts (Zürn 2000). In addition, the interplay between economic, political, scientific and civil society actors needs to be addressed when looking beyond just governmental or corporate actions.

It is the goal of IRGC to focus particularly on risk areas which have multidimensional and transnational implications, rather than revisiting classic areas of risk regulation by individual governments or routine risk handling by private corporations. In this focus, one major aspect of risk governance concerns political culture, i.e. *regulatory regimes or governmental styles*. Each country and, in many instances, different risk domains within a country pursue different pathways for dealing with risk. The multitude of risk classification documents and meta-analyses of risk taxonomies is obvious proof of the plurality of risk handling processes and conceptual approaches. It may thus be helpful to search for some underlying principles of these approaches and classify them accordingly.

This exercise of finding common denominators in cultural and national diversity is less of a challenge than one may assume at first glance. Most analysts agree that many of the cognitive factors that govern risk perception are similar throughout the world (Rohrmann and Renn 2000). In addition, risk management styles are also becoming increasingly homogenous as the world becomes more globalised (Löfstedt and Vogel 2001). In spite of the distinct cultural differences among nations and the variations with respect to educational systems, research organisations, and structures of scientific institutions, assessment and management of risks and concerns have become universal enterprises in which nationality, cultural background or institutional setting play a minor role only. This is particularly due to the role of science in proposing and justifying regulatory standards. Research establishments as well as universities have evolved into multinational and cosmopolitan institutions that speak identical or at least similar languages and exchange ideas on world-wide communication networks.[19]

Risk management depends, however, not only on scientific input. It rather rests on three components: *systematic knowledge*, *legally prescribed procedures* and *social values*. Even if the same knowledge is processed by different risk management authorities, the prescriptions for managing risk may differ in many aspects (e.g. with regard to inclusion and selection rules, interpretative frames, action plans for dealing with evidence, and others). National culture, political traditions, and social norms furthermore influence the mechanisms and institutions for integrating knowledge and expertise in the policy arenas. Policy analysts have developed a classification of governmental styles that address these aspects and mechanisms. While these styles have been labelled inconsistently in the literature, they refer to common procedures in different settings (O'Riordan and Wynne 1987). They are summarised in Table 7.

- The **'adversarial' approach** is characterised by an open forum in which different actors compete for social and political influence in the respective policy arena. The actors in such an arena use and need scientific evidence to support their position. Policy makers pay specific attention to formal proofs of evidence because their decisions can be challenged by social groups on the basis of insufficient use or negligence of scientific knowledge. Risk management and communication is essential for risk regulation in an adversarial setting because stakeholders demand to be informed and consulted. Within this socio-political context, stakeholder involvement is mandatory.
- In the **'fiduciary' approach**, the decision-making process is confined to a group of patrons who are obliged to make the 'common good' the guiding principle of their action. Public scrutiny and involvement of the affected public are alien to this approach. The public can provide input to and arguments for the patrons but is not allowed to be part of the negotiation or policy formulation process. The system relies on producing faith in the competence and the fairness of the patrons involved in the decision-making process. Advisors are selected

[19] Indeed, this tendency towards a universal understanding of risk problems and a common language to describe risks and risk reduction measures is one of the most relevant reasons for establishing the IRGC.

Table 7 Characteristics of policy making styles.

Style	Characteristics	Risk Management
1 *Adversarial approach*	• open to professional and public scrutiny • need for scientific justification of policy selection • precise procedural rules • oriented towards producing informed decisions by plural actors	• main emphasis on mutual agreements on scientific evidence and pragmatic knowledge • integration of adversarial positions through formal rules (due process) • little emphasis on personal judgement and reflection on the side of the risk managers • stakeholder involvement essential for reaching communication objectives
2 *Fiduciary approach (patronage)*	• closed circle of 'patrons' • no public control, but public input • hardly any procedural rules • oriented towards producing faith in the system	• main emphasis on enlightenment and background knowledge through experts • strong reliance on institutional in-house 'expertise' • emphasis on demonstrating trustworthiness • communication focused on institutional performance and 'good record'
3 *Consensual approach*	• open to members of the 'club' • negotiations behind closed doors • flexible procedural rules • oriented towards producing solidarity with the club	• reputation most important attribute • strong reliance on key social actors (also non-scientific experts) • emphasis on demonstrating social consensus • communication focused on support by key actors
4 *Corporatist approach*	• open to interest groups and experts • limited public control, but high visibility • strict procedural rules outside of negotiating table • oriented towards sustaining trust to the decision-making body	• main emphasis on expert judgement and demonstrating political prudence • strong reliance on impartiality of risk information and evaluation • integration by bargaining within scientifically determined limits • communication focused on fair representation of major societal interests

according to national prestige or personal affiliations. In this political context, stakeholder involvement may even be regarded as a sign of weakness or a diffusion of personal accountability.

- The **'consensual' approach** is based on a closed circle of influential actors who negotiate behind closed doors. Social groups and scientists work together to reach a predefined goal. Controversy is not present and conflicts are reconciled on a one-to-one basis before formal negotiations take place. Risk communication in this context serves two major goals: it is supposed to reassure the public that the 'club' acts in the best interest of the public good and to convey the feeling that the relevant voices have been heard and adequately considered. Stakeholder participation is only required to the extent that the club needs further insights from the affected groups or that the composition of the club is challenged.
- The **'corporatist' approach** is similar to the consensual approach, but is far more formalised. Well-known experts are invited to join a group of carefully selected policy makers representing the major forces in society (such as the employers, the unions, the churches, the professional associations, the environmentalists). Similar to the consensual approach, risk communication is mainly addressed to the outsiders: it has the goal of creating the impression that the club is open to all 'reasonable' public demands and that it tries to find a fair compromise between public protection and innovation. Often the groups represented within the club are asked to organise their own risk management and communication programmes as a means of enhancing the credibility of the whole management process.

Although these four styles cannot be found in pure form in any country, they form the backdrop of socio-political context variables against which specific risk governance structures are formed and operated. These structures, along with the individual actors' goals and the institutional perspectives they represent, would need more specific attention and, for the time being, are difficult to classify further.

Conclusions

One of the main mandates of IRGC is to assist risk/concern assessors and managers in exploring and handling risks and to promote effective and fair approaches for improving, and enhancing the visibility of, the present risk governance processes. IRGC's aim is to offer guidance and advice on how to approach the complexities, uncertainties and ambiguities of risk issues and to promote a wider understanding of their interconnectedness, particularly in relation to newly emerging systemic risks. To this end IRGC is developing an integrative framework that takes into account scientific, physical, economic, social and cultural aspects and includes effective and appropriate engagement of stakeholders – not least to ensure that both risk appraisal and risk management strategies command the widest possible acceptance and sup-

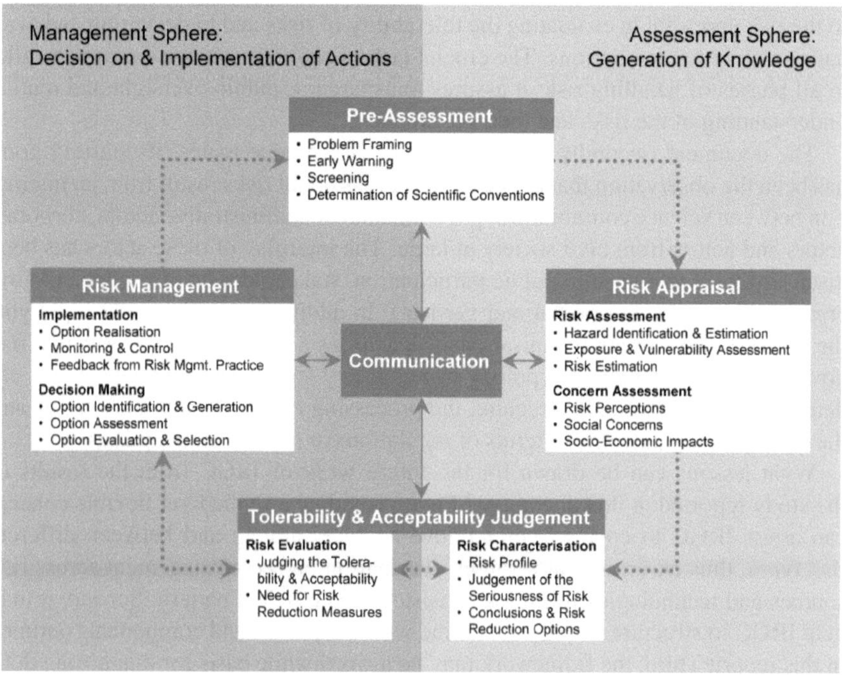

Fig. 5 IRGC Risk Governance Framework.

port. A prototype version of this framework is outlined in the present paper and summarised in Figure 5.

The framework has been designed, on one hand, to include enough flexibility to allow its users to do justice to the wide diversity of risk governance structures and, on the other hand, to provide sufficient clarity, consistency and unambiguous orientation across a range of different risk issues and countries.

This document, firstly, discussed a comprehensive risk handling chain, breaking down its various components into three main phases: 'pre-assessment', 'appraisal', and 'management'. The two intermediate and closely linked stages of risk characterisation and evaluation have been placed between the appraisal and management phases and can be assigned to either of them, depending on the circumstances: if the interpretation of evidence is the guiding principle for characterising risks, then risk and concern assessors are probably the most appropriate people to handle this task; if the interpretation of underlying values and the selection of yardsticks for judging acceptability are the key problems, then risk managers should be responsible. In an ideal setting, however, this task of determining a risk's acceptability should be performed in a joint effort by both assessors and managers. At any rate, a comprehensive, informed and value-sensitive risk management process requires a systematic compilation of results from risk assessment, risk perception studies and other context-related aspects as recommended and subsumed under the category of risk appraisal. Risk managers are thus well advised to include all the information related to the risk appraisal in evaluating the tolerability of risks and in designing and evaluating risk reduction options. The crucial task of risk communication runs parallel to all phases of handling risk: it assures transparency, public oversight and mutual understanding of the risks and their governance.

The document, secondly, addresses wider governance issues. Its starting point has been the observation that collective decisions about risks result from an interaction between science communities, governmental or administrative actors, corporate actors and actors from civil society at large. The interplay of these actors has been discussed with reference to public participation, stakeholder involvement and governance structures (horizontal and vertical). In addition, the document highlights the need for appropriate organisational capacity as a prerequisite for effective risk governance and provides a typology of regulatory styles. These variables also co-determine the institutional structure, the processing of information and values and the quality of the outcome in terms of regulations or management options.

What lessons can be drawn for the future work of IRGC from the results of the study reported in this document? First, providing a unified yet flexible concept can assist IRGC to conduct comparative analyses among and between different risk types, thus ensuring that resource distribution on risk management across risk sources and technologies follows a consistent and efficient pattern. Second, it may help IRGC to structure its projects in line with the phases and components outlined in this report. Third, the framework may be a worthwhile basis for diagnosing deficiencies in existing risk governance regimes around the world and provide suggestions for how to improve them. Lastly, the document may serve a heuristic function by adding to the worldwide efforts for harmonising risk governance approaches and finding some common denominators for risk governance that provide a credible and substantive response to both the globalisation of the planet and the need for a coherent approach to the risks faced by our increasingly interconnected populations.

Glossary of Terms

Acceptability: Risks are deemed to be acceptable if they are insignificant and adequately controlled. There is no pressure to reduce acceptable risks further, unless cost effective measures become available. In many ways, acceptable risks are equivalent to those everyday risks which people accept in their lives and take little action to avoid. (See also 'Intolerable Risks' and 'Tolerability'.)

Agent: In the context of risk a substance, energy, human activity or psychological belief that can cause harm.

ALARA: As Low As Reasonably Achievable.

ALARP: As Low As Reasonably Practicable. (Note: There is little or no difference in practice between ALARA and ALARP. 'Reasonably practicable' is defined in some countries through case law which says that a reduction in risk is 'reasonably practicable' unless the improvement achieved is grossly disproportionate to the cost of achieving that improvement.)

Ambiguity: Giving rise to several meaningful and legitimate interpretations of accepted risk assessments results. See also 'Interpretative Ambiguity' and 'Normative Ambiguity'. ('Ambiguity' is one of three major challenges confronting risk assessment; the others are 'complexity' and 'uncertainty'.)

Buffer Capacity: Capacity of a system to withstand a risk event (e.g. the failure of a component) through the incorporation of additional protective measures.

Complexity: Complexity refers to the difficulty of identifying and quantifying causal links between a multitude of potential causal agents and specific observed effects. ('Complexity' is one of three major challenges confronting risk assessment; the others are 'uncertainty' and 'ambiguity'.)

Coping Capacity: Building into systems, society, organisations or individuals measures to reduce the impact of a risk if it is realised. For example, measures to improve the ability of a building to resist earthquakes. (See also 'Resilience'.)

Design discourse: A form of deliberation for defining and specifying the most appropriate route for assessment and management of a given risk.

Dose-Response Relationship: The relationship between the amount of exposure (dose) to a substance (or other hazard) and the resulting changes in health or body function. (Note: Usually applied to human beings but can be applied more widely in the environment.)

Early warning: Institutional arrangement for (systematically) looking for indicators of potentially damaging events or their precursors.

Epistemological: Concerning the nature, origin and scope of knowledge. So an 'epistemological discourse' is about the scope and the quality (validity, reliability and relevance) of the information available and is aimed at finding the best estimates for characterising the risk.

Exposure: Contact of a risk target (humans, ecosystems) with a hazard.

Flexibility: One of the skills essential to tackling modern risk situations. The ability to look for new ways to make sense of a dynamic situation, if necessary to fight against traditional practices and institutional inertia, and to find novel solutions.

Framing: The initial analysis of a risk problem looking at what the major actors, e.g. governments, companies, the scientific community and the general public, select as risks and what types of problems they label as risk problems. This defines the scope of subsequent work.

Governance: At the national level, the structure and processes for collective decision making involving governmental and non-governmental actors (Nye and Donahue 2000). At the global level, governance embodies a horizontally organised structure of functional self-regulation encompassing state and non-state actors bringing about collectively binding decision without superior authority (cf. Rosenau 1992; Wolf 2000).

Hazard: A source of potential harm or a situation with the potential to causes loss. (Australian/New Zealand risk management standard)

Horizontal Governance: This involves all the relevant actors including government, industry, NGOs and social groups in decision-making processes with a defined geographical or functional segment, such as a community or region.

Indeterminacy: See 'Stochastic Effects'.

Instrumental [discourse]: Used in the case of 'simple risks'. It is aimed at finding the most cost-effective measures to make the risk acceptable or at least tolerable.

Interpretative Ambiguity: Different interpretations of an identical assessment result: e.g. as an adverse or non-adverse effect.

Intolerable Risks (alternatively 'Unacceptable Risks'): A risk that society deems to be unacceptable, no matter what benefits arise from the activity giving rise to the risk.

Justification: The case for undertaking an activity that carries an element of risk. In effect, some kind of risk/benefit analysis which demonstrates the case for the activity.

Latency: Concealed or dormant risks; latency refers to those risks where the harm emerges some considerable time after exposure (e.g. effects of exposure to radiation).

Normative Ambiguity: Different concepts of criteria or yardsticks that help to determine what can be regarded as tolerable referring e.g. to ethics, quality of life parameters, risk-benefit balance, distribution of risks and benefits, etc.

Organisational Capacity: The ability of organisations and individuals within organisations to fulfil their role in the risk governance process.

Participative [decision making/discourse]: Open to public input; possibly including new forms of deliberation. Examples of participative discourse include citizens' juries, consensus conferences, etc.

Probabilistic Risk Assessment (PRA): Methods for calculating probability-loss functions based on statistical, experimental and/or theoretically derived data (such as event treed or fault trees). PRA is often used in the context of engineered systems.

Reflective [discourse]: Collective reflection on the course of action to take e.g. balancing possibilities of over- and under-protection in the case of large remaining uncertainties about probabilities and/or magnitude of damage(s). Examples of reflective discourse include round tables, open space forums and negotiated rule making.

Resilience: A protective strategy to build in defences to the whole system against the impact of the realisation of an unknown or highly uncertain risk. Instruments for resilience include strengthening the immune system, designing systems with flexible response options, improving emergency management, etc.

Risk: An uncertain consequence of an event or an activity with respect to something that humans value (definition originally in Kates et al. 1985: 21). Such consequences can be positive or negative, depending on the values that people associate with them.

Risk Analysis: Some organisations, e.g. Codex Alimentarius, use risk analysis as a collective term which covers risk assessment, risk management and risk communication.

Risk Appraisal: The process of bringing together all knowledge elements necessary for risk characterisation, evaluation and management. This includes not just the results of (scientific) risk assessment but also information about risk perceptions and economic and social implications of the risk consequences.

Risk Assessment: The task of identifying and exploring, preferably in quantified terms, the types, intensities and likelihood of the (normally undesired) consequences related to a risk. Risk assessment comprises hazard identification and estimation, exposure and vulnerability assessment and risk estimation.

Risk Characterisation: The process of determining the evidence based elements necessary for making judgements on the tolerability or acceptability of a risk. (See also 'Risk Evaluation'.)

Risk Estimation: The third component of risk assessment, following hazard identification and estimation, and exposure/vulnerability assessment. This can be quantitative (e.g. a probability distribution of adverse effects) or qualitative (e.g. a scenario construction).

Risk Evaluation: The process of determining the value-based components of making a judgement on risk. This includes risk-benefit balancing or incorporation of quality of life implications and may also involve looking at such issues as the potential for social mobilisation or at pre-risk issues such as choice of technology and the social need of the particular operation giving rise to the risk. (See 'Justification'.)

Risk Governance: Includes the totality of actors, rules, conventions, processes, and mechanisms concerned with how relevant risk information is collected, analysed and communicated and management decisions are taken. Encompassing the combined risk-relevant decisions and actions of both governmental and private actors, risk governance is of particular importance in, but not restricted to, situations where there is no single authority to take a binding risk management decision but where instead the nature of the risk requires the collaboration and co-ordination between a range of different stakeholders. Risk governance however not only includes a multifaceted, multiactor risk process but also calls for the consideration of contextual factors such as institutional arrangements (e.g. the regulatory and legal framework that determines the relationship, roles and responsibilities of the actors and co-ordination mechanisms such as markets, incentives or selfimposed norms) and political culture including different perceptions of risk.

Risk Management: The creation and evaluation of options for initiating or changing human activities or (natural and artificial) structures with the objective of increasing the net benefit to human society and preventing harm to humans and what they value; and the implementation of chosen options and the monitoring of their effectiveness.

Risk Mitigation: Measures to reduce the impact of a realised risk; for example, design features in a chemical plant to direct any explosive failure in a particular direction away from sensitive parts of the plant.

Risk Perception: The outcome of the processing, assimilation and evaluation of personal experiences or information about risk by individuals or groups in society.

Risk Prevention: Measures to stop a risk being realised. This often means stopping the activity giving rise to the risk. But this, because of the need for substitution, can often give rise to other risks in the substituted activity.

Risk Reduction: Measures to reduce the level of risk, for example by reducing the likelihood of the risk being realised or reducing the impact of the risk.

Risk Screening: The process of sifting and selecting information about risk in order to allocate the risk to a particular category or to a particular control regime; the process needs to be done in a manner that avoids unnecessary compartmentalisation of a risk.

Risk Trade-Offs (or Risk-Risk Trade-Offs): The phenomenon that interventions to reduce one risk can increase other risks, or shift risk to a new population.

Risk Transfer: Passing on some or all of the consequences of a risk to a third party. In some cases, this may be part of legitimate risk management e.g. to an insurance company; in other cases, for example, where those benefiting from the risk generating activity are not those who suffer from the risk (e.g. those suffering pollution down stream from a chemical plant), risk governance needs to ensure that such transfers are dealt with fully and equitably.

Robustness: This concerns primarily the insensitivity (or resistance) of parts of systems to small changes within well defined ranges of the risk consequences (contrast with 'resilience' which more concerns whole systems).

Semantic Risk Patterns: Classes of risk that reflect certain perceptive or psychological approaches to risk. For example, one such class concerns risks posing an immediate threat such as nuclear energy; another concerns activities where an individual's perception of their vulnerability is underestimated because they believe they are 'in charge', e.g. when driving a car.

Social Amplification of Risk: An overestimation or underestimation of the seriousness of a risk caused by public concern about the risk or an activity contributing to the risk.

Social Mobilisation: Social opposition or protest that feeds into collective actions (such as voting behaviour, demonstration or other forms of public protest).

Stakeholder: Socially organised groups that are or will be affected by the outcome of the event or the activity from which the risk originates and/or by the risk management options taken to counter the risk.

Stochastic Effects: Effects due to random events.

Systemic Risk: Those risks that affect the systems on which society depends – health, transport, energy, telecommunications, etc. Systemic risks are at the crossroads between natural events (partially altered and amplified by human action such as the emission of greenhouse gases), economic, social and technological developments and policy-driven actions, both at the domestic and the international level.

Taxonomy: A structure for classifying risks and approaches to methods of dealing with risks.

Tolerability: An activity that is seen as worth pursuing (for the benefit it carries) yet requires additional efforts for risk reduction within reasonable limits. (See also 'Acceptability' and 'Intolerable Risks'/ 'Unacceptable Risks'.)

Ubiquity: In the context of risk, one for which the impact of the risk being realised is widespread, usually geographically.

Unacceptable Risks: See 'Intolerable Risks'.

Uncertainty: A state of knowledge in which, although the factors influencing the issues are identified, the likelihood of any adverse effect or the effects themselves cannot be precisely described. (Note: this is different from ignorance about the effects or their likelihood. 'Uncertainty' is one of three major challenges confronting risk assessment; the others are 'complexity' and 'ambiguity'.)

Vertical Governance: This concerns the links between the various segments which may have an interest in an issue, e.g. between local, regional and state levels (whereas 'horizontal governance' concerns the links within those segments).

Vulnerability: The extent to which the target can experience harm or damage as a result of the exposure (for example, immune system of target population, vulnerable groups, structural deficiencies in buildings, etc.).

References

Amy, D.J., 1983, Environmental mediation: An alternative approach to policy stalemates, *Policy Sciences* **15**, 345–365.

Applegate, J., 1998, Beyond the usual suspects: The use of citizens advisory boards in environmental decisionmaking, *Indiana Law Journal* **73**, 903.

Armour, A., 1995, The citizen's jury model of public participation, in: O. Renn, T. Webler and P. Wiedemann (eds.), *Fairness and Competence in Citizen Participation. Evaluating New Models for Environmental Discourse*, Kluwer Academic Publishers, Dordrecht, pp. 175–188.

Baram, M., 1984, The right to know and the duty to disclose hazard information, *American Journal of Public Health* **74**(4), 385–390.

Barber, B., 1984, *Strong Democracy. Participatory Politics for a New Age*, University of California Press, Berkeley, CA.

Beck, U., 1994, The reinvention of politics: Towards a theory of reflexive modernization, in: U. Beck, A. Giddens and S. Lash (eds.), *Reflexive Modernization. Politics, Tradition and Aesthetics in the Modern Social Order*, Stanford University Press, Stanford, CA, pp. 1–55.

Beierle, T.C. and Cayford, J., 2002, *Democracy in Practice. Public Participation in Environmental Decisions*, Resources for the Future, Washington, DC.

Benz, A. and Eberlein, B., 1999, The Europeanization of regional policies: Patterns of multi-level governance, *Journal of European Public Policy* **6**(2), 329–348.

Boholm, A., 1998, Comparative studies of risk perception: A review of twenty years of research, *Journal of Risk Research* **1**(2), 135–163.

Bradbury, J.A., 1989, The policy implications of differing concepts of risk, *Science, Technology, and Human Values* **14**(4), Fall, 380–399.

Brehmer B., 1987, The psychology of risk, in W.T. Singleton and J. Howden (eds.), *Risk and Decisions*, Wiley, New York, pp. 25–39.

British Broadcasting Company (BBC), 2002, Chronology of rail crashes, 10 May 2002. Online on Internet: http://news.bbc.co.uk/1/hi/uk/465475.stm (accessed on 2005-08-11).

Brown, H. and Goble, R., 1990, The role of scientists in risk assessment, *Risk: Issues in Health and Safety* **VI**, 283–311.

Bruijn, J.A. and ten Heuvelhof, E.F., 1999, Scientific expertise in complex decision-making processes, *Science and Public Policy* **26**(3), 151–161.

Burns, W.J., Slovic, P., Kasperson, R.E., Kasperson, J.X., Renn, O. and Emani, S., 1993, Incorporating structural models into research on the social amplification of risk: Implications for theory construction and decision making, *Risk Analysis* **13**(6), 611–623.

Chess, C., Dietz, T. and Shannon, M., 1998, Who should deliberate when?, *Human Ecology Review* **5**(1), 60–68.

Clark, W., 2001, Research systems for a transition toward sustainability, *GAIA* **10**(4), 264–266.

Codex Alimentarius Commission, 2001, *Procedural Manual: Twelfth Edition*, UN Food and Agriculture Organisation, FAO, Rome. Online on Internet: http://www.fao.org/documents/show_cdr.asp?url_file=/DOCREP/005/Y2200E/y2200e07.htm (accessed on 2005-08-11).

Coglianese, C., 1999, The limits of consensus, *Environment* **41**(28), 28–33.

Coglianese, C. and Lazer, D., 2003, Management-based regulation: Prescribing private management to achieve public goals, *Law and Society* **37**, 691–730.

Cooke, R.M., 1991, *Experts in Uncertainty: Opinion and Subjective Probability in Science*, Oxford Press, Oxford.

Covello, V.T., 1983, The perception of technological risks: A literature review, *Technological Forecasting and Social Change* **23**, 285–297.

Cross, F.B., 1998, Facts and values in risk assessment, *Reliability Engineering and Systems Safety* **59**, 27–45.

Dienel, P.C., 1989, Contributing to social decision methodology: Citizen reports on technological projects, in: C. Vlek and G. Cvetkovich (eds.), *Social Decision Methodology for Technological Projects*, Kluwer Academic Publishers, Dordrecht, pp. 133–151.
Douglas, M., 1990, Risk as a forensic resource, *DEADALUS* **119**(4), 1–16.
Drottz-Sjöberg, B.M., 1991, *Perception of Risk. Studies of Risk Attitudes, Perceptions, and Definitions*, Center for Risk Research, Stockholm.
Dryzek, J.S., 1994, *Discursive Democracy. Politics, Policy, and Political Science*, Second Edition, Cambridge University Press, Cambridge.
Durant, J. and Joss, S., 1995, *Public Participation in Science*, Science Museum, London.
Environment Agency, 1998, *Strategic Risk Assessment. Further Developments and Trials*, R&D Report E70, Environment Agency, London.
European Commission/Health & and Consumer Protection Directorate General, Directorate C, 2000, *Scientific Opinions: First Report on the Harmonisation of Risk Assessment Procedures*, EU, Brussels.
European Commission, 2001a, *European Governance. A White Paper*, EU, Brussels.
European Commission, 2001b, *European Transport Policy for 2010: Time to Decide. White Paper*, EU, Luxembourg.
European Commission, 2003, *Final Report on Setting the Scientific Frame for the Inclusion of New Quality of Life Concerns in the Risk Assessment Process*, EU, Brussels.
Fiorino, D.J., 1990, Citizen participation and environmental risk: A survey of institutional mechanisms, *Science, Technology, and Human Values* **15**(2), 226–243.
Fischhoff, B., 1985, Managing risk perceptions, *Issues in Science and Technology* **2**(1), 83–96.
Fischhoff, B., 1995, Risk perception and communication unplugged: Twenty years of process, *Risk Analysis* **15**(2), 137–145.
Fischhoff, B., Slovic, P., Lichtenstein, S., Read, S. and Combs, B., 1978, How safe is safe enough? A psychometric study of attitudes toward technological risks and benefits, *Policy Sciences* **9**, 127–152.
Functowicz, S.O. and Ravetz, J.R., 1992, Three types of risk assessment and the emergence of post-normal science, in: S. Krimsky, S. and D. Golding (eds.), *Social Theories of Risk*, Praeger, Westport, CT, 251–273.
Gigerenzer, G. and Selten, R., 2001, Rethinking rationality, in: G. Gigerenzer and R. Selten (eds.), *Bounded Rationality: The Adaptive Toolbox. Dahlem Workshop Report*,.(MIT Press, Cambridge, MA, pp. 1–12.
Goodwin, P. and Wright, G., 2004, *Decision Analysis for Management Judgement*, Wiley, London.
Gosh, D. and Ray, M.R., 1997, Risk, ambiguity and decision choice: Some additional evidence, *Decision Sciences* **28**(1), Winter, 81–104.
Graham, J.D. and Rhomberg, L., 1996, How risks are identified and assessed, in: H. Kunreuther and P. Slovic (eds.), *Challenges in Risk Assessment and Risk Management*, The Annals of the American Academy of Political and Social Science, Sage, Thousand Oaks, CA, pp. 15–24.
Graham, J.D. and Wiener, J.B., 1995, *Risk vs. Risk*, Harvard University Press, Cambridge, MA.
Greeno, J.L. and Wilson, J.S., 1995, New frontiers in environmental, health and safety management, in: R. Kolluru, S. Bartell, R. Pitblade and S. Stricoff (eds.), *Risk Assessment and Management Handbook. For Environmental, Health, and Safety Professionals*, McGraw-Hill, New York, pp. 3.1–2.17.
Gregory, R.S., 2004, Valuing risk management choices, in: T. McDaniels and M.J. Small (eds.), *Risk Analysis and Society. An Interdisciplinary Characterization of the Field*, Cambridge University Press, Cambridge, pp. 213–250.
Gregory, R., McDaniels, T. and Fields, D., 2001, Decision aiding, not dispute resolution: A new perspective for environmental negotiation, *Journal of Policy Analysis and Management* **20**(3), 415–432.
Grossi, P. and Kunreuther, H. (eds.), 2005, *Catastrophe Modeling: A New Approach to Managing Risk*, Springer, New York.
Hajer, M. und Wagenaar, H., 2003, *Deliberative Policy Analysis: Understanding Governance in the Network Society*, Cambridge University Press, Boston, MA.

Hammond, J., Keeney, R. and Raiffa, H., 1999, *Smart Choices: A Practical Guide to Making Better Decisions*, Havard Business School Press, Cambridge, MA.

Hampel, J. and Renn, O. (eds.), 2000, *Gentechnik in der Öffentlichkeit. Wahrnehmung und Bewertung einer umstrittenen Technologie*, Second Edition, Campus, Frankfurt/Main.

Hance, B.J., Chess, C. and Sandman, P.M., 1988, *Improving Dialogue with Communities: A Risk Communication Manual for Government*, Environmental Communication Research Programme, Rutgers University, New Brunswick, NJ.

Hattis, D., 2004, The conception of variability in risk analyses: Developments since 1980, in: T. McDaniels and M.J. Small (eds.), *Risk Analysis and Society. An Interdisciplinary Characterization of the Field*, Cambridge University Press, Cambridge, pp. 15–45.

Hattis, D. and Kennedy, D., 1990, Assessing risks from health hazards: An imperfect science, in: T.S. Glickman and M. Gough (eds.), *Readings in Risk*, Resources for the Future, Washington, DC, pp. 156–163.

Ho, J.L.L., Keller, R. and Keltyka, P., 2002, Effects of probabilistic and outcome ambiguity on managerial choices, *Journal of Risk and Uncertainty* **24**(1), 47–74.

Hohenemser, C., Kates, R.W. and Slovic, P., 1983, The nature of technological hazard, *Science* **220**, 378–384.

HSE, 2001, *Reducing Risk – Protecting People*, Health and Safety Executive, London.

Hsee, C. and Kunreuther, H., 2000, The affection effect in insurance decisions, *Journal of Risk and Uncertainty* **20**, 141–159.

IAEA, 1995, *Guidelines for Integrated Risk Assessment and Management in Large Industrial Areas*, Technical Document: IAEA-TECDOC PGVI-CIJV, International Atomic Energy Agency, Vienna.

IEC, 1993, *Guidelines for Risk Analysis of Technological Systems*, Report IEC-CD (Sec) 381 issued by the Technical Committee QMS/23, European Community, Brussels.

IPCS and WHO, 2004, *Risk Assessment Terminology*, World Health Organization, Geneva.

Jaeger, C., Renn, O., Rosa, E. and Webler, T., 2001, *Risk, Uncertainty and Rational Action*, Earthscan, London.

Jasanoff, S., 1986, *Risk Management and Political Culture*, Russell Sage Foundation, New York.

Jasanoff, S., 2004, Ordering knowledge, ordering society, in: S. Jasanoff (ed.), *States of Knowledge: The Co-Production of Science and Social Order*, Routledge, London, pp. 31–54.

Kahneman, D. and Tversky, A., 1979, Prospect theory: An analysis of decision under risk, *Econometrica* **47**(2), 263–291.

Kasperson, J.X., Kasperson, R.E., Pidgeon, N.F. and Slovic, P., 2003, The social amplification of risk: Assessing fifteen years of research and theory, in: N.F. Pidgeon, R.K. Kasperson and P. Slovic (eds.), *The Social Amplification of Risk*, Cambridge University Press, Cambridge, pp. 13–46.

Kasperson, R.E., Renn, O., Slovic, P., Brown, H.S., Emel, J., Goble, R., Kasperson, J.X. and Ratick, S., 1988, The social amplification of risk. A conceptual framework, *Risk Analysis* **8**(2), 177–187.

Kasperson, R.E., Golding, D. and Kasperson, J.X., 1999, Risk, trust and democratic theory, in: Cvetkovich, G. and Löfstedt, R. (eds.), *Social Trust and the Management of Risk*, Earthscan, London, pp. 22–41.

Kasperson, R.E., Jhaveri, N. and Kasperson, J.X., 2001, Stigma and the social amplification of risk: Toward a framework of analysis, in: J. Flynn, P. Slovic and H. Kunreuther (eds.), *Risk Media and Stigma*, Earthscan, London, pp. 9–27.

Kates, R.W., Hohenemser, C. and Kasperson, J., 1985, *Perilous Progress: Managing the Hazards of Technology*, Westview Press, Boulder, CO.

Keeney, R., 1992, *Value-Focused Thinking. A Path to Creative Decision Making*, Harvard University Press, Cambridge, MA.

Keeney, R. and McDaniels, T., 2001, A framework to guide thinking and analysis regarding climate change policies, *Risk Analysis* **6**, December, 989–1000.

Kemp, R., 1998, Modern strategies of risk communication: Reflections on recent experience, in: R. Matthes, J. Bernhardt and M. Repacholi (eds.), *Risk Perception, Risk Communication and*

Its Application to EMF Exposure, ICNRP 5/98, International Commission on Non-Ionising Radiation Protection and World Health Organisation, Geneva, pp. 117–125.

Kemp, R. and Greulich, T., 2004, *Communication, Consultation, Community: MCF Site Deployment Consultation Handbook*, Mobile Carriers Forum, Melbourne.

Klinke, A. and Renn, O., 2002, A new approach to risk evaluation and management: Risk-based, precaution-based and discourse-based management, *Risk Analysis* **22**(6), December, 1071–1094.

Kolluru, R.V., 1995, Risk assessment and management: A unified approach, in: R. Kolluru, S. Bartell, R. Pitblade and S. Stricoff (eds.), *Risk Assessment and Management Handbook. For Environmental, Health, and Safety Professionals*, McGraw-Hill, New York, pp. 1.3–1.41.

Kunreuther, H. and Heal, G., 2003, Interdependent security, *Journal of Risk and Uncertainty*, Special Issue on Terrorist Risks, **26**(2/3), March/May, 231–249.

Kunreuther, H., Novemsky, N. and Kahneman, D., 2001, Making low probabilities useful, *Journal of Risk and Uncertainty* **23**, 103–120.

Laudan, L., 1996, The pseudo-science of science? The demise of the demarcation problem, in: L. Laudan (ed.), *Beyond Positivism and Relativism. Theory, Method and Evidence*, Westview Press, Boulder, CO, pp. 166–192.

Lave, L., 1987, Health and safety risk analyses: Information for better decisions, *Science* **236**, 291–295.

Leiss, W., 1996, Three phases in risk communication practice, in *Annals of the American Academy of Political and Social Science*, Special Issue, H. Kunreuther and P. Slovic (eds.), *Challenges in Risk Assessment and Risk Management*, Sage, Thousand Oaks, CA, pp. 85–94.

Liberatore, A. und Funtowicz, S., 2003, Democratizing expertise, expertising democracy: What does this mean, and why bother?, *Science and Public Policy* **30**(3), 146–150.

Löfstedt, R.E., 1997, *Risk Evaluation in the United Kingdom: Legal Requirements, Conceptual Foundations, and Practical Experiences with Special Emphasis on Energy Systems*, Working Paper No. 92, Akademie für Technikfolgenabschätzung, Stuttgart.

Löfstedt, R.E. and Vogel, D., 2001, The changing character of regulation. A comparison of Europe and the United States, *Risk Analysis* **21**(3), 393–402.

Loewenstein, G., Weber, E., Hsee, C. and Welch, E., 2001, Risk as feelings, *Psychological Bulletin* **127**, 267–286.

Lundgren, R.E., 1994, *Risk Communication: A Handbook for Communicating Environmental, Safety, and Health Risks*, Battelle Press, Columbus, OH.

Lyall, C. and Tait, J., 2004, Shifting policy debates and the implications for governance, in: C. Lyall and J. Tait (eds.), *New Modes of Governance. Developing an Integrated Policy Approach to Science, Technology, Risk and the Environment*, Ashgate, Aldershot, pp. 3–17.

Mayo, D.G. and Hollander, R.D. (eds.), 1991, *Acceptable Evidence: Science and Values in Risk Management*, Oxford University Press, Oxford.

Morgan, M.G., 1990, Choosing and managing technology-induced risk, in: T.S. Glickman and M. Gough (eds.), *Readings in Risk*, Resources for the Future, Washington, DC, pp. 17–28.

Morgan, M.G. and Henrion, M., 1990, *Uncertainty: A Guide to Dealing with Uncertainty in Quantitative Risk and Policy Analysis*, Cambridge University Press, Cambridge.

Morgan, M.G., Fischhoff, B., Bostrom, A., Lave, L. and Atman, C., 1992, Communicating risk to the public, *Environmental Science and Technology* **26**(11), 2049–2056.

Morgan, M.G., Florig, K., DeKay, M., Fischbeck, P., Morgan, K., Jenni, K. and Fischhoff, B., 2000, Categorizing risks for risk ranking, *Risk Analysis* **20**(1), 49–58.

Morgan, M.G., Fischhoff, B., Bostrom, A. and Atman, C.J., 2002, *Risk Communication: A Mental Models Approach*, Cambridge University Press, Boston, MA.

National Research Council, Committee on Risk and Decision Making: Risk and Decision Making, 1982, *Perspectives and Research*, National Academy Press, Washington, DC.

National Research Council, Committee on the Institutional Means for Assessment of Risks to Public Health, 1983, *Risk Assessment in the Federal Government: Managing the Process, National Academy of Sciences*, National Academy Press, Washington, DC.

Nye, J.S. and Donahue, J. (eds.), 2000, *Governance in a Globalising World*, Brookings Institution, Washington, DC.

OECD, 2002, *Guidance Document on Risk Communication for Chemical Risk Management*, OECD, Paris.

OECD, 2003, *Emerging Systemic Risks*, Final Report to the OECD Futures Project, OECD, Paris.

OECD, 2004, *Large-Scale Disasters – Lessons Learned*, OECD, Paris.

Olin, S., Farland, W., Park, C., Rhomberg, L., Scheuplein, R., Starr, T. and Wilson, J., 1995, *Low Dose Extrapolation of Cancer Risks: Issues and Perspectives*, ILSI Press, Washington, DC.

O'Riordan, T. and Wynne, B., 1987, Regulating environmental risks: A comparative perspective, in: P.R. Kleindorfer and H.C. Kunreuther (eds.), *Insuring and Managing Hazardous Risks: From Seveso to Bhopal and Beyond*, Springer, Berlin, pp. 389–410.

Paquet, G., 2001, The new governance, subsidiarity, and the strategic state, in: OECD (ed.), *Governance in the 21st Century*, OECD, Paris, pp. 183–215.

Perritt, H.H., 1986, Negotiated rulemaking in practice, *Journal of Policy Analysis and Management* **5**, Spring, 482–95.

Petts, J., 1997, The public-expert interface in local waste management decisions: Expertise, credibility, and process, *Public Understanding of Science* **6**(4), 359–381.

Pidgeon, N.F., 1998, Risk assessment, risk values and the social science programme: Why we do need risk perception research, *Reliability Engineering and System Safety* **59**, 5–15.

Pidgeon, N.F. and Gregory, R., 2004, Judgment, decision making and public policy, in: D. Koehler and N. Harvey (eds.), *Blackwell Handbook of Judgment and Decision Making*, Blackwell, Oxford, pp. 604–623.

Pidgeon, N.F, Hood, C.C., Jones, D.K.C., Turner, B.A. and Gibson, R., 1992, Risk perception, in: Royal Society Study Group (eds.), *Risk Analysis, Perception and Management*, The Royal Society, London, pp. 89–134.

Pinkau, K. and Renn, O., 1998, *Environmental Standards. Scientific Foundations and Rational Procedures of Regulation with Emphasis on Radiological Risk Management*, Kluwer Academic Publishers, Dordrecht.

Plough, A. and Krimsky, S., 1987, The emergence of risk communication studies: Social and political context, *Science, Technology, and Human Values* **12**, 78–85.

Pollard, S.J.T., Duarte-Davidson, R., Yearsley, R., Twigger-Ross, C., Fisher, J., Willows, R. and Irwin, J., 2000, *A Strategic Approach to the Consideration of 'Environmental Harm'*, The Environment Agency, Bristol.

Ravetz, J., 1999, What is post-normal science, *Futures* **31**(7), 647–653.

Renn, O., 1998, Three decades of risk research: Accomplishments and new challenges, *Journal of Risk Research* **1**(1), 49–71.

Renn, O., 2004a, Perception of risks, *The Geneva Papers on Risk and Insurance* **29**(1), 102–114.

Renn, O., 2004b, The challenge of integrating deliberation and expertise: Participation and discourse in risk management, in: T.L. MacDaniels and M.J. Small (eds.), *Risk Analysis and Society: An Interdisciplinary Characterization of the Field*, Cambridge University Press, Cambridge, pp. 289–366.

Rhodes, R.A.W., 1996, The new governance: Governing without government, *Political Studies* **44**(4), 652–667.

Rhodes, R.A.W., 1997, *Understanding Governance: Policy Networks, Governance, Reflexivity and Accountability*, Open University Press, Buckingham.

RISKO, 2000, Mitteilungen für Kommission für Risikobewertung des Kantons Basel-Stadt: Seit 10 Jahren beurteilt die RISKO die Tragbarkeit von Risiken, *Bulletin* **3**, June, 2–3.

Rohrmann, B. and Renn, O., 2000, Risk perception research – An introduction, in: O. Renn and B. Rohrmann (eds.), *Cross-Cultural Risk Perception. A Survey of Empirical Studies*, Kluwer Academic Publishers, Dordrecht, pp. 11–54.

Rosa, E.A., 1998, Metatheoretical foundations for post-normal risk, *Journal of Risk Research* **1**, 15–44.

Rosenau, J.N., 1992, Governance, order, and change in world politics, in: J.N. Rosenau and E.O. Czempiel (eds.), *Governance without Government. Order and Change in World Politics*, Cambridge University Press, Cambridge, pp. 1–29.

Ross, L.D., 1977, The intuitive psychologist and his shortcomings: Distortions in the attribution process, in: L. Berkowitz (ed.), *Advances in Experimental Social Psychology*, Vol. 10, Random House. New York, pp. 173–220.

Rowe, G. and Frewer, L., 2000, Public participation methods: An evaluative review of the literature, *Science, Technology and Human Values* **25**, 3–29.

Shome, N., Cornell, C.A., Bazzurro, P. and Carballo, J.E., 1998, Earthquakes, records and nonlinear responses, *Earthquake Spectra* **14**(3), August, 469–500.

Shrader-Frechette, K.S., 1991a, *Risk and Rationality. Philosophical Foundations for Populist Reforms*, University of California Press, Berkeley, CA.

Shrader-Frechette, K.S., 1991b, Reductionist approaches to risk, in: D.G. Mayo and R.D. Hollander (eds.), *Acceptable Evidence: Science and Values in Risk Management*, Oxford University Press, Oxford, pp. 218–248.

Shrader-Frechette, K.S., 1995, Evaluating the expertise of experts, *Risk: Health, Safety & Environment*, **6**, 115–126.

Shubik, M., 1991, Risk, society, politicians, scientists, and people, in: M. Shubik (ed.), *Risk, Organizations, and Society*, Kluwer Academic Publishers, Dordrecht, pp. 7–30.

Sjöberg, L., 1999, Risk perception in Western Europe, *Ambio* **28**(6), 543–549.

Skinner, D., 1999, *Introduction to Decision Analysis*, Second Edition, Probabilistic Publishers, London.

Slovic, P., 1987, Perception of risk, *Science* **236**, 280–285.

Slovic, P., 1992, Perception of risk: Reflections on the psychometric paradigm, in: S. Krimsky and D. Golding (eds.), *Social Theories of Risk*, Praeger, Westport, CT, pp. 153–178.

Slovic, P., Fischhoff, B. and Lichtenstein, S., 1982, Why study risk perception?, *Risk Analysis* **2**, June, 83–94.

Slovic, P., Finucane, E., Peters, D. and MacGregor, R., 2002, The affect heuristic, in: T. Gilovich, D. Griffin and D. Kahneman (eds.), *Intuitive Judgment: Heuristics and Biases*, Cambridge University Press, Boston, MA, pp. 397–420.

Stein, R.S., 2003, Earthquake conversations, *Scientific American* **288**(1), 72–79.

Stern, P.C. and Fineberg, V., 1996, *Understanding Risk: Informing Decisions in a Democratic Society*, National Research Council, Committee on Risk Characterization, National Academy Press, Washington, DC.

Stirling, A., 1998, Risk at a turning point?, *Journal of Risk Research* **1**(2), 97–109.

Stirling, A., 1999, *On 'Science' and 'Precaution' in the Management of Technological Risk, Volume I: Synthesis Study*, Report to the EU Forward Studies Unit by European Science and Technology Observatory (ESTO), EUR19056 EN, IPTS, Sevilla. Online on Internet: ftp://ftp.jrc.es/pub/EURdoc/eur19056IIen.pdf (accessed on 2005-08-11).

Stirling A., 2003, Risk, uncertainty and precaution: Some instrumental implications from the social sciences, in: F. Berkhout, M. Leach and I. Scoones (eds.), *Negotiating Change*, Edward Elgar, London, pp. 33–76.

Stirling, A., 2004, Opening up or closing down: Analysis, participation and power in the social appraisal of technology, in M. Leach, I. Scoones and B. Wynne (eds.), *Science and Citizens Globalization and the Challenge of Engagement*, Zed, pp. 218–231.

Streffer, C., Bücker, J., Cansier, A., Cansier, D., Gethmann, C.F., Guderian, R., Hanekamp, G., Henschler, D., Pöch, G., Rehbinder, E., Renn, O., Slesina, M. and Wuttke, K., 2003, *Environmental Standards. Combined Exposures and Their Effects on Human Beings and Their Environment*, Springer, Berlin.

Stricoff, R.S., 1995, Safety risk analysis and process safety management: Principles and practices, in: R. Kolluru, S. Bartell, R. Pitblade and S. Stricoff (eds.), *Risk Assessment and Management Handbook. For Environmental, Health, and Safety Professionals*, McGraw-Hill, New York, pp. 8.3–8.53.

Swiss Re, 1995, *The Great Hanshin Earthquake: Trial, Error, Success*, Swiss Reinsurance Company, Zurich.

Thompson, M., Ellis, W. and Wildavsky, A., 1990, *Cultural Theory*, Westview Press, Boulder, CO.

Trustnet, 1999, *A New Perspective on Risk Governance*, Document of the Trustnet Network, EU, Paris. Online on Internet: www.trustnetgovernance.com (accessed on 2005-08-11).

Turner, B.L., Clark, W.C., Kates, R.W., Richards, J.F., Mathews, J.T. and Meyer, W.B., 1990, *The Earth as Transformed by Human Action*, Cambridge University Press, Cambridge.

Tversky, A. and Kahneman, D., 1974, Judgement under uncertainty. Heuristics and biases, *Science* **85**, 1124–1131.

Tversky, A. and Kahneman, D., 1981, The framing of decisions and the psychology of choice, *Science* **211**, 453–458.

UK Treasury Department, 2004, *Managing Risks to the Public: Appraisal Guidance*, Draft for Consultation, HM Treasury Press, London. Online on Internet: www.hm-treasury.gov.uk (accessed on 2005-08-11).

USEPA Environmental Protection Agency, 1997, *Exposure Factors Handbook*, NTIS PB98-124217, EPA, Washington, DC. Online on Internet:
http://cfpub.epa.gov/ncea/cfm/recordisplay.cfm?deid=12464 (accessed on 2005-08-11).

Uff, J., Cullen, W.D. and HSC, 2001, *The Southall and Ladbroke Grove Joint Inquiry into Train Protection Systems*, HSE Books, London.

Van Asselt, M.B.A., 2000, *Perspectives on Uncertainty and Risk*, Kluwer Academic Publishers, Dordrecht.

Van Asselt, M.B.A., 2005, The complex significance of uncertainty in a risk area, *International Journal of Risk Assessment and Management* **5**(2/3/4), 125–158.

Van der Sluijs, J.P., Risbey, J.S., Kloprogge, P., Ravetz, J.R., Funtowicz, S.O., Corral Quintana, S., Guimaraes Pereira, A., De Marchi, B., Petersen, A.C., Janssen, P.H.M., Hoppe, R. and Huijs, S.W.F., 2003, *RIVM/MNP Guidance for Uncertainty Assessment and Communication*, Report No. NWS-E-2003-163, Copernicus Institute for Sustainable Development and Innovation and Netherlands Environmental Assessment Agency, Utrecht and Bilthoven.

Van der Sluijs, J.P., Janssen, P.H.M., Petersen, A.C., Kloprogge, P., Risbey, J.S., Tuinstra, W. and Ravetz, J.R., 2004, *RIVM/MNP Guidance for Uncertainty Assessment and Communication: Tool Catalogue for Uncertainty Assessment*, Report No. NWS-E-2004-37, Copernicus Institute for Sustainable Development and Innovation and Netherlands Environmental Assessment Agency, Utrecht and Bilthoven.

Viklund, M., 2002, *Risk Policy: Trust, Risk Perception, and Attitudes*, Stockholm School of Economics, Stockholm.

Viscusi, W.K., 1994, Risk-risk analysis, *Journal of Risk and Uncertainty* **8**, 5–18.

Vogel, D., 2003, Risk regulation in Europe and in the United States, in: H. Somsen (ed.), *Yearbook of European Environmental Law*, Volume 3, Oxford University Press, Oxford.

Von Winterfeldt, D. and Edwards, W., 1984, Patterns of conflict about risk debates, *Risk Analysis* **4**, 55–68.

WBGU (Wissenschaftlicher Beirat der Bundesregierung Globale Umweltveränderungen), 2000, *World in Transition: Strategies for Managing Global Environmental Risks*, Springer, Berlin.

Webler, T., 1995, Right discourse in citizen participation. An evaluative yardstick, in: O. Renn, T. Webler and P. Wiedemann (eds.), *Fairness and Competence in Citizen Participation. Evaluating New Models for Environmental Discourse*, Kluwer Academic Publishers, Dordrecht, pp. 35–86.

Webler, T., 1999, The craft and theory of public participation: A dialectical process, *Risk Research* **2**(1), 55–71.

Webler, T., Levine, D., Rakel, H. and Renn, O., 1991, The Group Delphi: A novel attempt at reducing uncertainty, *Technological Forecasting and Social Change* **39**, 253–263.

Wiener, J.B., 1998, Managing the iatrogenic risks of risk management, *Risk: Health, Safety & Environment* **9**, 39–83.

Wisdon, J. and Willis, R., 2004, *See-Through Science. Why Public Engagement Needs to Move Upstream*, Demos, London.

Wolf, K.D., 2000, *Die neue Staatsräson – zwischenstaatliche Kooperation als Demokratieproblem in der Weltgesellschaft. Plädoyer für eine geordnete Entstaatlichung des Regierens jenseits des Staates*, Nomos, Baden-Baden.
Wolf, K.D., 2002, Contextualizing normative standards for legitimate governance beyond the state, in: J.R. Grote and B. Gbikpi (eds.), *Participatory Governance. Political and Societal Implications*, Leske und Budrich, Opladen, pp. 35–50.
Wolf, K.D., 2005, Private actors and the legitimacy of governance beyond the state. Conceptional outlines and empirical explorations, in: A. Benz and I. Papadopoulos (eds.), *Governance and Democratic Legitimacy: Transnational, European, and Multi-Level-Issues*, Routledge, London, pp. 200–227.
Wynne, B., 1992, Risk and social learning: Reification to engagement, in: S. Krimsky and D. Golding (eds.), *Social Theories of Risk*, Praeger, Westport, CT, pp. 275–297.
Wynne, B., 2002, Risk and environment as legitimatory discourses of technology: Reflexivity inside out?, *Current Sociology* **50**(30), 459–477.
Zürn, M., 2000, Democratic governance beyond the nation-state: The EU and other international institutions, *European Journal of International Relations* **6**(2), 183–221.

Part 2. A Framework for Risk Governance: Critical Reviews

Chapter 2
A Framework for Risk Governance Revisited

Ragnar Löfstedt[1] and Marjolein van Asselt[2]
[1]*King's Centre for Risk Management, King's College London, UK*
[2]*Faculty of Arts and Social Sciences, Maastricht University, The Netherlands*

Introduction

Risk managers like models; they use them to compartmentalise the world they operate in. Good models offer ways to better assess and cope with risks. Over the past decades, a multitude of risk management models has been proposed, some more useful than others. Among the most well known are the National Research Council's 1983 report *Federal Risk Assessment-Managing the Process*, which calls for separation of risk assessment from risk management (NRC 1983), the National Research Council's 1996 report *Understanding Risk: Informing Decisions in a Democratic Society*, which argues for a risk characterisation process to be implemented throughout the entire risk analysis process (NRC 1996) and the UK Royal Commission for Environmental Pollution 1998 report *Setting Environmental Standards*, which argues that the articulation of people's values should more or less be incorporated in risk management processes (RCEP 1998). Models for risk management are neither easy to develop, nor are they by definition useful for policy makers. Good risk management models need to be easily understood in order to be usable by risk management practitioners, while at the same time such models should reflect the scholarly state of the art. They are simplified versions of a complex body of knowledge. In other words, sound frameworks for risk management are both as sophisticated and as simple as possible. Developing a risk management model thus requires synthesis of available insights as well as simplification and translation into easy to understand schemes and recommendations. This is the challenge the International Risk Governance Council (IRGC) took on in developing its risk governance framework.

O. Renn and K. Walker (eds.), Global Risk Governance: Concept and Practice Using the IRGC Framework, 77–86.

The IRGC Framework for Risk Governance

The IRGC's framework for risk governance (IRGC 2005) is both a progression and extension from both the NRC 1996 and RCEP 1998 models in that it calls for inclusion of the societal context and categorisation of risk-related knowledge.[1] With the attention to social context, it is recognised and emphasised that various publics, stakeholders and regulators perceive risks differently (Slovic 2000) and that such differences in risk perceptions are not reducible to a 'knowledge deficit' on the part of non-experts (Wynne 1989, 1996; Irwin and Wynne 1996).

With the categorisation of risk-related knowledge, the IRGC framework recognises that not all risks are the same. Some are better understood than others: potential negative consequences are obvious, applied values are non-controversial and uncertainties can be captured in probability distributions. Other risks are inherently uncertain, complex and/or ambiguous. These risks should be treated differently. In line with the state of the art in the field of risk research, the IRGC framework calls attention to such classes of risks which cannot be addressed with traditional assessment and decision-making tools. In this regard, the IRGC categorisation resembles Funtowicz and Ravetz's Post Normal Science model (Funtowicz and Ravetz 1990, 1992; Ravetz 1999; see also van Asselt 2000). By categorising risks via the 'risk escalator', the IRGC framework attempts to deal with management issues in a proactive way so as reduce chances of surprise, or worse, a crisis.

The IRGC model brings together both the assessment and management spheres and calls on communication to be part of pre-assessment, risk management, risk evaluation and risk assessment. The plea for communication is particularly important in the European context, where there has been a spate of scandals over recent years leading to a 'post-trust society' (Löfstedt 2005). These events have made humble, honest and proactive risk communication a necessary, if not the most crucial, element in risk management.

Reception

To date, the IRGC framework has gained considerable attention. It served as a topic for an initial discussion in a well-attended meeting arranged by IRGC in Switzerland in June 2004 and a follow up meeting in the fall the same year. In 2005, the framework was presented at a number of workshops including one at Her Majesty's Treasury in London, leading a reference to the IRGC's 'risk escalator' in one of the report's appendices to the Treasury guidelines on communicating risks to the pub-

[1] The risk governance model described in the IRGC White Paper is clearly rooted in Ortwin Renn's earlier work on risk governance funded by the European Commission ("Precautionary Strategies and Research Needs to Compose and Specify a European Policy on the Application of the Precautionary Principle") and the German Advisory Council on Global Change (WBGU), entitled "World in Transition. Strategies for Managing Global Environmental Risks" (Klinke and Renn 2001, 2002; Klinke et al. 2006; Renn et al. 2003; WBGU 2000).

lic (HM Treasury 2005). The report was also extensively discussed at the Society for Risk Analysis in Orlando, Florida in December 2005. It served as a background both for the keynote speech given by Dr. Renn, as a well as for an entire session at the meeting. Since then, IRGC has attempted to operationalise the framework in the areas of nanotechnology (IRGC 2006) and food (Stirling et al. 2006). At the Society for Risk Analysis meeting in Baltimore, Maryland in 2006, a number of case studies ranging from acrylamide to genetically modified crops (GMOs) were introduced to explore application of the IRGC model (see Part 3 of this book). Furthermore, the Dutch Health Council has employed the IRGC framework in its recent assessment on potential health effects associated with nanotechnology (Health Council of the Netherlands 2006). Other research bodies, most notably the University of Michigan, have also expressed interest in employing the IRGC framework for risk governance.

Strengths

The IRGC framework offers a new model that is conceptually more sophisticated than the ones developed in the past and at the same time it is much more proactive. It does recognise that risk managers need to manage risks in instances of 'complexity', 'uncertainty' and/or 'ambiguity'. We are convinced that the IRGC framework can help risk managers to avoid unnecessary pitfalls and aid them in applying the 12 'late' lessons so ably summarised in the European Environment Agency report (Harremoes et al. 2002). As IRGC argues, the framework can be of use in structuring risk governance processes, in diagnosing deficiencies and understanding problems in risk controversies, and it can serve as a framework for comparison. Furthermore, it offers vocabulary to talk about risks which cannot be addressed by traditional assessment and decision-making tools.

The major strengths of the model reside in the following factors:

- risk governance as a *dynamic process* with many iterative loops and the emphasis on *flexibility* in applying the model in order to do justice to the particular societal context and contingencies;
- *communication* as the core of risk governance;
- *pre-assessment*, e.g. framing, as well as *monitoring* as important, integral parts of risk governance;
- the ambition to blend knowledge with perceptions in *concern assessment* and *risk appraisal*;
- the importance of *organisational capacity*.

Furthermore, IRGC should be commended for its plea to integrate this kind of thinking about risk into education, as opposed to 'compartmentalised specialisation' (p. 61) fostered by most current education systems (compare the pleas for uncertainty awareness in education in Pollack 2003; and van Asselt and Petersen 2003). Finally, we think that the distinction between risk management strategies directed to risk agents (i.e. the sources of risks) and strategies directed to risk absorbing systems (i.e. the potential 'risk victims') is also useful for risk managers. We are convinced that if the IRGC framework is applied as it is intended, it can facilitate responsible, ethical, efficient and effective management of the important risks facing 'post-trust' societies today.

Critique

Any risk governance model or framework can be easily criticised. It is either too simple or too complex, or too pragmatic or too scholarly. Criticising the IRGC framework therefore may seem a cheap shot. However, models should not be taken at face value either. In order to understand whether and how a new model is better than its predecessors, it is necessary that they be constructively reviewed. IRGC has explicitly presented the framework as a 'prototype', which needs further development and testing.

The model, as noted, has many strengths so it already renders previous models outdated. Nevertheless, in the remainder of this commentary, we would like to discuss some of the features which we believe to be noteworthy weaknesses.[2] With our commentary, we aim to contribute to further development of a sound framework for risk governance of complex, uncertain and/or ambiguous risks.

The Need for Further Simplification

As noted, a good risk governance model is a simplified version of a complex body of knowledge translated into easy to understand schemes and recommendations. IRGC's endeavour is impressive because of its sophisticated, highly interdisciplinary coverage and as such, it is an important and highly welcome contribution to the field of risk research. We have no doubt that this White Paper will develop into a frequently consulted reference work. On the other hand, the impressive and quite comprehensive overview can be at times overwhelming to someone unfamiliar with the scholarly debate surrounding this issue making it difficult to discern 'the forest for the trees'.

First, the document gives the appearance of having been drafted by an academic largely for an academic audience. The paper is not easy to follow, in particular for non-native, non-risk policy specialists. It is far too long; the executive summary should be limited to a page at the most, and the report itself should be no more than 20 pages in length. The language should also be made much simpler. Terms such as 'aleatory', concepts like 'ambiguity', technical jargon such as 'exceedance probability', are difficult even for well versed risk professionals to understand, let alone to apply. Many of the schemes and tables are not self-explanatory. Although

[2] It is actually quite difficult to do so, as Renn has already stolen our thunder with the many self-critical questions and caveats he already brought to the fore.

examples are usually useful, the examples on high speed trains, earthquakes and nuclear energy do not really add much to the framework.

The ultimate challenge to developing a risk governance framework lies in communicating points of agreement in the scholarly state of the debate. To do so requires ignoring many academic nuances, interesting subtleties and theoretical subdivisions. In our opinion, the IRGC framework did not adequately withstand the scholar's inclination to detail details and to differentiate differentiations.

Nonetheless, we would consider it highly unfortunate if the insights and utility of IRGC's risk governance framework were to be lost in the style and scholarly density of the current report. We therefore advise IRGC to commission a science journalist or another professional experienced in popularising scientific knowledge to a broader public, to produce a version of the risk governance framework that can be 'sold' to policy-making communities. The current White Paper can still serve as a scientific background document for the academic audience or academically oriented risk managers, but it needs a 'quick scan' complement.

The Need for Adequate Positioning

We regret that IRGC could not resist the temptation to portray their risk governance framework as comprehensive, i.e. applicable in all instances of risk management. A careful reading of the document suggests that IRGC was motivated to develop the framework primarily for human-induced globally relevant risks (transboundary, international, ubiquitous, large scale effects), which involve many actors ('horizontal governance') and/or different levels of decision-making ('vertical governance'). We would argue that the IRGC risk governance framework is especially targeted towards those global risks which are not 'simple' or 'known', i.e. risks which cannot be (fully) estimated in terms of probability and effect, because science is inconclusive, historical experience is limited or lacking, and causalities are complex and interdependent (compare van Asselt and Vos 2006). We fully agree with IRGC that some risks may seem simple at first sight, but are in fact complex, uncertain and/or ambiguous. Nevertheless, we are convinced that it would help, both in terms of practical and academic value, if the IRGC framework were to be portrayed and described as a 'risk governance framework for complex, uncertain and/or ambiguous risks'.

The Need for Rethinking

The IRGC framework calls for greater public and stakeholder involvement in risk management process.[3] In many cases this is indeed important. This is especially the case in a post-trust society, where deliberation addressing fairness issues is crucial

[3] The framework builds on previous work on fairness and public participation (e.g. Renn 2004; Renn et al. 1995).

to (re)build public trust (Renn and Levine 1991; Löfstedt 2005). The public and stakeholders need to feel listened to. It leads to a sense of ownership of the outcome (Schelling 1960), and helps to reassure the public and stakeholders that regulators and policy makers have their best interest at heart and are not captured by industry (Löfstedt 2005). An up-to-date risk governance model thus needs to emphasise that risk managers should seriously think about deliberation as an option, rather than rely too reflexively on technocratic and bureaucratic solutions.

Although deliberation involving the public and stakeholders can positively contribute to risk management, it should not be treated as a panacea (Barber 1984; Chess et al. 1995; Fiorino 1989; Petts 1995; Rowe and Frewer 2000; Wynne 1989, 1996). For each risk management issue, the issue of whether and how deliberation may or may not be adequate needs to be critically evaluated. There are numerous issues to consider:

First, deliberation alone does not guarantee public trust in risk producers, risk managers and regulators. In many cases, the public distrusts regulators, policy makers and/or industry for reasons other than fairness. Issues of lack of competence and lack of efficiency can also be important factors in public distrust (Löfstedt 2005), which cannot be addressed by deliberation. If distrust is rooted in concerns about competence and efficiency, deliberation may even increase distrust and hamper responsible dealing with risk. In cases where public trust is otherwise high, deliberation may result in growing inefficiency, which may actually diminish trust in industry and public authorities (Löfstedt 2001).

Second, deliberation is usually very expensive and time consuming. Many governments cannot afford to spend scarce resources on deliberative processes.

Third, it is difficult to assure representativeness of those participating in the deliberative process. There is often the self-selection problem, i.e. most members of the public who are asked to participate in the policy-making process do not wish to do so (response rates often vary between 1–20%) (Löfstedt 1999). Even in cases where deliberative processes do not aim to be representative, it is nevertheless problematic when the participants are primarily retired, unemployed and/or lonely people, or merely members of an active, concerned, and risk averse minority (Löfstedt 2004, 2005).

Fourth, in cases of highly contested risks, it may be especially problematic to get relevant stakeholders and publics willing to participate. Some risks may be so contentious that organising a deliberative process is a controversial act in itself. The Swiss Institute for Environmental Decisions (IIED) at the ETH recently had this experience when they tried to organise a transdisciplinary process pertaining to the siting of low-level radioactive waste in a canton which had a history of being a repository candidate and which presented fierce opposition from civil society (personal communication of van Asselt with IIED staff, November 2006).

Despite our reservations about deliberation as the 'be all and end all' of risk management, we want to support and confirm that communication is at the heart of risk governance. We therefore fully agree with the central position of communication in the IRGC's framework. Industry, governments, experts, governmental and non-governmental organisations should communicate about risks amongst themselves and to stakeholders and the public. Risk scandals have made crystal clear that any risk management which does not involve honest, humble and transparent communication is doomed to failure.

We make the distinction, however, that communication and the serious consideration of public and stakeholder concerns does not necessarily require deliberation, i.e. active participation of the public and stakeholders in deliberative risk management processes. We fear what appear to be naïve pleas for deliberation as a standard model for risk communication in the current version of the IRGC's framework. The framework appears to advocate deliberation as a central feature of the model: 'an obligation to ensure the early and meaningful involvement of all stakeholders and, in particular, civil society'; the lengthy section devoted to stakeholder involvement and participation as prelude to the section on risk communication; and in the section on stakeholder involvement where it is explicitly argued that 'the (...) framework advocates the notion of inclusive governance (...) this means that (...) political, business, scientific and civil society players should jointly engaged (...) and [come] to a joint conclusion'.

IRGC might argue that deliberation is 'just' needed in cases of ambiguous risks, those risks where fundamental differences exist amongst different actors with respect to in the interpretation of information or values associated with particular outcomes. While we do agree that not all ambiguous risks are by definition uncertain or complex, we find it difficult to imagine complex and/or uncertain risks that are not ambiguous or contested in some way (compare van Asselt 2005). If so, the majority of complex and/or uncertain risks are also ambiguous, with the consequence that deliberation appears to be the de facto model for governance of any risk that is not strictly 'simple' in the IRGC framework.

Given the strong preference for public and stakeholder involvement generally in the framework, we were surprised that IRGC treats public and stakeholder participation so asymmetrically in the management and risk assessment components of the framework. While on the one hand, deliberation is strongly advocated for the risk *management* part of risk governance, public and stakeholder participation is rather absent as part of risk *assessment*. Stakeholder and public input are portrayed in terms of 'emotions, hopes, fears and apprehension' and much less in terms of experiential knowledge and particular local context expertise. This approach appears to ignore literature advocating public and stakeholder participation in the knowledge gathering process, also referred to as transdisciplinary science (among others, Wynne 1982; van Asselt and Rijkens-Klomp 2002; Kasemir et al. 2003; van de Kerkhof 2004). Rather, the IRGC framework appears to argue that the 'factual basis' of risk assessment should come from experts (compare Breyer 1993; Cross 1994; Graham and Hartwell 1997) and that 'the basis of risk assessment' is 'the systematic use of analytical – largely probability-based – methods'. IRGC seems to ignore that in case of complex, uncertain and/or ambiguous risks, risk assessment is not just a matter of knowledge gathering ('what is known?'), but that in those cases the delineation of so-called 'uncertainty information' ('what is unknown?' and 'which unknowns are relevant for public decision making?') (van Asselt and Petersen 2003; Harremoes et al. 2002) is equally, if not more, important and definitely more challenging.

Stakeholder and public involvement, in one way or the other, might contribute to the demanding production of uncertainty information.

So on the one hand, we would like to see a more critical treatment – and some toning down – of deliberation in risk *management.* On the other hand, we would like to challenge IRGC to pay some serious attention to the potential value of stakeholder and public participation in risk *assessment.*

Conclusions

We feel that this IRGC framework is a very important document for the wider risk management community. It offers a new model that is conceptually more sophisticated and more proactive than the ones developed in the past. The framework could be used by risk managers to improve policy-making processes in ways that may prevent unpleasant surprises and pre-empt societal crises. Nevertheless, we have argued that serious simplification is needed in order to develop the IRGC framework into a usable tool risk management tool. Furthermore, we have recommended that IRGC explicitly position the framework as a risk governance model primarily for complex, uncertain and/or ambiguous risks, rather than for simple risks. Last but not least, we sketched issues that need to be rethought by the scholarly risk research community at large. The IRGC White Paper is not only a very valuable reference work for the academic community, but it also sets a research agenda which is as challenging and interdisciplinary as it is important for today's societies.

Acknowledgement

We are grateful to Chris Bunting at IRGC for giving us an opportunity to comment on the IRGC framework.

References

Barber, B.R., 1984, *Strong Democracy: Participatory Politics for a New Age*, University of California Press, Berkeley, CA.

Breyer, S., 1993, *Breaking the Vicious Circle: Toward Effective Risk Regulation*, Harvard University Press, Cambridge, MA.

Chess, C., Salamone, K.L., Hance, B.J. and Saville, A., 1995, Results of national symposium on risk communication: Next steps for government agencies, *Risk Analysis* **15**, 115–125.

Cross, F.B., 1994, The public in risk control, *Environmental Law* **24**, 888–969.

Fiorino, D., 1989, Environmental risk and democratic process: A critical review, *Columbia Journal of Environmental Law* **14**, 501–547.

Funtowicz, S.O. and Ravetz, J., 1990, *Uncertainty and Quality in Science for Policy*, Kluwer Academic Publishers, Dordrecht.
Funtowicz, S.O. and Ravetz, J., 1992, Three types of risk assessment and the emergence of post-normal science, in: S. Krimsky and D. Golding (eds.), *Societal Theories of Risk*, Praeger Publishers, Westport, CT.
Graham, J. and Hartwell, J.K. (eds.), 1997, *The Greening of Industry: A Risk Management Approach*, Harvard University Press, Cambridge, MA.
Harremoes, P., Gee, D., MacGarvin, M., Stirling, A., Keys, J., Wynne, B. and Guedes Vaz, S., 2002, *The Precautionary Principle in the 20th Century: Late Lessons from Early Warnings*, Earthscan, London.
Health Council of the Netherlands, 2006, *Health Significance of Nanotechnologies*, Health Council of the Netherlands, Publication No. 2006/06, The Hague.
HM Treasury, 2005, *Managing Risks to the Public: Appraisal Guidance*, HM Treasury, London.
IRGC, 2005, *White Paper on Risk Governance: Towards an Integrative Approach*, IRGC, Geneva.
IRGC, 2006, *White Paper on Nanotechnology: Risk Governance*, IRGC, Geneva.
Irwin, A. and Wynne, B., 1996, *Misunderstanding Science? The Public Reconstruction of Science and Technology*, Cambridge University Press, Cambridge, UK.
Kasemir, B., Jäger, J., Jaeger, C. and Gardner, M.T., 2003, *Public Participation in Sustainability Science*, Cambridge University Press, Cambridge, UK.
Klinke, A. and Renn, O., 2001, Precautionary principle and discursive strategies: Classifying and managing risks, *Journal of Risk Research* **4**, 159–174.
Klinke, A. and Renn, O., 2002, A new approach to risk evaluation and management: Risk-based, precaution based and discourse-based strategies, *Risk Analysis* **22**(6), 1071–1094.
Klinke, A., Dreyer, M., Renn, O., Stirling, A. and van Zwanenberg, P., 2006. Precautionary risk regulation in European governance, *Journal of Risk Research* **9**(4), 373–392.
Löfstedt, R.E., 1999, The role of trust in the North Blackforest: An evaluation of a citizen panel project, *Risk, Health, Safety and Environment* **10**, 7–30.
Löfstedt, R.E., 2001, Risk and regulation: Boat owners' perceptions to recent antifouling legislation, *Risk Management* **3**(3), 33–46.
Löfstedt, R.E., 2004, Risk communication in the 21st century, *International Public Management Journal* **7**(3), 335–346.
Löfstedt, R.E., 2005, *Risk Management in Post Trust Societies*, Palgrave/MacMillan, Basingstoke, UK.
National Research Council, 1983, *Risk Assessment in the Federal Government: Managing the Process*, National Academy Press, Washington, DC.
National Research Council, 1996, *Understanding Risk: Informing Decisions in a Democratic Society*, National Academy Press, Washington, DC.
Petts, J., 1995, Waste management strategy development: A case study of community involvement and consensus building in Hampshire, *Journal of Applied Social Psychology* **26**, 1427–1453.
Pollack, H.N., 2003, *Uncertain Science … Uncertain World*, Cambridge University Press, Cambridge, UK.
Ravetz, J., 1999, What is post-normal science?, *Futures* **31**, 647–653.
RCEP, UK Royal Commission for Environmental Pollution, 1998, *Setting Environmental Standards*, RCEP, London.
Renn, O., 2004, The challenge of integrating deliberation and expertise: Participation and discourse in risk management, in: T.L. McDaniels and M.J. Small (eds.), *Risk Analysis and Society: An Interdisciplinary Characterization of the Field*, Cambridge University Press, Cambridge, UK.
Renn, O. and Levine, D., 1991, Credibility and trust in risk communication, in: R. Kasperson and P.J. Stallen (eds.), *Communicating Risks to the Public: International Perspectives*, Kluwer, Amsterdam.
Renn, O., Webler, T. and Wiedemann, P., 1995, *Fairness and Competence in Citizen Participation. Evaluation Models for Eenvironmental Discourse*, Kluwer Academic Publishers, Dordrecht.
Renn, O., Dreyer, M., Klinke, A. et al., 2003, *The Application of the Precautionary Principle in the European Union. Regulatory Strategies and Research Needs to Compose and Specify*

a European Policy on the Application of the Precautionary Principle, Final Report to the European Commission, Dialogik, Stuttgart (unpublished).

Rowe, G. and Frewer, L.J., 2000, Public participation methods: A framework for evaluation, *Science, Technology and Human Values* **225**, 3–29.

Schelling, T., 1960, *The Strategy of Conflict*, Harvard University Press, Cambridge, MA.

Slovic, P., 2000, *Perception of Risk*, Earthscan, London.

Stirling, A., Ely, A., Renn, O. et al., 2006, *A General Framework for the Precautionary and Inclusive Governance of Food Safety in Europe*, Dialogik, Stuttgart.

van Asselt, M.B.A., 2000, *Perspectives on Uncertainty and Risk: The PRIMA Approach to Decision Support*, Kluwer Academic Publishers, Dordrecht.

van Asselt, M.B.A., 2005, The complex significance of uncertainty in a risk era: Logics, manners and strategies in use, *International Journal for Risk Assessment and Management* **5**(2/3/4), 125–158.

van Asselt, M.B.A. and Petersen, A.P., 2003, *Not Afraid of Uncertainty*, Lemma/RMNO, The Hague, the Netherlands [in Dutch].

van Asselt, M.B.A. and Rijkens-Klomp, N., 2002, A look in the mirror: Reflection on participation in integrated assessment from a methodological perspective, *Global Environmental Change* **12**(3), 167–184.

van Asselt, M.B.A. and Vos, E., 2006, The precautionary principle and the uncertainty paradox, *Journal of Risk Research* **9**(4), 313–336.

van de Kerkhof, M., 2004, *Debating Climate Change. A Study on Stakeholder Participation in an Integrated Assessment of Long-Term Climate Policy in the Netherlands*, Lemma, Utrecht, the Netherlands.

WBGU, German Advisory Council on Global Change, 2000, *World in Transition; Strategies for Managing Global Environmental Risks*, Annual Report, Springer, Berlin.

Wynne, B., 1982, *Rationality and Ritual: The Windscale Inquiry and Nuclear Decisions in Britain*, British Society for the History of Science, Chalfont, St. Giles.

Wynne, B., 1989, Sheepfarming after Chernobyl: A case study in communicating scientific information, *Environment* **31**(2), 10–15, 33–39.

Wynne, B., 1996, May the sheep safely graze?, in: S. Lash, B. Szerszynski and B. Wynne (eds.), *Risk, Environment and Modernity: Toward a New Ecology*, Sage Publications, London.

Chapter 3
Enterprise Risk Management Perspectives on Risk Governance

Robin Cantor
Navigant Consulting, Inc., Washington, DC, USA

Most of my professional life is spent addressing issues related to the unlawful, unintended, or completely unanticipated consequences of product, production or market activities. Many of my projects involve analysis of product liability and the interaction of the various private and public institutions that regulate, resolve disputes, compensate, or punish the relevant parties. The problems can be quite limited as in a simple insurance dispute regarding coverage for a single company's environmental damage at one site, or highly complex as in the long-term forecast of liability for asbestos risks at the national level over a 50 year time horizon. My consulting experiences have taught me that risk governance, even for a 'simple' risk problem, not only occurs on many different levels of society, but also must contend with the widely varying interests of the participants who influence choices and outcomes. That any single framework could begin to tame the messy assessment, appraisal and management tasks of complex and/or global risk problems with the hope of offering effective guidance for governance is certainly a bold claim. Nonetheless, the IRGC risk governance framework, in my view, moves us towards that goal by its attention to integrating key insights about risk from the social, behavioural, natural, and engineering sciences. At the very least, the integrated messages of the framework are important contributions to directing a high quality analysis to support risk governance planning, perhaps even more so at non-governmental levels, where risk governance is a relatively novel concept.

My comments on the framework, therefore, will largely focus on the efficacy of such a framework for risk governance planning at levels below the national authority. By way of background, I note that in recent years there has been a renewed interest in Enterprise Risk Management (ERM), both in the corporate and financial worlds.[1] Some of this interest is motivated by better measurement, data collection,

[1] See, for example, *Overview of Enterprise Risk Management*, Casualty Actuarial Society, May 2003.

O. Renn and K. Walker (eds.), Global Risk Governance: Concept and Practice Using the IRGC Framework, 87–91.

analysis, and forecasting methods regarding a wide range of risks – environmental, product, operational, market, and financial. There are also greater demands from regulatory and financial oversight institutions for the corporate world to analyse, disclose, and actively manage existing or emerging risk problems. Many of the topics and insights of the IRGC framework should be a great value to corporate decision makers as they plan for and respond to increasing ERM responsibilities.

The framework as proposed emphasises 'a common *analytical structure for investigating and supporting the treatment of risk issues* by the relevant actors in society' (emphasis in original). As developed, the framework seems more consistent with public decision making in a context less driven by litigation than the typical high-profile risk experience of the US. To realise its potential for the corporate context, however, the framework needs to expand its treatment of (1) accountability and liability, (2) retrospective applications, and (3) *ex post* as opposed to *ex ante* risk governance activities.

The concepts accountability and liability are critical in the world of corporate decision making because accountability often drives trust, and liability often drives financial survival. A retrospective phase of the framework would be important for its credibility with corporate decision makers, because it corresponds to the institutionalised business focus on historical case studies and 'lessons learned'. Moreover, retrospective applications provide essential information for calibrating the framework to real-world experiences. Retrospective analyses will also further our understanding of *ex post* as opposed to *ex ante* risk governance and important distinctions between the two for the interaction of industry and government. In the *ex ante* phase of this relationship, the focus is often risk assessment and public acceptance. In the *ex post* phase, the focus is more likely, 'What did you know and when did you know it?'.

For example, a (brief) retrospective evaluation of the recent history of risk governance related to asbestos in the US clearly reveals the important role of liability and *ex post* governance reactions.

Beginning in the 1970s, and for the first decade of personal injury tort claims in the US, seriously ill individuals – those with cancer or severe asbestosis – filed almost all of the claims, and much of the litigation targeted the Johns-Manville Corporation.[2] Under the weight of the litigation, Manville declared bankruptcy in 1982, resulting in the formation of the Manville Personal Injury Settlement Trust (the 'Manville Trust').

In the years following the formation of the Manville Trust, the focus of asbestos litigation shifted towards diseases with weaker relationships to asbestos exposure and towards claimants alleging non-malignant conditions such as pleural plaques that had not caused impairment. Specifically, beginning in the mid-1980s there was a surge in filings of asbestos-related claims, of which a growing proportion were claims alleging non-malignant conditions. To date, well over 700,000 claims have been filed, more than $70 billion has been spent on asbestos litigation, and over

[2] 'Recent Developments in Assessing Future Asbestos Claims under the Fair Act', hearing before the Senate Committee on the Judiciary: Prepared Statement of Testimony of Mark Lederer, November 17, 2005, 109th Cong.

70 companies have petitioned for bankruptcy protection in the US due to asbestos personal injury liabilities (Carroll et al. 2005).

Importantly, risk governance of asbestos was almost entirely an *ex post* problem. Notwithstanding the health and safety regulations, product bans and voluntary product withdrawals introduced by public and private decisions beginning in the 1970s,[3] most risk governance of asbestos occurred once asbestos-containing products were ubiquitous in the workplace. As a result, estimates indicated that 27.5 million workers in the US had some occupational exposure to asbestos between 1935 and 1979 (Nicholson et al. 1982). For the hundreds of thousands of bodily injury and property damage claims that were filed, attention quickly focused on a defendant company's liability for 'defective' products and 'failure to warn' exposed claimants about the dangers of its products. In total, more than 8000 companies have been named as defendants in these lawsuits.

Not surprisingly, there have been several attempts over the years to bring some level of control to the liability problem through federal legislation of various designs. The most recent is the Fairness in Asbestos Injury Resolution (FAIR) Act, which has been before the US Congress in various forms since 2003. However, the FAIR Act appears to have met the fate of previous Congressional efforts to devise a legislative solution. A key reason for failure was that no one could predict with reasonable certainty how much the new system would cost, and this amplified the concerns that the funding would fall short and the government would ultimately end up with the liability.

In contrast, there is mounting evidence in the US that other reforms to control the number and cost of asbestos claims are succeeding. Several states, including those which have drawn large numbers of claims, have enacted general or asbestos-specific tort reform legislation. These bills require plaintiffs with non-malignant conditions to provide evidence of impairment that meets strict criteria, and often require those with malignancies to meet medical evidence standards as well, in order to have an actionable claim. Importantly, medical criteria controls have now been enacted in most of the states with the highest historical claim filings and in the claims procedures for payment under the Manville Trust.

Figure 1 shows claim count data from the Manville Trust and the dramatic decline in claims filed after the full implementation of the stricter medical criteria. The pattern observed in the trust claims can also be seen in the claims filed against many US defendants and other bankruptcy settlement trusts.

Empirical evidence that these non-federal changes[4] are reducing claim counts and liability includes a more than 70% reduction since 2002 in the claims filed

[3] The first US Occupational Safety & Health Administration standards setting permissible exposure levels for workers exposed to asbestos fibers was issued in 1971. In the late 1970s, the US Consumer Product Safety Commission (CPSC) banned the use of asbestos in many products used for home construction. In 1989, the US Environmental Protection Agency issued a final rule banning most asbestos-containing products. While particular aspects of the regulation were overturned by a Court of Appeals, the remaining components of the rule did ban the use of asbestos in products that did not historically contain asbestos, otherwise referred to as 'new uses' of asbestos.

[4] Other provisions enacted in several states include venue reform to restrict 'forum shopping'; limits on consolidation of cases; restrictions on non-product claims and claims against innocent

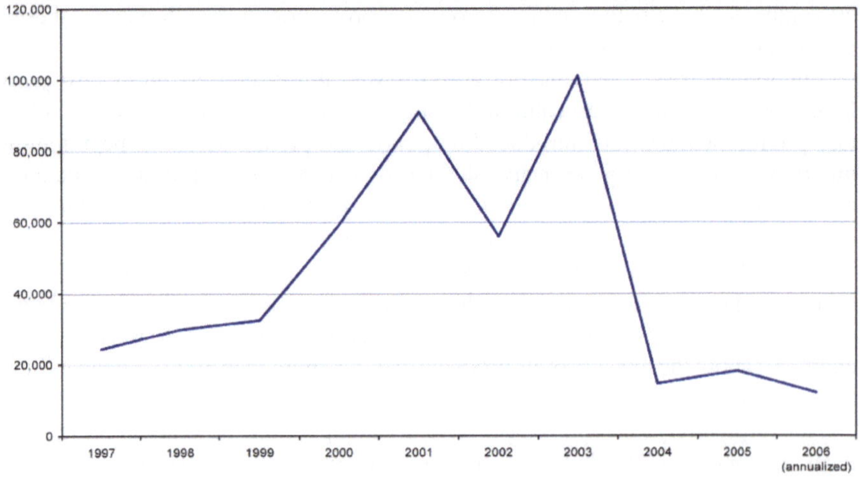

Fig. 1 Manville Trust annual claims fillings. Source: Manville Trust quarterly reports.

against large defendants publicly reporting such data, and reported dismissal rates that often exceed 75%. Public filings also show that some companies are experiencing lower average settlement values even for malignant claims or lower average resolution values across all diseases, or have reduced their liability estimates for pending and future claims.

In light of the recent dramatic changes in asbestos claims, one might believe that the liability crisis is over. Whether or not effective *ex post* governance has arrived for asbestos risks, it is important to understand that the asbestos experience in the US has fundamentally altered many of the options available to corporations to manage and govern risk. Following the experiences with asbestos and general liability policies for defendant companies, insurance policies now routinely include complex exclusion terms to limit insurance/reinsurance exposure to product risks. With or without insurance coverage, a history of substantial product liability has been shown to affect a company's ability to raise capital, which in turn increases the costs of borrowing funds for productive investments.

Retrospective application of the framework to other mass tort problems like asbestos would be of great value specifically to sort out the accountability and liability raised here and the lessons learned. The integration of resulting insights into the framework is likely to be critical for developing effective risk governance solutions for widely dispersed products. To have any measure of success, such solutions must be designed to balance the needs of diverse public *and* private stakeholders who often may find their expectations at odds regarding innovation, market expansion, transnational development, and unintended consequences.

---

sellers; limits on successor liability; and elimination or restriction of joint and several liability, non-economic damages and punitive damages.

References

Casualty Actuarial Society, 2003, *Overview of Enterprise Risk Management*, May 2003.

Carroll, S.J. et al., 2005, *Asbestos Litigation*, RAND Institute for Civil Justice.

Nicholson, W.J., Perkel, G. and Selikoff, I.J., 1982, Occupational exposure to asbestos: Population at risk and projected mortality, 1980–2030, *American Journal of Industrial Medicine* **3**, 259–311.

Chapter 4
Comments on the IRGC Framework for Risk Governance[1]

D. Warner North
NorthWorks, Inc., Belmont, and Department of Management Science and Engineering, Stanford University, California, USA

Overview

My overall reaction to the IRGC risk governance framework is extremely positive. However, I have concerns that it may be misinterpreted as a 'how-to-do-it' manual, a check list, or a catalogue of what IRGC considers the 'approved' set of analytical tools and methodology. Such interpretations could be restrictive and impede progress in improving risk analysis and risk governance. The IRGC framework should be viewed as a broad summary of what risk analysis is, and why risk analysis is needed in support of improving risk governance. It provides a useful taxonomy and an overview of how risk analysis can be effectively accomplished and communicated, through an iterative process of work by trained analysts and dialogue with political leaders and the interested/affected parties on risk issues. The framework should not be regarded as rigid, that is, defining a set of methods for risk analysis practice, but rather it should be regarded as an overview of the potential for risk analysis to assist in improving risk governance. In particular, it should not be interpreted as restricting methods and analytical tools for dealing with uncertainty, complexity, and ambiguity. Selection of such methods should be consistent with best practices from the scientific and engineering disciplines supporting risk analysis, as these practices evolve through further research and further experience.

---

[1] At the Society for Risk Analysis Annual Meeting in Orlando, Florida in December of 2005, I delivered prepared comments on the *IRGC White Paper on Risk Governance* (hereafter, the White Paper) at a session in which it was presented and discussed. I have since become a member of the IRGC Science and Technology Council. My thinking on the White Paper has continued to evolve since December. This version of my comments was written prior to attending my first S&T Council meeting May 19–20, 2006. Some of these comments may become outdated as I learn more about IRGC and the risk governance problems that IRGC is addressing in its activities. But I suspect much of what I am writing here will not change, as the words reflect my deeply-held beliefs about the role of analysis and how it can help decision making. It is my hope that these comments will promote useful discussion within IRGC of the issues involved and will assist IRGC in making further revisions to the risk governance framework.

O. Renn and K. Walker (eds.), Global Risk Governance: Concept and Practice Using the IRGC Framework, 93–99.

Context and Purpose

I particularly liked the discussion of the motivation and purpose of the IRGC risk governance framework as set forth in its Preface:

> The establishment of IRGC was the direct result of widespread concern within the public sector, the corporate world, academia, the media and society at large that the complexity and interdependence of an increasingly large number of risk issues was making it ever more difficult for risk managers to develop and implement adequate risk governance strategies. Consequently, IRGC is committed to promoting a multidisciplinary, multi-sectoral, and multi-regional approach to risk governance.
>
> The IRGC risk governance framework represents a fundamental step towards the achievement of our mission – the development of an integrated, holistic, and structured approach, a framework, by which we can investigate risk issues and the governance processes and structures pertaining to them. (IRGC 2005: 5)

I think it is really important to recognise that the task being addressed is risk governance, meaning decision making on the management of risk by a variety of interacting governmental and non-governmental organisations (NGOs), for-profit businesses, and individual citizens. This risk governance task does not require that political and business leaders be educated about the myriad complexities and uncertainties on emerging and inherently dangerous technologies, natural disasters, or other sources of risk. Rather, it requires that planning and analysis be carried out in a fashion that is credible and trustworthy to all the involved and interested parties.

Good risk management decision making should be informed by the best available scientific knowledge – including acknowledgement of uncertainty in the consequences of following alternative risk management policies. Individuals and societies differ in judgements about overall goals, objectives, and values. Such differences generate ambiguities for decision making. Separation of these issues of judgement on overall goals, objectives, and values from the complex and uncertain science issues is one of the most important aspects of good risk analysis. Political leaders and the citizens who elect them are the ultimate authorities for making these value judgements in modern democratic societies. Therefore, a risk governance challenge is how to enable and encourage that the citizens and their leaders be properly informed. The third US President, Thomas Jefferson wrote in the last decade of his life:

> I know of no safe depository of the ultimate powers of society but the people themselves, and if we think them not enlightened enough to exercise their control with a wholesome discretion, the remedy is not to take it from them, but to inform their discretion. (Thomas Jefferson, letter to William Charles Jarvis, September 28, 1820)

Good risk governance requires understanding of uncertainty, complexity, and ambiguity, and communication of that understanding. This understanding – perhaps in the form of a collection of insights on what we know, what we can do, and what we want – will allow risks to be managed based on best available scientific knowledge, and on shared values among the diverse individuals and organisations that constitute society, on a regional, national, or international basis. IRGC should address all levels but particularly, international risk governance.

At the same time, scholarly research in economics and political theory (cf. Kenneth Arrow's Impossibility Theorem, Condorcet's Paradox: see www.econlib.org/library/ENC/bios/Arrow.html), and current events such as those involving terrorism and religious fundamentalism, tell us that when values differ greatly, social consensus on goals and objectives, and therefore consensus on how decision alternatives should be evaluated, may not be possible. We must recognise the limits of risk analysis, that it may not be capable of creating consensus on values when strong differences in values make concensus impossible. But risk analysis may help leaders to understand this reality, and help them to work toward resolve these differences by increasing the extent of shared values.

The activities of the IRGC, and the case studies in particular, should avoid becoming compendiums of scientific information relevant to a risk issue or sets of issues. Rather, the focus should be on risk governance, especially the decision making and consensus-building aspects. Support to good risk governance and to good decision making must depend on judgement and communication – about uncertainty, complexity, and ambiguity. I believe these are the key aspects of the framework set forth by IRGC.

Good decision making is not a technical process that can be reduced to a cookbook-style recipe or formula. The IRGC risk governance framework must provide an overview of principles and process – similar in character to how one might foster appreciation of great music, art, or literature. It cannot be a 'manual' that will enable a beginner to do excellent risk analysis, just as a book presenting an overview of classical music will not enable a beginner to write a great symphony. A book on classical music explaining the elements and traditions of the symphonic form may greatly assist listeners in understanding and appreciating the symphonies of Mozart, Beethoven, Brahms, Mahler, Shostakovich, and other composers. The IRGC risk governance framework can similarly assist decision makers and citizen leaders in understanding and appreciating risk analysis and its proper role in improving risk governance.

The IRGC framework appropriately focuses on how risk analysis can help to achieve better risk management and reduce 'gridlock' and controversy. It was written primarily for political leaders and citizens who do NOT have extensive training in physical science, social science, mathematics, statistics, probability, etc. The IRGC risk governance framework is appropriately aimed at bringing outsiders into risk analysis, to enable them to understand risk issues better, and to prepare them to engage in dialogue and with risk analysis specialists and to consider institutional and process reforms to improve risk governance.

It is not a textbook for risk analysts. It appropriately avoids going into extensive detail on the tools and techniques of how to do risk analysis. To some in the technical community, the framework may seem to lack rigor and needed technical detail. But I think this would be based on a misunderstanding of the goal and the audience for the IRGC risk governance framework. It was not written primarily for the technical community. This technical community has its own very extensive network of communications and stored knowledge via textbooks and scholarly journals – similar to the scholarly writings and the teachings at music conservatories about the symphonies of the great composers. The details of performance technique will be of interest to those with extensive experience and understanding, but NOT for the general public – even those who often attend and enjoy orchestra performances.

Comments on Strengths and Weaknesses

I view *strengths of the current version* of the IRGC risk governance framework as:

- *Provision of an overview of risk concepts and ideas*, bringing together much experience from Europe and North America.
- *Use of practical examples of how the framework might be used.* The applications provided in the annexes are not intended as deep or thorough expositions of complex applications, but rather as illustrations. They provide a good introduction, and they can be expanded and improved as IRGC gains more case study experience.
- *An emphasis on problem framing and dialogue* among the concerned parties. The details of the science and the risk numerology are not highlighted, and in my judgement, this is appropriate in view of the goal and the audience.
- *The recommendation of analysis, particularly use of decision analysis.* At the same time, the details of how to do decision analysis and the mathematical and statistical tools involved are not described, again, in my judgement, appropriately for the goal and the audience.
- *The concept of the risk management 'escalator'* (Figure 4, Chapter 1). I like this idea. Risk issues can move up as the frame expands – and move down as understanding improves and ambiguity is resolved; Table 6 similarly. It is useful for concerned parties to understand that risk issues should be addressed differently, depending on the uncertainty, complexity, and ambiguity. However, please note concerns about this taxonomy of risks becoming too rigid below.

Areas of weakness and need for improvement:

- *The IRGC risk governance framework needs more balance of references and discussion of traditional science and engineering.* Three examples are provided: train wrecks, earthquakes, nuclear power/waste. There is little referencing of the large professional literature on these problems! Normative analysis by scientists is complex. It requires hard work and it takes advanced training. Its importance should not be minimised. Although extensive detail is not needed, the text should do more to indicate the importance of the scientific and engineering literature, the experience of the scientific and engineering community, and to centers of outstanding practice and scholarship. The methods of science and engineering should be briefly explained and endorsed: use logic, check the logic. These are essential parts of the scientific method. Laypeople who are an important part of the audience for the IRGC risk governance framework often have very limited understanding of probability theory/statistics and mathematics, which form the basic language for the process of doing science. We need our audience to understand that science uses logic, mainly through use of mathematics and probability/statistics and a careful process of acquiring, selecting, and using appropriate data all of which should be subject to external peer review.

- *How much analysis and deliberation are enough?* Risk analysis must balance 'paralysis by analysis' against 'extinction by instinct'. Decisions must often be made under time pressure, with much less information than decision makers and scientists would like to have regarding the uncertain consequences of the decision alternatives. A good risk analysis process needs the right mix of iterative deliberation and analysis, guided by sensitivity analysis and value-of-information analysis to come to closure on choices that must be made in the near term among decision alternatives. Adaptive management techniques may be used to revise choices based on further information and experience as these are obtained over time.
- *Most risk issues are ongoing, and risk management should be ongoing.* The 1997 US Presidential/Congressional Commission on Risk Assessment and Risk Management (www.riskworld.com/riskcommission/Default.html) and the 1996 National Research Council report *Understanding Risk: Informing Decisions in a Democratic Society* (www.nap.edu/catalog/5138.html) are useful references on these aspects, which are very important from the risk governance perspective.

Other weaknesses that I consider less critical but deserving attention:

- *Language and the concepts for some terms like aleatory uncertainty need better explanation and clarification.* For example: '... aleatory uncertainty will remain fuzzy no matter how much research is invested in the subject'. This discussion is too technical (and misleading!) for a lay audience and insufficient for a sophisticated technical audience.
- *The simple example on damage to homes from earthquakes in Seattle is potentially misleading.* The probability of damage to a structure in an earthquake depends on the construction, the location – particularly type of soil – and a lot of other details well known to civil engineers who specialise in assessing potential earthquake damage to structures. Simple exceedance potential charts that do not include such details can be misleading, because these details are frequently important!
- *Separation of evidence and values is a key principle* – but this idea seems buried in the text. It should be moved up and made more prominent.
- *Watch out for the taxonomy becoming overly rigid!* While I like the idea of the 'escalator', the degree of uncertainty, complexity, and ambiguity can differ from problem to problem, and each can evolve over time for a given problem.
- *Traffic Light Model* (Figure 2 in Chapter 1). I find that this simple triage is a useful first step in thinking how a large number of similar risk issues, such as regulation of carcinogenic chemicals, might be sorted out. For some, the risk is clearly unacceptable, and little or no further analysis is needed. For others, the risk appears minimal and little further investigation is needed. For others, where there has existed longstanding and extensive human exposure and the potential

for a moderate level of elevated cancer risk (but not an observable epidemic in the large exposed population), a great deal of expensive, detailed, and careful analysis may be needed. However, the context of the risk management decision is very important, and simple generalisations may be as misleading. A simple sorting based on a cursory inspection may be an excellent first step. But the IRGC framework should not mislead readers into thinking that a three-bin sort is always the way to do it! Examples of difficult decision choices include Stephen Schneiders decisions about cancer treatment in *The Patient from Hell* and Donald Kennedy's editorial, 'Risks and risks', published in *Science* in 2005, which compares risks from climate change to risks from civilian nuclear power.

Next Steps and Outreach

Ortwin Renn, the primary author of the IRGC's framework on risk governance is an outstanding ambassador for risk analysis, both within the technical community and to the audience of leaders and decision makers. I think he has done an excellent job in drafting this document and deserves hearty commendation for the accomplishment. His perspective from the social sciences side of the risk analysis community is really enlightening, especially on issues of improving communication and achieving shared understanding as a way toward improving governance. This perspective now embodied in the IRGC framework will greatly benefit those in the engineering/physical sciences/biological sciences community who do not yet understand why so many risk management decisions in their areas of interest are bogged down in controversy.

> The testing of the framework's efficacy will involve its application in a number of areas where the risks appear not fully understood or where there is a desire or need to improve risk governance. ... Upon completion of the testing phase, we will make any necessary revisions to the framework and use it as our core mechanism for looking into risk issues. (IRGC 2005)

I find the language in the paragraph above bit formal, and maybe the wording appears to promise too much from the IRGC framework. I do not think this framework will serve as a 'core mechanism' for how to do outstanding risk analysis; the 'core mechanism' for outstanding work will always involve high levels of training and talent in science and engineering, perhaps in the same manner as musical talent and training for symphony composers. The IRGC framework may enable more outstanding risk analysis by helping to persuade our world's leaders that they need it, and that they should be willing to pay for it.

The IRGC risk governance framework describes risk analysis. Let us consider an analogy, that of describing an elephant. A group of blind men may give descriptions of various parts of an elephant. To understand the whole, their individual descriptions must be appropriately combined. The IRGC framework should give an overview of the whole of risk analysis, made up of many parts that are subjects for study in different disciplines of physical science, engineering, and social science. A holistic description should be a useful aid for a decision – in the elephant analogy, on whether to learn more about elephants, or whether a community should have elephants in its local zoo. The IRGC risk governance framework is not intended, nor is it adequate, to be a manual for elephant care in routine decision situations, such as zoo keepers might face daily. It is certainly not a manual for how to deal with emergency situations, when an elephant is sick or injured! One might need a highly trained veterinarian with a specialisation on elephants, supported by technical tools and the experience of colleagues.

Concluding Quote

I concluded my comments at the SRA meeting with a favorite quote, which I learned from my Ph.D. thesis advisor, Stanford Professor Ronald A. Howard, more than 40 years ago. The quote is now over 400 years old, and still very applicable to the task of improving risk governance:

> If a man will begin with certainties he shall end in doubts, but if he will be content to begin with doubts, he shall end in certainties.
>
> – Francis Bacon, *The Advancement of Learning*, 1605.

We in IRGC have the potential to help decision makers who must deal with uncertainty, complexity, and ambiguity. The IRGC risk governance framework sets forth a set of concepts and a taxonomy for how this can be done, and how it can be done in a way that helps build trust and confidence in the decision making process. We at IRGC need to move from framework into case studies that demonstrate how we at IRGC can help decision makers and interested/affected parties on the risks in facing up to these difficult aspects of uncertainty, complexity, and ambiguity. In my judgement, we will make progress most rapidly by (1) demonstrating how with current tools and methods we can provide assistance to those with risk management/risk governance responsibility, and by (2) building an understanding among those who allocate the funding that further investment in tools, training, research, and education in the relevant scientific and engineering disciplines underlying risk analysis can enable greatly improved decision making and risk governance by our society's leadership.

References

IRGC, 2005, *White Paper on Risk Governance: Toward an Integrative Framework*, IRGC, Geneva.
Kennedy, D., 2005, Risks and risks, *Science* **309**, 2137.
Schneider, S. and Lane, J., 2005. *Patient from Hell: How I Worked with My Doctors to Get the Best of Modern Medicine and How You Can Too*, Da Capo Press, Cambridge, MA.

Chapter 5
White, Black, and Gray: Critical Dialogue with the International Risk Governance Council's Framework for Risk Governance[1]

Eugene A. Rosa
Edward R. Meyer Professor of Natural Resource and Environmental Policy, Washington State University, Pullman, Washington, USA

> Scientists tend to ignore academic philosophy
> as an empty pursuit. Surely, any intelligent
> person can think straight by intuition ... I deplore
> the unwillingness of scientists to explore
> seriously the logical structure of arguments ...
> Many great theories are held together by chains
> of dubious metaphor and analogy.
>
> Stephen Jay Gould
> *Ever Since Darwin: Reflections in Natural History*

Introduction

There has been a growing need among governments and a wide variety of policymakers and managers at all levels for a comparative risk framework. Needed in particular is a framework that integrates the cumulative knowledge on the analytic domain of risk research with the policy and regulatory domain of risk management and with the democratic processes of informed citizen choices. The momentum of globalising processes has accelerated this need. The inevitable risks accompanying global processes rapidly expand in variety, scale, and in disrespect for national borders. Expanding daily are concerns about global terrorism, the expansion of nuclear weapons, the emergence and spread of new diseases, the potential threats of new technologies, such as nanotechnology, and others. The International Risk Governance Council (IRGC) has responded to this need with its first White Paper, *Risk Governance: Toward an Integrative Approach* (IRGC 2005), principally authored by IRGC Scientific and Technology Council member Ortwin Renn. There is little doubt that this tome is a scholarly and practical landmark that markedly advances

[1] Portions of these comments were presented at the annual meetings of the Society for Risk Analysis, Orlando, Florida (4–7 December 2005).

O. Renn and K. Walker (eds.), Global Risk Governance: Concept and Practice Using the IRGC Framework, 101–118.

our understanding of risk as an analytic concept and as a management tool. The White Paper is hereafter referred to as the IRGC risk governance framework or IRGC framework.

The goal of the IRGC framework is stated in the first sentence of the Executive Summary:

> ... [to put] forward an integrated framework for risk governance which provides guidance for the development of comprehensive assessment and management strategies to cope with risks, in particular at the global level. (IRGC 2005: 11)

The IRGC framework comprises three phrases: pre-assessment, risk appraisal, and risk management. Pre-assessment focuses on identifying risk problems and framing how they will be analysed and eventually, managed. Risk appraisal, the core activity in determining the likelihood, exposure, and extent or consequences of risks realised provides the risk estimates used in risk evaluation and characterisation – and ultimately the assessment of the tolerability or acceptability of risks. Management comprises the decision making about and implementation of management strategies to attain tolerable or acceptable risks.

Presented here is a set of critical comments that focus on the first two phases, with particular attention to the definitional[2] and analytic features of the IRGC risk governance framework. Ignored is the management phase, though the first two phases inevitably shade into it. The goal of these comments is not to cast fatal doubt on the worthiness of the project, but to offer guidance for a tightening of the logical structure and coherence of the IRGC's framework, and for promoting a dialogue for its improvement as it evolves. Because these criticisms will be centered on the analytic side of the framework (especially the pre-assessment and appraisal components), they will not only ignore the management side but also the intermediate components between the two sides. While the policy and management domains are integral parts of the IRGC risk governance framework, they are beyond the scope of criticism here.

Presuppositional and Scope Issues

The IRGC (2005) White Paper on risk governance is explicitly devoted to the development of an analytic framework, in contrast to an interpretative, adumbrating, or other form of general framework. This goal, as noted above, is explicitly stated in the first sentence of the Executive Summary.

Hence, it is around this self-proclaimed identity as an analytic framework that the following definitional issues and critical comments will be directed. The orienting perspective adopted here, from the outset and throughout, is the presupposition that

[2] In preparing this dialogue I discovered that one principal task of analytic philosophy, according to Princeton philosopher Harry G. Frankfurt, is to clarify the meaning of words that are widely used but little understood (2006). Hence, because of its emphasis on definitional issues this dialogue inadvertently is a synthesis of analytic philosophy, social science, and political pragmatism.

the analytic domain of risk is a scientific enterprise and, as such, should conform to the best practices of scientific investigation. The scope of the critical comments that follow will address these issues:

- Definitional clarity.
- General coherence of framework and internal consistency of terms.
- Uncertainty in risk estimation.
- Political implications and unintended consequences.

Definitional Clarity: Defining Risk

> Everything has been said before, but
> since nobody listens we have to keep
> going back and beginning all over again.
>
> André Gide

Unlike other frameworks, analytic ones typically require, as a first step, a pre-analytic vision – one that precedes even the pre-assessment phase specified in the IRGC framework – accompanied by a set of foundational premises. Often the keystone of those premises is a definition, or set of definitions, of fundamental concepts. Thus, a concern with definitional issues is neither merely a Talmudic preoccupation nor a game of semantic tag for intellectual amusement – despite the attractiveness of this pursuit for some academic scholars. Rather definitions are pivotal to the entire analytic enterprise as they hold in place the foundation upon which an analytic framework rests. Weak foundations ensure shaky frameworks. In short, foundations provide one safeguard against the kinds of dubious chains of thinking by metaphor and analogy that Gould refers in the epigram introducing this chapter.

The IRGC framework, relying on an OECD[3] definition, defines risk as

> ... mental constructions.[4] They are not real phenomenon but originate in the human mind.[5] Actors, however, creatively arrange and reassemble signals that they get from the 'real world' providing structure and guidance to an ongoing process of reality enactment. (IRGC 2005: 23)

There are a number of problems with this definition that challenge its meeting of the criteria specified above – clarity, consistency, coherence, politics, and consequences. For example, if risk originates in the human mind and is not a real phenomenon, how can 'real world' signals be part of that origination?

[3] While it seems fully appropriate to adopt the definition of a policymaking agency to guide risk management, it is curious that it would pass academic scrutiny as an analytic definition.

[4] Ortwin Renn, principal author of the IRGC's White Paper on risk governance, explicitly rejects the idea that the adopted definition is a socially constructed one (IRGC 2005).

[5] Renn's rejection of a social construction orientation leaves open the question of whether the emphasis on 'not real phenomenon ... [but] in the human mind' is tantamount to the idea that risk is a psychological construction, in service to a reductionist methodology. The IRGC's framework is silent on this fundamental point, as it is about a social constructivist stance on risk.

General Coherence of Framework

Internal Consistency Criterion: Part I

We first ask whether the logic is internally consistent. The consistency criterion is challenged by the earlier definition in the IRGC risk governance framework stating that

> Risk is understood in this document as an uncertain consequence of an event or an activity with respect to something that humans value[6] ... Risks always refer to a combination of two components: the likelihood or chance of potential consequences and the severity of consequences of human activities, natural events, or a combination of both. (IRGC 2005: 19)

It seems clear here that the meaning of risk is something more than a 'mental construct', that it encompasses an actual state of the world, a state where a human or a natural event has some probability of leading to untoward outcomes. And here nothing is said about whether a percipient human recognises the existence of this state of the world – a state of risk – or not. This perspective says the world is 'out there' before it is 'in here'. In the extreme case there are risks 'out there' that not only do we not know, but that we don't know that we do not know.

Examples abound. A flaw in a safety system of a complex technology, like a space shuttle, is a risk regardless of whether anyone has detected it.[7] Similarly, an undetected Tsunami on its way or an undetected asteroid about to collide with the earth are both risks to those in its path. The history of atmospheric chemistry offers a final example. Sophisticated satellite images failed to detect the Antarctic ozone hole, but eventually British scientists on the ground did (their findings were later confirmed by satellite as well). In the meantime the ozone hole, though undetected, posed health and other risks especially for occupants of the Southern Hemisphere. Both human and natural events have, in the words of Kates and Kasperson, some 'likelihood or chance of potential consequences'. Hence, we can ask is there even a need to invoke the notion of mentality or of construct in defining risk? Yes, but not in the Platonic way conceived in the IRGC framework. That invocation is necessary, not in the risk condition of the world, but in human judgement about the stakes embedded in the risk – a point returned to in the alternative definition of risk recommended below.

Clarity Criterion

Returning to the IRGC's mental construction definition itself, the clarity criterion is challenged on a variety of fronts. We can unveil a lack of clarity from either of

[6] This definition is derived from Kates et al. (1985), which is a restatement of Kates and Kasperson, 1983: 'A hazard, in our parlance, is a threat to people and to what they value (property, environment, future generations, etc.) and a risk is a measure of hazard. Specifically, risks are measures of the likelihood of specific hazardous events leading to certain adverse consequences' (1983: 7029).

[7] Detected too late in the case of the Columbia Shuttle.

two diametrically opposite directions. From one direction, we can begin by taking the stated meaning at face value – that risk is nothing more than a mental construct – without invoking second order meanings. At face value, the definition leaves us with few objections. Of course, a conceptualisation, an idea, a definition, like risk, is categorically a mental construct. But therein lies the very problem. The statement is either a tautology; that is, it is true by definition and therefore violates the presupposition stated above that we follow best scientific practices.[8] Scientific investigation must contain a step in the process of its logic where it breaks out of the circularity of tautology in conceptual and mathematical definitions and permits itself to be threatened by disconfirming empirical evidence. The testability of ideas, a principle originating with logical positivism, remains a central criterion via the principle of 'falsifiability' in scientific practice even as logical positivism has been superceded by other versions of scientific logic – such as logical empiricism or instrumental realism (Ihde 1991).

Or, taken at face value, the definition, even if not a tautology, is empty of meaning. It is similar to saying that the sun rises in the east. With a simple thought experiment we can imagine the sun following many other paths, and these imagined outcomes would not violate the laws of cosmology. But the sun follows a regularity that is referred to by analytic philosophers as probabilistic certainty. So, while it is possible for the sun to traverse alternative orbs the likelihood of it doing so is so infinitesimal that the practical probability of this is effectively zero. Hence, in practice, the statement that the sun rises in the east remains unequivocally true empirically, but it does not tell us anything we do not already know. Similarly for a claim that risk is a mental construct. Such a definition, therefore, begs the very question it is supposed to answer: What is risk? What do we mean by this mental construction?

There is virtually no limit to the forms of mental constructions that humans can create.[9] For simplicity and without loss of generalisation, let us reduce the vast range of possibilities down to two categories: Mental Constructions One (MC_1) that refer to conceptualisations that are based upon some verifiable empirical referent in the real world[10] and Mental Constructions Two (MC_2) that are conceptualisations derived purely from imagination and do not have a verifiable referent in the real world. The fact that mental constructions can take two forms (at least) renders the IRGC framework's definition, and the OECD one upon which it is based, unclear. We are in no position to choose whether risk is grounded in the real world or in imagination.

---

[8] There is, perhaps, no better exemplar and expositor of best practices in science than Stephen Jay Gould who observes the following about tautologies. 'Tautologies are fine as definitions, but not as testable scientific statements – there can be nothing to test in a statement true by definition' (1977: 40).

[9] Furthermore, in unconditionalised form the mental concept definition of risk can become a taffy term, stretchable to an infinity of interpretations – a sharp blow to scientific investigation.

[10] This real world assumption is not universally accepted, but does not seem to pose a problem here since the extended definition of the IRGC framework states: '... Actors, however, creatively arrange and reassemble signals that they get from the '*real world*' (IRGC 2005: 23, emphasis added).

This ambiguity goes to the heart of a key point. Let us develop the point further by comparing examples of each category, MC_1 and MC_2. One unchallenged example of the first category is energy, the capacity for doing work. An example of the second category is the 'green eggs' in the children's book by Dr. Seuss titled *Green Eggs and Ham* (Geisel 1960).

Literally no one would question the ontological status of energy, or the four laws that govern it (especially the second law of thermodynamics of which Einstein said is the one law in nature least resistant to being overthrown). In its status as an MC_1, energy has correspondence to the real world. We can identify energy in its various forms and we can see the consequences – work – when it is applied. The mental construction, 'green eggs', does not enjoy a similar status; as an MC_2 it enjoys no intrinsic ontological status.[11] Green eggs correspond not to a real world, but to the imaginary outcome of a fertile mind.

In view of the foregoing considerations, we can now ask: To which of the two categories of mental construction does the idea of risk belong? If we answer that it belongs in MC_1, as does energy, then risk is something more than a mental construct; it is a state of the world with identifiable features and parameters – just as is energy. What features of risk make it similar enough to energy to piggyback it on energy into the MC_1 category? To answer this question it is important to note that energy is always, until realised in work, a potential. Indeed, that is how energy is defined: it is the potential for doing work. Similarly, risk is, until consequences are realised, always a potential, too; the potential for the realisation of certain outcomes – either wanted or unwanted. Hence, there is a definitional symmetry between energy and risk.

If the appropriate placement of risk is, like energy, in the MC_1 category it challenges the key example in the IRGC risk governance framework designed to affirm that the mental construction definition of risk is valid. To illustrate the affirmation the IRGC framework asks us to reflect on the difference between risk and tangible features of our environment: trees and houses. 'Unlike trees or houses, one cannot scan the environment, identify the objects of interest, and count them' (IRGC 2005: 23). One problem with this conclusion is that it is about the outcome of actions – of biota or human agency – not, like risk, the potential for certain outcomes. In other words it asks that we compare the actual outcomes of biological processes or of human effort – trees and houses – with, not actual outcomes, but with potential outcomes. Hence, the comparison is asymmetrical and misplaced. Using the principle of symmetry, we should compare the outcome of biological or physical processes (number of downed trees, number of damaged houses) with the consequences of realised risk where we have tangible outcomes to count (number of sick or injured people or animals, number of dead bodies or animals, etc.). Therefore, IRGC's presumed demarcation between these two classes of phenomena axiomatically dissolves.

---

[11] We could create green eggs by colouring real eggs with food dye, thereby producing a construction of ontological reality. But this action only converts an imaginary construct into a real one. In no way does it alter the original, imaginary status of green eggs as an MC_2.

Another problem is that trees and houses are mental constructions, too. 'Tree' is a word we give to certain of nature's assets and not to others. 'House' is a word we give to a mental construction of another type of construction, a physically circumscribed space that provides a collection point, a shelter, a place of comfort, and a variety of cultural attributes. In each case the signifier (tree or house) is constructed, the signified is not.

Recognition that trees and houses are mental constructions, too, raises the question whether they differ from risk. And, if so, how? Or putting the question the other way around, why do we name some mental constructions 'risks', but not others? Answering these questions leads to two possibilities: (1) either there is no difference between the anticipation of trees, houses and risk – as they are all mentally constructed; or (2) fundamental differences exist between them. If the correct answer is (1), it leads to the further question of whether the claim that all these things – trees, houses, and risks – are the same would ever attract the support of a majority of independent, even-handed observers. It is doubtful that it would; the family of resemblances between trees and houses are too intuitively different from those between trees, for example, and risks. Or, if the correct answer is (2), that they are different, why have they been classified with the same term, that is, mental construct? Classification initiates the process of pattern recognition and organised cognition, which is based upon distinguishing the dominant, recognizable features of phenomena.

What happens to the logic if we switch the symmetry from the consequences themselves to the pre-conditions for the consequences or outcomes? It would seem that in case of the growth of trees and the building of houses, the outcomes are certain or nearly so, not as a matter of chance, but by well-understood processes of nature or of human intention. Of course, for risk the same cannot be true. Risk outcomes, by definition, are uncertain. Vanishingly few thinkers would challenge that verity.

Internal Consistency Criterion: Part II

The IRGC framework's definition, as noted above, is elaborated beyond the original statement that risks are 'mental constructs'. It goes on to point out that:

> Actors, however, creatively arrange and reassemble signals that they get from the 'real world' providing structure and guidance to an ongoing process of reality enactment. So risks represent what people observe in reality and what they experience. The link between risk as a mental concept and reality is forged through the experience of actual harm (the consequence of risk) in the sense that human lives are lost, health impacts can be observed, the environment is damaged or buildings collapse. (IRGC 2005: 23).

This elaboration of the original definition reaffirms a commitment to risk as the product of mental processes, but now permits that those processes are shaped by signals from the real world. Tying risk as a mental process to signals from the real world is an essential step in maintaining consistency within an analytic framework. At the same time, it raises a core question. What is to prevent those signals from including a 'sense' or fear that there is some probability, if the dice come up in a particular way, of 'actual harm' to the individual or to society? After all humans have experienced harm both indirectly and directly since we evolved into sentient beings. On what basis can we reject the idea that the mental construct that we use to organise those experiences, which we call risk, is not shaped by the experience of harm? If we cannot logically make such a rejection, it follows that the idea of risk is formed from real world observations and experiences – from observations about a state of the world independent of the percipient observer. It exists in the external world, the world 'out there'. The mental construction is, therefore, an emergent property arising out of experience – not out of a circumscribed self-referential mind.

Alternative Conceptualisations

These critical observations would lead to a dead-end if they did not offer alternative ways of thinking about risk. An important emphasis in metamethodological work on risk – seldom found in the contemporary risk literature[12] – has been to investigate the logical and epistemological status of the idea of risk (Shrader-Frechette 1985, 1991; Krimsky and Golding 1992). One key goal of my investigation in this domain has been to develop a definition that can serve as the foundation of that status. The definition has been introduced (Rosa 1998), refined only slightly (Jaeger et al. 2001), and reaffirmed in the literature (Rosa 2003). It asserts that risk is an ontological state of the world defined as:

> A situation or event where something of human value (including humans themselves) is at stake and where the outcome is uncertain. (Rosa 1998)

The definition presupposes the classic distinction in analytic philosophy between ontology and epistemology. Ontology refers to the state of the world independent of any observers. Epistemology refers to the interpretation or knowledge claims about the world by percipient observers, such as humans. The key features[13] of this definition are: (1) it defines risk as real, not imagined; (2) it situates risk as an ontological state of the world not as a state of our minds alone; and (3) it intentionally leaves open the issue of whether any given risk will attract commensurate mental constructions that lead to epistemic agreement. Indeed, it redefines the ideational domain of risk to be an issue of epistemology (what risks do we choose to concern ourselves with and what is our understanding of those risks?) where, indeed, mental constructs reign.

The definition leaves open the pivotal question: What is the connection between the ontology and the epistemology of risk? The answer lies in the unavoidable truth that there is never isomorphic congruency between the world and our understanding of it. For risk, there is rarely a one-to-one correspondence between the ontology (existence) of risk and the epistemology (our understanding) of risk because of the intervening role of human perception and knowledge (and or interpretation of that

[12] A recent exception is Althaus (2005).

[13] A more thorough discussion of these features and their logical implications is presented in Rosa (1998) and will, therefore, not be repeated here.

knowledge) of the real world of risk. Virtually all conceptions of risk and all knowledge claims about risk, despite its ontological state of realism, are off the mark.

But it is crucial to note that not all risk estimates are equally off the mark. Some knowledge claims are very close to the mark while others are remarkably distant from it. In short, not all claims to knowledge of risk are of equal credibility. Instead, claims of knowledge about risk form a hierarchy where some claims are highly credible (such as the connection between smoking and ill-health) while others are far more dubious (such as a claimed connection between global climate change and the mating patterns of domestic pets). We can name this ordered continuum *Epistemological Hierarchicalism* (EH).

How can we judge where a given knowledge claim about risk lies on this hierarchical, 'close' to 'distant' continuum? We can develop a principle based on the consistency of signals from the external world; that is, from the world itself. In particular, we can apply a principle grounded in two criteria: *ostensibility* (we can point to examples of the risk) and *repeatability* (over some period of time the risk will repeat and show itself).[14]

Combining a hierarchical epistemology (HE for short) with the presupposition that risk enjoys a status in realist ontology (RO for short) produces the convenient acronym HERO. But HERO is much more than an acronym. It provides a coherent framework for risk that rests on a foundation of clarity. That same coherence accommodates the status of risk with human interpretation and knowledge (construction, if you wish) within the same framework. Risk is not simply what any individual or group says it is, but begins with what the external world says about the conditions of risk. It ends with interpretation about the conditions and about the stakes humans have in them. Hence, this way of looking at risk does not require *ad hoc* adjustments to connect constructions with 'reality enactment', as does the IRGC framework.

Uncertainty in Risk Estimation

Correspondence Rules

Risk estimation, the procedures for assigning numerical values to probabilities and for anticipating consequences, unavoidably embeds a distinct form of uncertainty, that is, a second-order uncertainty. The first-order uncertainty in risk, of course, is the uncertainty in the idea of probability itself. A quantitative probability is simply a numerical mapping of uncertainty. The second-order uncertainty is over how true our knowledge is of that probability; is it precise or fuzzy?

---

[14] The criteria were developed in Rosa (1998) from the universal examples of physical gravity and the 365-day calendar. The ostensibility criterion asks: 'Do you see what I see?' If the answer is 'yes' we have inter-subjective agreement. If the answer is 'no' the repeatability criterion responds: 'Just wait and you will have other opportunities to see what I see'.

Procedural Issues in Estimating Probabilities

Risk estimation (risk assessment in the IRGC framework) is the pivotal link between the conceptualisation of a given risk, the characterisation of its seriousness and evaluation of its tolerability or acceptability, presumably by stakeholders and ultimate management of that risk. The goal of estimation is to specify what is at stake and with what likelihood it will occur. In practice, this means the application of a variety of analytic or judgemental tools for estimating the probability of any event with untoward or unwanted consequences, the magnitude or impact of those consequences, and for aggregating the probabilities and consequences if the event should occur.

A key historical fact is often lost, or papered over, in debates about appropriate procedures for estimating probabilities. From the very beginnings of its evolution, the idea of probability – during the era of advances in our understanding of gambling and of mathematical uncertainty – was conceived as a duality.

The IRGC's framework explicitly recognises this duality when it states:

> Risk analysts consequently distinguish between *aleatory* and *epistemic uncertainty*: epistemic uncertainty can be reduced by more scientific research while aleatory uncertainty will remain fuzzy regardless of how much research is invested in the subject. (IRGC 2005: 28)[15]

However, it inverts, mistakenly in my view, the timeworn duality of what we mean by probability. As explained below, the IRGC framework's distinction contradicts the widely accepted conventional interpretation of these two forms of uncertainty. There are clear advantages to the inversion, such as the coherence between the emphasis on epistemic uncertainty and the definition of risk as a mental construct. If risk is principally a product of our mind, then it follows that improvements in estimates of it should emanate from the mind, too. But this coherence advantage is trumped by the many costs that accrue as one travels the logic of this inversion.

Ian Hacking, in his classic *The Emergence of Probability Theory* (1975), describes the conventionally accepted duality much differently:

> There is hardly any history to record before Pascal (p. 1)[16] ... the probability emerging at the time of Pascal is essentially dual. It is, as I shall put it, both aleatory and epistemological[17] (p. 10) ... [probability] is Janus-faced. On the one side it is statistical, concerning Itself with stochastic laws of chance processes. On the other side it is epistemological, ded-

[15] The source for this distinction between the two types of second-order uncertainty in the IRGC's framework is Shome et al. (1998), a technical article devoted to earthquake prediction. Its relevance to the epistemic/aleatory duality is somewhat cryptic.

[16] Hacking reminds us, too, that in applying probabilistic reasoning to real problems, rather than just games of chance, Pascal thereby invented decision theory.

[17] The source for the idea of 'aleatory' is traceable to the first book on probability, *Liber de Ludo Aleae* (*Book on Games of Chance*) by the Italian physician Girolamo Cardano (or Jerome Cardan), written sometime in the mid-1500s but not published until 1663 in Latin. The meaning of the word 'aleae' in the title and developed in this treatise refers to games of dice, while its extension, Aleatorius, refers to games of chance in general (Bernstein 1996).

icated to assessing reasonable degrees of belief in propositions quite devoid of statistical background (p. 12).[18]

This duality, this distinction between a statistically based probability, the aleatory, and a belief-based probability, the epistemological, has been a feature of the statistical canon ever since. In modern parlance, the aleatory emphasises actuarial data and stable frequencies, while the latter emphasises degrees of belief.[19]

Underscoring the dual meaning of probability is neither a misplaced preoccupation, nor a misplaced fastidiousness over trivial issues. Instead, it speaks to the very logic of correspondence between concept and the empirical world, between the meaning of risk and its measurement properties. It addresses the basic estimation question: by what means can we judge the likelihood that a risk will be realised and, if it is, what the consequences will be? And since estimation is pivotal to judgement about risk acceptability its importance is critical.

It is useful to recall that probability is a representation of uncertainty. When the probability is expressed as a numerical fraction it represents the mapping of uncertainty into a mathematical continuum with the dimension $0 < p < 1$. There are four fundamental forms of probability: (1) propositional, (2) *a priori*, (3) empirical (or actuarial), and (4) subjective. The propositional form, virtually absent from the risk literature, defines probability as a purely theoretical concept and locates it within a system of logic, sometimes expressed in formal or symbolic forms. *A priori* probability, typically associated with games of chance, is probability where the logical chances of all outcomes is known beforehand because they are inherent in the instrument generating the outcomes – e.g. a fair coin. Empirical probability or actuarial probability, the gold standard of probability estimation, is derived from actual frequency data. Subjective probability is an exercise of judgement, especially under conditions where actuarial data is unavailable or meager.[20]

It is only with the first two definitions of probability – propositional, and *a priori* – that our knowledge of this probability is certain. We can either define probability with logical certainty or, like games of chance, specify the probability with certainty because all outcomes and their likelihood are known. Such a level of certainty over

[18] Bernstein, too, underscores the dual meaning of probability. 'Probability has always carried [a] double meaning, one looking into the future, the other interpreting the past, one concerned with our opinions, the other concerned with what we know ... In the first sense, probability means the degree of belief or approvability of an opinion – the gut view of probability. Scholars use the term 'epistemological' to convey this meaning; epistemological refers to the limits of human knowledge not fully analyzable" (1996: 48–49).

[19] Hacking avers: 'Chances, odds, 'hazards' (the stock in trade of aleatory probability) are basically quantitative. There is no way to understand odds without understanding numerical ratios. Epistemic probability is not like this. You can compare the degree to which evidence warrants several propositions without recourse to numbers' (1975: 73). Bernstein, using remarkably more dramatic, exaggerating prose, concurs: 'Without numbers, there are no odds and no probabilities; without odds and probabilities, the only way to deal with risk is to appeal to the gods and the fates. Without numbers risk is wholly a matter of gut' (1996: 23).

[20] Another putatively subjective approach is the Bayesian one. However, in the context of the four-part typology it can best be viewed as a hybrid, where the subjective judgement is formed from actuarial data, however meager or complete those data.

the value of a probability is not available to the second two definitions of probability – the actuarial and the subjective. For each, an estimation procedure becomes necessary. But, as already noted the procedure introduces a second-order uncertainty, summarised in the question: Is the estimate accurate or inaccurate, that is, is it crisp or fuzzy?

Thus, in virtually all instances of practical risk estimation the four-part probability typology boils down to just two types: the empirical and the subjective. In short, the risk estimator's toolbox contains but two generic tools. That fact begs the question: 'under which conditions is one tool more appropriate than the other?' The answer is pre-framed by the duality of meaning in risk itself. With the conventional interpretation of the duality – aleatory as data driven and epistemic as logic driven – it is clear that the actuarial definition of probability is matched to the aleatory and the subjective definition of probability is matched to the epistemic. The question that follows, expressed in the IRGC risk governance framework itself, is: Which side of the risk duality offers more hope for a reduction in the fuzziness over the probability representing the side's uncertainty? The answer is straightforward: it is the data-laden, empirical side. Why?

The answer lies with the development of the idea of the law of large numbers and the central limit theorem.[21] The law eventually established itself as a fundamental concept in statistics and probability. Succinctly put, the larger the volume of actuarial, empirical data (essentially a virtual sample for many applications, and a literal sample for infinite populations) used to estimate a risk probability the closer the sample data will be to the true probability. It follows that estimates of probability will become more precise (less fuzzy) with the availability of more empirical evidence, either from natural conditions or experimentation. This is why the aleatory, not the epistemic, is the gold standard of probability – *qua* risk – estimation.

Statistical theorems, such as the central limit theorem, seem to play a compelling role in the argument that actuarial-based risk is the gold standard of risk estimation. However, those theorems embed unexamined presuppositions. One is particularly relevant here. Statistical theorems presuppose realism; that actuarial data are real data and, that, ineluctably, they reflects a world that is real, not just constructed. Further, the larger the sample the more precise the estimate of some feature of that world. The first problem for the IRGC framework is that, by accepting a data-based type of probability it neglects to examine this critical pre-supposition. Second, it leaves open the question of how a reduction in epistemic uncertainty can outpace the reduction in data-driven uncertainty.

These issues are explicitly addressed with the alternative definition explicated above and with the HERO framework that incorporates it. Because that definition is grounded in realist ontology, it explicitly anticipates and accommodates the presupposed realism of statistical theorems. And because it defines the degree of epi-

---

[21] Attributable to the 19th century mathematician Siméon-Denis Poisson. Poisson's insights were later codified by the 19th century Russain mathematician, Tchebycheff (also spelled Chebyshev), known as Tchebycheff's inequality. It states that as the sample size – i.e. number of data points – grows larger, the difference between the sample expected value (mean of a normal distribution) and the population expected value (a parameter) will approach zero (Hacking 1990).

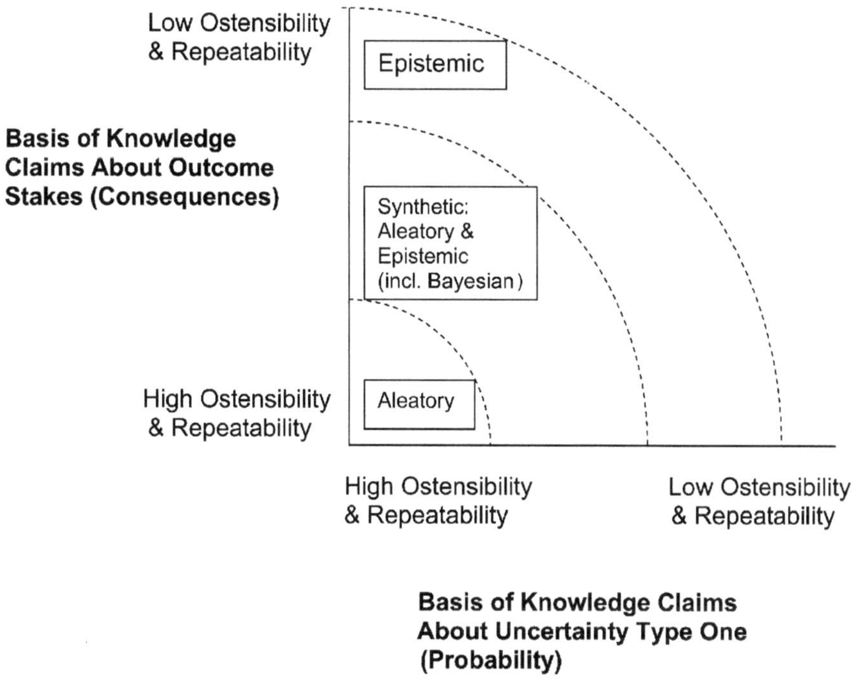

Fig. 1 The OR principal applied to the two sides of risk, the estimation of probability and the estimation of consequences. Note: Because the diagram compresses four variables-ostensibility, Repeatability, uncertainty (probability), and outcome stakes (consequences)-into two dimensions, the orientation of the axes is high to low, rather than in the typical Cartesian style.

stemological certainty according to the two criteria, ostensibility and repeatability, it connects directly to aleatory processes and actuarial based evidence in estimating risks – both of which assume the same principle.

We can summarise the variety of elements and the criteria of the alternative conceptualisation of HERO with Figure 1.

There are five key features to note in Figure 1.[22] First, the ostensibility and repeatability (OR) principle maps directly onto the aleatory form of risk. The closer the evidence underlying risk estimation is to the origin on either axis, the closer the estimate is to a fully aleatory, evidence-based estimation. Second, the higher the ostensibility and repeatability of data, the greater the availability of actuarial data to estimate risk. The salutary consequence is that the values in this finite sample are a more precise estimate of the signals about risk coming from the world.

Third, the OR principle provides demarcation points indicating the types of evidence available for risk estimation and, consequently maps a continuum between aleatory and epistemic risk. The demarcation points also delineate the degree of judgement (or construction) in risk estimates. The further away from the origin,

[22] Developed independently is a similar figure that highlights the management or 'problem-solving' strategies available to risk managers by Funtowicz and Ravetz (1991).

the more miserly the evidence, and the greater the amount judgement contained in knowledge claims about risk. Fourth, the axes along with the demarcation arcs concretise and provide a visual version of the hierarchical epistemology of HERO (Hierarchical Epistemology/Realist Ontology). Fifth and finally, the demarcations are intentionally dashed to indicate their dynamic relationship with each other. If a selected risk becomes more ostensible and if it realised in multiple instances, the more amenable it becomes to an aleatory-based estimate. Hence, with a growth in actuarial evidence an entirely epistemic risk can become a synthetic risk, (i.e. a synthesis of judgement and evidence), and, in turn, a synthetic risk can become an aleatory risk.

Political Implications and Unintended Consequences

We now come to the final issue for critical evaluation: the political implications of epistemological choice. Risk governance ineluctably embeds a variety of political issues. A critical one emerges with constructivist definitions of risk. Where risk is defined as a 'mental construct' we can then ask how consensus is possible. How can experts and laypersons alike make reasonable decisions if all of them operate from a position where there is no 'state of the world' to discipline their perceptions, only products of mental processes? Where can we expect traction in an arena of presumed mental anarchy?

Technological and environmental policy, perhaps more than virtually any other form of public policy, is dependent upon expert knowledge and analytic support – often including risk assessments. Taking the IRGC risk governance framework definition of risk at face value leaves the meaning of risk open to an entirely relativistic interpretation; an individual or institution is entirely free to choose and emphasise, without real world (or ontological) discipline, whatever interpretation of risk they please. This would seem to be an advantage to democratic process since from a governance point of view, it would apparently make the process more democratic; each point of view is as valid as any other. All voices are in the arena of democratic process. All flowers bloom.

At first blush, this arena would appear to be an exemplar of Habermas' idea of the 'ideal speech community', a community where all members can engage in free and open discussion (1984 [1981]). The community functions as a forum for open discourse where conflicts are resolved in an equitable manner and where agreed upon norms and values are established. The goal and outcome of these debates is 'communicative rationality' that, in turn, is, for Habermas, rationality itself. Rationality and truth are not ideals, independent of social actors, but is the product of communicative interaction.

But, as pointed out elsewhere (Rosa 1998, 2006), to render all knowledge claims about risk equally valid opens the way for the abuse of democracy. In Habermasian terms the discourse arena is devoid of the key pre-conditions of the ideal speech community – safeguards against coercion and against time and other constraints.

Thus, by equalising knowledge claims, one consequence of a definition of risk devoid of a real world component, is that it permits the powerful to say with impunity that 'my knowledge is just as good as anyone else's, so I am justified in using my power since it is grounded in that knowledge'. Habermas would find this not an ideal speech community, but a manipulated speech community.

I warned of this danger in a widely cited paper (Rosa 1998), but whose message was just as widely ignored.[23] Ignored, that is, until recently when Bruno Latour (2004a, 2004b, 2004c) was compelled to admit that it was this very type of reasoning, of which he was culpable, that granted the George W. Bush Administration the impunity to deny the threat of global warming and therefore choose a policy that ignored the Kyoto Treaty. Latour's project, as pied piper of the field of the Social Studies of Scientific Knowledge (SSK), has been to problematise the knowledge claims of establishment science. One unfortunate, and likely unintended consequence of the project was to redefine science as a cultural activity in the 'science wars'. More disturbing to democracy was its disarming of the underclass of one of their key weapons for speaking truth to power – the weapon of scientific findings.

But the problem goes even deeper than letting political priorities trump scientific evidence over an unwanted policy. It opens the way to the institutionalised practice of subordinating scientific knowledge to political expediency. Mooney (2005), for example, has convincingly shown how the political right[24] in the United States has systematically ignored, disparaged, or misrepresented scientific knowledge in service to its political agenda. Brought into question by this salvo of political righteousness is the scientific evidence on acid rain, global warming, the efficacy of condoms in preventing sexually transmitted diseases, the status of endangered species, the potential therapeutic benefits of stem cell research, and others – subjects all that involve questions of risk.

In some cases, scientists and scientific administrators have been harassed via burdensome and frivolous inquiries. In others, established scientific knowledge has been challenged or obfuscated by spotlighting the work of outlier, if not crackpot scientists (often in the gray literature, outside the peer-review process). The entire project turns on exaggerating the uncertainty inherent in science and on playing to ambiguous or multiple interpretations of key concepts. For the field of risk analysis, such practices violate the view of risk analysis as a scientific enterprise complying with its best practices. This is why further clarification of the key concepts put forth in the IRGC risk governance framework is critical. While it may not guarantee impunity from political manipulation, it should make it a more challenging activity.

---

[23] While the foundations of my critical comments are grounded in analytic philosophy and empirical science the delineation of its political implications are consonant with a wide range of views, from George Orwell to, remarkably enough, the postmodern thought of Foucault (1980) who argue that words, the elements of knowledge claims and discourse, are a form of power.

[24] While Mooney's work identifies the US Republican Party and conservatives on the right in the US as the guilty leader of this practice, it should be emphasised that the practice, itself, is open to any political faction. One clear instance of this was the Lysenko era in the Soviet Union where the science of genetics was preframed by the discredited Lamarckian theory of evolution in service to the ideological presuppositions of communism.

As such, it should be of greater service to a democratic process, not to a power 'free-for-all'.

Conclusions

The IRGC's first White Paper, *Risk Governance: Towards an Integrative Approach* (IRGC 2005) is doubtless a landmark event in risk analysis, management, and democracy. There is little question that the integrated framework it proposes will prove to be effective guidance for risk policymakers and managers at all levels of society and government. It will inform risk policy and practice. Quite apart from its underlying foundations there is much in the IRGC risk governance framework to appeal to a broad range of risk practitioners.

Despite its practical appeal, the ultimate resilience of the framework lies with the logical soundness of its analytic framework. Judgement and management of risk are fully dependent upon a sound explication of what is meant by risk and by sound procedures for estimating risks. The goal of the critical dialogue developed here was to pinpoint flaws in the explication of risk and in the procedures for estimating risk. While practitioners may choose to ignore the finer features of the analytic framework and its flaws, the resilience and legitimacy of their practices will ultimately rise or fall on the logical soundness of all parts of the framework.

The critical dialogue was organised around four criteria in logic and political process: definitional clarity and internal consistency; correspondence between concepts and measures; general coherence; and political pragmatism. The criteria were addressed in four parts. The first part, Definitional Clarity, developed a number of key questions about the IRGC risk governance framework definitions. The second, General Coherence of the Framework elaborated problems of clarity in the framework's definitions and logic, and unpacked two principal indiscretions in internal consistency: the underdeveloped connection between what risk is and what we understand about it; and the apparent contradiction between two versions of the IRGC framework's definition of risk. Further, it provided an alternative conceptualisation and definition of risk.

The third section, Uncertainty in Risk Estimation, focused on the correspondence rules between risk as a concept and risk as an estimated value. It elaborated the problem with the IRGC framework's inversion of the conventional meaning of the duality of probability, between the aleatory and the epistemic. It went beyond this elaboration to specify a principle for connecting the alternative conceptualisation of risk in the previous part with operational estimation procedures. This extra step illustrated the greater coherence of the alternative version over the IRGC framework's version in the mapping of the conventional distinctions between aleatory and epistemic risk to empirical evidence. The fourth part, Political Implications and Unintended Consequences, excavated a deeply troubling political problem inherent in constructivist definitions of risk, the potential for granting decision impunity to those in political power. This potential represents a direct threat to the global momentum for furthering the democratisation of risk decisions.

The analytic foundation of the IRGC's integrative framework risk governance will be strengthened by attention to these indiscretions in definition, logic, and political pragmatism. Unaddressed, they leave not only the analytic foundation embedded with an incoherent logic open to continuous scrutiny and vulnerability, but also leave the entire framework open to misuse by the politically powerful.

Acknowledgement

I appreciate the comments and criticisms of Kennon Kuykendall, Lauren Richter, and Robert (Mick) McKinnon Wilkes.

References

Althaus, C.E., 2005, A disciplinary perspective on the epistemological status of risk, *Risk Analysis* **25**, 567–588.

Bernstein, P.L., 1996, *Against the Gods: The Remarkable Story of Risk*, John Wiley and Sons, New York.

Foucault, M., 1980, *Power/Knowledge: Selected Interviews and Other Writings 1972–1977*, edited by Colin Gordon, Pantheon Books, New York.

Frankfurt, H.G., 2006, Potter Memorial Lecture in Philosophy, Washington State University, Pullman, Washington, USA (October).

Funtowicz, S.O. and Ravetz, J.R., 1991, A new scientific methodology for global environmental issues, in: R. Costanza (ed.), *Ecological Economics: The Science and Management of Sustainability*, Columbia University Press, New York, pp. 137–152.

Geisel, T. Seuss (Dr. Seuss), 1960, *Green Eggs and Ham*, Random House, New York.

Gould, S.J., 1977, *Ever Since Darwin: Reflections in Natural History*, W.W. Norton, New York.

Habermas, J., 1984 [1981], *The Theory of Communicative Action, Volume 1* Polity Press, Cambridge [*Theorie des kommunikativen Handelns. Band 1: Handlungsrationalitat und gesellschaftliche Rationalisierung*, Suhrkamp Verlag, Frankfurt am Main].

Hacking, I., 1975, *The Emergence of Probability*, Cambridge University Press, Cambridge.

Hacking, I., 1990, *The Taming of Chance*, Cambridge University Press, Cambridge.

Ihde, D., 1991, *Instrumental Realism*, Indiana University Press, Bloomington, IN.

IRGC, 2005, *White Paper on Risk Governance: Toward an Integrative Approach*. The International Risk Governance Council, Geneva.

Jaeger, C., Renn, O., Rosa, E.A. and Webler, T., 2001, *Risk, Uncertainty, and Rational Action*, Earthscan, London.

Kates, R.W. and Kasperson, J.X., 1983, Comparative risk analysis of technological hazards (A review), *Proceedings of the National Academy of Sciences* **80**, 7027–7038.

Kates, R.W., Hohenemser, C. and Kasperson, J.X., 1985, *Perilous Progress: Managing the Hazards of Technology*, Westview Press, Boulder, CO.

Krimsky, S. and Golding, D., 1992, The role of theory in risk studies, in: S. Krimsky and D. Golding (eds.), *Social Theories of Risk*, Praeger, Westport, CT, pp. 3–22.

Latour, B., 2004a, Why has the critique run out of steam?, *Critical Inquiry*, Winter.

Latour, B., 2004b, The last critique, *Harpers*, April.

Latour, B., 2004c, *Politics of Nature: How to Bring the Sciences into Democracy*, Harvard University Press, Cambridge, MA.

Mooney, C., 2005, *The Republican War on Science*, Basic Books, New York.

Renn, O., 2005, Comments made at the annual meetings of the Society for Risk Analysis, Orlando, Florida (4–7 December).

Rosa, E.A., 1998, Metatheoretical foundations for postnormal risk, *Journal of Risk Research* **1**, 15–44.

Rosa, E.A., 2003, The logical structure of the social amplification of risk framework (SARF): Metatheoretical foundations and policy implications, in: N. Pidgeon, R.E. Kasperson and P. Slovic (eds.), *The Social Amplification of Risk*, Cambridge University Press, Cambridge, pp. 47–79.

Rosa, E.A., 2006, The place of knowledge in decision making, *Natures, Sciences, Societes Environment and Technology*, Winter [in French].

Shome, N., Cornell, C.A., M.EERI, Bazzurro, P. and Carballo, J.E., 1998, Earthquakes, records, and nonlinear responses, *Earthquake Spectre* **14**, 469–500.

Shrader-Frechette, K.S., 1985, *Risk Analyis and Scientific Method*, D. Reidel, Dordrecht.

Shrader-Frechette, K.S., 1991, *Risk and Rationality: Philosophical Foundations for Populist Reforms*, University of California Press, Berkeley, CA.

Chapter 6
Synopsis of Critical Comments on the IRGC Risk Governance Framework

Ortwin Renn[1] and Alexander Jäger[2]
[1]*University of Stuttgart, Stuttgart, Germany and DIALOGIK gGmbH, Stuttgart, Germany*
[2]*Interdisciplinary Research Unit on Risk Governance and Sustainable Technology Development (ZIRN), Stuttgart, Germany*

Introduction

This chapter provides a synthesis of informal comments, remarks, and statements by as many as 42 individuals received by IRGC following the publication of the risk governance framework.[1] Many comments were given to IRGC in the form of e-mails and short correspondence. More input was received from transcribed oral statements from participants during the discussion periods at numerous conferences and symposia where the framework had been presented. Our objectives for the chapter were to put them in a consistent format, to organise them by topic and to link them to the four phases of the risk governance framework. In so doing, we want both to express our appreciation for the thoughtful comments received and to begin to identify the recurring issues that may require work in subsequent versions of the framework.

This chapter has three major sections. The first presents general comments associated with the conceptual basis and structure for the framework. The second summarises the more specific comments and suggestions with respect to each governance phase. The final section summarises the major recurring themes among the all comments received, both formal and informal.

General Comments

Conceptual/Theoretical Issues

Many comments pointed out conceptual issues that the authors felt raised concerns about the usefulness or the limits of the framework for risk governance.

[1] Many contributors asked IRGC to treat their comments in confidence (i.e. they did not want their remarks to be attributed to their names), a request we have therefore honoured for all contributors.

O. Renn and K. Walker (eds.), Global Risk Governance: Concept and Practice Using the IRGC Framework, 119–130.

Beginning with some fundamental theoretical issues, several commentators argued that the document did not delineate a clear distinction between an objectivist and a constructivist perceptive on risk. They had the impression that the document at times takes a more realist (objectivist) position, that is, advocating that risks are real.[2] Some commentators also felt that the discussion of uncertainty was not clear; that the document meanders between uncertainty as a qualifier of risks and uncertainty as a description of knowledge about risks. A final settlement of the 'true dimensions of risk and knowledge' was perceived to be missing.

A major theme was the issue of compatibility of the framework with other existing taxonomies or manuals. As one observer puts it, 'The challenge for the IRGC terminology is to be in line with other important international documents and clear about the concepts used'. Many contributors mentioned that the framework provides a useful comparison of terminologies of different international standards but does not follow the terminology used in other important international sources such as the ISO or CSA. One commentator suggested, for example, that 'tolerability/acceptability judgement' be renamed as 'risk evaluation' to be consistent with the terms used in the CODEX Alimentarius, WHO and IEA. This commentator acknowledged, however, that the existing taxonomies and frameworks vary from one risk area to the next and that a clear consensus about the use of terms and concepts does not yet exist.

A last general point about the overall framework was that it was too rigid. Many criticised that all activities were forged into neat boxes disregarding the many interactions between the phases and the need for flexible adjustments. They believed it understates the changing dynamics of public preferences and systematic knowledge. Feedback systems and real case tests were recommended to bring the model in line with real risk governance processes.

Purpose of the Framework

Referring to the section on purpose and objectives of the paper, several commentators expressed the opinion that certain objectives were promised but not met. According to these critics, the framework does not provide an 'analytic' tool but rather a conceptual outline and set of definitions that allow describing risk governance processes. They felt that the document falls short of being an operative framework; it lacks clear instructions about how the suggested distinctions of knowledge might be used to select assessment methods and to specify protocol of how to evaluate and manage risks. The reference to Integrated Disaster Risk Management (IDRM) was perceived to be neither justified nor adequately developed in the document.

It was also mentioned that the language and the style in which the document has been written undermines the purpose of the framework. Many critics claimed that the document at times was rather difficult to understand and that it appeared more

---

[2] In reality, the framework encompasses both; it is based on the idea that risks are 'knowledge constructs' in which available knowledge plays a central role.

suited to an intellectual or academic audience than the targeted parties involved in assessing or managing concrete risks.

Scope of the Framework

In regard to the scope of the paper, risks originating from intentional human action and/or from malicious agents were seen as either absent or addressed insufficiently by the framework. In general, a clearer concept of what constitute hazardous agents (including human action) was demanded. Commentators did not feel that it was plausible to place climate change, mass hysteria and terrorists into the same hazard category. As a consequence, the aspects of intentional destruction or sabotage appear not adequately treated in the framework. Yet, given the prominence of terrorism and other malicious acts (cyber-attacks) one should expect a more detailed analysis of these risks. Furthermore, an analysis of critical infrastructure or modern large-scale technologies would be incomplete without due consideration to malicious acts. Ultimately, multiple risks combining technical failures and malicious acts would be of special interest to the international community and should be incorporated into any future versions of the framework

Several contributors regretted that only little attention was given to other situations in which multiple risks might occur at the same time. Events may happen synchronously – e.g. a hurricane takes of power at the same day as a nuclear accident takes place – or progressively in relation to the same hazardous agent – e.g. a hurricane takes of power and water and health problems occur. Looking at the risk of such synchronous events would be particularly helpful for addressing the question of organisational capacities.

Exploring Risk Governance

Several academic commentators discussed the use of the terms governance and risk governance. Since the document is titled 'risk governance' they expected a more in-depth review of the meaning, concept and structure of governance. They felt it should include a clearer understanding and definition of governance, regulation and regimes, a more detailed explication of how these terms relates to risk and a more elaborate review of the actors, processes, and structures that constitute risk governance. For example, one commentator noted that there are groups and institutions that drive changes, those who impede changes and others who just sail along with changes. Each of these actor groups were seen as having different needs in terms of the governance process.

Another point that was mentioned frequently was the importance of distinguishing between horizontal and vertical governance. Vertical governance refers to the various levels from the local to the international scene, whereas horizontal governance describes the variety of actors on each vertical governance level, including governments, scientific experts, NGOs, industry and other actors within civil society.

The willingness of actor groups and stakeholders to be part of a governance process is crucial for having meaningful stakeholder involvement. Some groups, however, might have limited interest in being part in an open governance process. The authors of the framework 'shy away from the important issue of power and power distribution' commented one individual. Unless the subject of unequal distribution of power among stakeholders is adequately addressed, there can be no meaningful concept of governance. Commentators believed evolving power structures to be more complex and chaotic than anticipated in the well-ordered model of participation that is being developed in the IRGC document.

One of the major shortcomings of the framework relating to risk governance was perceived to be its lack of theoretical as well as practical elaboration on best governance practices. One comment referred to risk governance as a process where actors make concrete decisions, in a context defined by institutional and organisational constraints. In this case, actors and decision makers would need effective principles and behavioural rules of 'good risk governance'. The framework falls short of this demand because it describes all the governance phases, and their key attributes, but does not provide concrete suggestions for what actors and decision makers should do in order to be more effective and efficient at each phase.

Categorisation and Quality of Risk-Related Knowledge

One of the important basic concepts introduced in the IRGC framework is the classification of risk problems according to the state and quality of knowledge available: complexity, uncertainty and ambiguity. The knowledge categories were viewed as being useful but also confusing. Often commentators had difficulty understanding the distinction between the three terms or felt that they needed to be explained in clearer and simpler terms. Many commentators opted for other terminologies. One suggestion was to replace complexity with degree of sophistication, and ambiguity with degree of expected controversy. Other comments referred to what they believed was the anti-science message hidden in IRGC's classification. They preferred a classification according to objective, inter-subjective and subjective knowledge. Finally, one person found the classification too rigid and suggested using different categories in accordance with risk under investigation and the corresponding knowledge available to assess it, rather than using a fixed categorisation scheme.

Other contributors suggested that the categories of complexity, uncertainty and ambiguity should not been seen as a linear chain of stages but as interrelated aspects of a risk problem. A particular risk might display aspects of complexity, uncertainty and ambiguity at the same time. At different stages, a particular type of risk may be characterised best as being complex but, at others, as uncertain or ambiguous.

Seeing complexity, uncertainty and ambiguity as dynamic and interrelated aspects of a risk would make the framework more realistic.

This need for a more dynamic interpretation of the framework was also echoed in the comment that, over the course of time, new knowledge is being generated leading to a re-assessment of complexity, uncertainty and ambiguity characterising a particular risk. Some commentators suggested making the framework more explicit in that risks might move from one category to another as available knowledge changes. For example, simple risks might appear more complex or ambiguous in light of new facts or new interpretations of existing knowledge. Therefore, the hard lines in the framework between the knowledge related characteristics should be modelled as being permeable.

Several contributors noted that the classification of a particular risk is not obvious but depends on the judgements of the stakeholders involved. It was noted that whereas one group might evaluate a risk as being ambiguous another group may feel the risk is only complex.

Benefits and Costs

An issue raised by commentators from different stakeholder groups was that the IRGC risk governance framework should emphasis both risk *and* benefits. Many contributors advised IRGC to give benefits a more prominent role in the risk governance framework. 'Risks are not taken for their own sake but for reaching desirable goals' was one typical comment. In particular, individuals from industry and insurance companies stressed the need for a symmetrical treatment of risks and benefits. One person suggested renaming the appraisal phase into 'impact assessment phase' in which both risks and benefits are systematically identified and estimated. Almost all comments that addressed this issue felt that in the characterisation and evaluation phases, benefits should be treated with the same care as risks. A problem with tools like the traffic light model is that they do not emphasis the fact that risks are taken to attain particular benefits and therefore do not portray decision making as a true balancing act.

Alternatives to the approach presented in the IRGC framework would be to focus on the hazardous agent and the benefits from the risks taken. The risk of missing economic opportunities and benefits should be seen as a real risk that needs to get the same amount of consideration as the potential negative side effects. Ways to incorporate this perspective could include technology life cycle assessments, cost-benefit analysis, and other forms of comprehensive analysis.

One individual criticised that financial costs and benefits were not adequately treated in the document. Although the framework refers to cost-effectiveness as a criterion for selecting risk management options, it does not elaborate how cost-effectiveness analysis should be done and in which way financial risks should be taken into account.

Vulnerability and Resilience

Several commentators noted that the distinctions, as well as the similarities, between concepts such as risk, vulnerability or resilience should be described more explicitly. They also noted that each term would likely require a better clarification in the context of each area of application. The question was raised whether the IRGC framework addresses the issue of vulnerability in an adequate manner. Vulnerability was perceived to be of high importance in connection with natural hazards, critical infrastructures, and systemic risks. The same commentators also asked for a stronger focus on mitigation, in its various forms, as important elements of risk management.

The question of how to deal with 'surprises' or unexpected outcomes/events for industrialised as well as developing countries was raised in several comments. The emphasis of the IRGC framework for promoting resilience as an appropriate response to uncertainty was welcomed but commentators felt it needed more specification in terms of mitigating measures. Commentators suggested that a link to capacity building be made more explicitly. Other commentators emphasised that the strong focus on pre-assessment would strengthen the framework's capacity to cope with the issue of uncertainty and future surprises.

Comments about the Phases of the IRGC Risk Governance Framework

Overall View of the Four Phases of Risk Governance

Most comments supported the framework's distinction between four phases of the risk governance process: pre-assessment, appraisal, tolerability and acceptability judgement, and management. In particular, the idea of including a pre-assessment phase in which the problem is framed and the terms of reference specified was perceived as being a very helpful contribution to a productive exchange of views and perspectives. More controversial was the inclusion of the phase 'characterisation and tolerability judgement' as an interim stage between appraisal and management. Several commentators wanted this phase to be merged with risk management, others with risk appraisal. One contributor warned IRGC not to make tolerability and acceptability judgement an explicit step in the risk governance process as this would invite political pressure groups to interfere with the scientific process of risk analysis.

Pre-Assessment

The pre-assessment phase proposed in the framework received much support especially for its emphasis on the development of a proactive approach to management of potential risks before serious hazards evolve. Consistent with this view of the relevance of framing, some comments recommended that IRGC add activities in the pre-assessment phase that would support shared taxonomies of terms, mutual knowledge building processes, and the establishment of common rules and objectives at an early stage. These activities were seen as basic requirements for stakeholders to enter into a productive exchange of views and ideas with risk assessors and risk managers. In the pre-assessment phase, facilitating institutions that could support the collection and dissemination of concerns and perspectives to all actors involved were seen as being essential. One person noted that the term pre- assessment may be problematic; it means one thing for existing risks or hazards and another for newly emerging or innovative processes.

The comments on the pre-assessment phase noted that framing of problems cannot be performed in a neutral, unbiased way. Framing always prioritises viewpoints and creates a biased context where people evaluate what issues they would or would not like to see addressed in the governance process. One commentator noted: 'When you create a frame for the problem you create the context where people evaluate whether the benefits are desirable or not, so just the process of beginning to ask the questions influences how people will view the risk issue'. The pre-assessment of an issue therefore can influence subsequent phases of the risk governance process including later assessments about 'ambiguity' within the affected communities.

Risk Appraisal: Risk Assessment and Concern Assessment

Risk Assessment

With respect to risk assessment, the distinction between hazards, vulnerabilities and risks appeared to be inadequately addressed. Many commentators suggested a more elaborate conceptual distinction at this point. In particular, they requested a more thorough description of the differences between vulnerability analysis and risk assessment. Some contributors suggested a graphic illustration of the differences would be helpful for this purpose. One person suggested that the document should position risk assessment as the top activity providing the guiding principle for integrating hazard assessment, exposure assessment, and vulnerability assessment.

Concern Assessment

Overall, the public perception of risk was regarded as an important factor in driving risk policies and regulation and for pressing policymakers, company managers and other individual actors into making decisions. Although most contributors found the section on risk perception useful and informative, there was occasional criticism that the psychological review on risk perception in the White Paper, in particular use of terminology like 'immediate threat', 'insidious threat', would not apply to critical infrastructures. The problem perceived by one of the commentators was that critical infrastructures would pose risks to collective entities such as whole societies rather than simply to number of single individuals which this commentator felt to be the predominant focus of the framework. Threats to infrastructure were also seen to represent multiple risks – e.g. natural, system-related, malicious, and technological – that are closely interrelated. They also fall into several categories across both dimensions of the characteristics of risk knowledge as well as the patterns of risk perception. Risk perception for critical infrastructure was seen to depend on particular prevailing and constantly changing circumstances, e.g. fear of terrorism, attention to natural hazards, perception of market inequalities, among others. Using the example of critical infrastructures, one suggestion was to incorporate different risk perception patterns referring to different risks and different stakeholder groups into the IRGC document in order to highlight the value of multiple perspectives and perceptions.

Tolerability and Acceptability Judgement

Risk Characterisation and Evaluation

The 'traffic light' model introduced in the IRGC framework was seen as useful but at times very simplistic (see Chapter 1, Figure 2). It was recommended that IRGC further elaborates on the different scenarios behind the categories. The framework's reference to risk assessors and risk managers as separate categories and its discussion on 'evidence and values' was also seen to be at odds with the daily experiences of the commentators. They noted that in practical decision-making situations, facts and values cannot be separated and, that in any case, such a distinction would not contribute to solving risk problems. Other contributors argued that facts and values are impossible to distinguish and that therefore any attempt to divide the analytical from the normative aspect of risk governance would support what they argue is a biased view of an artificial juxtaposition between an allegedly value-free science and a value-laden political sphere.

Risk Management

Several comments focused on the risk management 'escalator' depicted in Figure 4, Chapter 1. It was noted that the four risk problems described (simple, complex, uncertain, and ambiguous) cannot be seen as mutually independent routes to risk management. Those different routes may change in the course of time as the predominant knowledge characteristics of the risk problem shifts for example, from uncertain to complex.

Concern was also expressed about whether the risk management escalator, as proposed, would impose too many responsibilities and restrictions on actors and stakeholders. In many instances, the comments suggested, it should be left to the actors themselves whether they want to be involved and in what form. If participation is rigidly structured and participants are assigned a specific role, the purpose of participation may be missed. It should be left to the social preferences of the actors in which way they want to be involved and how they want to contribute to the governance process.

Another important comment on the Risk Management section was that the link between the activities described and the overall goal of risk governance was missing. To some commentators, risk management was presented in a rather traditional view (decision analytic model) and lacked integration into the new context of risk governance.

Regarding the question of managing interdependencies, e.g. interactions between different risks and multiple input from stakeholder groups, it was commented that the document does not specify a concrete way of implementing this noble goal except for the recommendation of a 'well balanced mix of consensual, coercive strategies'. IRGC was encouraged to elaborate on how this 'mix' could be achieved in real case examples.

Stakeholder Involvement and Sharing Knowledge

The comments on the role of stakeholders in risk governance often spanned several phases of IRGC's risk governance framework. In general, the issue of sharing knowledge was perceived as being of major importance. Several contributors, particularly from NGOs, complained about the lack of opportunities for stakeholders to influence the risk management process. Others regretted that there was no reliable and approved platform where scientists, companies, NGOs and governments can exchange views and discuss available knowledge. They welcomed the IRGC initiative to provide more access for stakeholders to enter the risk governance process. However, they felt that the roles portrayed in the section on stakeholder involvement did not adequately address the influence of consumers and their position in governance systems. After all, the consumers have to be empowered to make informed choices within a governance system. The roles for insurance companies and insurance mechanisms in the IRGC framework were also perceived as being unclear. Further research on the composition of stakeholder involvement processes was therefore recommended.

In contrast to the overall positive views expressed above, some contributors felt that the stakeholders were given too much power in the IRGC framework. The multiple access points for stakeholder participation were seen as too sophisticated and cumbersome to be of any value to the process. Moreover, stakeholder involvement was criticised as compromising the scientific nature of risk analysis in particular if participation is extended to risk assessment. Reference to 'sound science' or 'evidence-based analysis' was made several times in the context of the need for rigorous scientific review and exclusion of political influences and social values.

One of the commentators noted that the framework assumes that all relevant actors can be identified. In the case of critical infrastructures and other systemic risks the ability to identify all relevant actors may be the exception rather than the rule. Who would represent the interests of those who benefit from highways, gas pipelines, and internet access or electricity grids? Selecting the stakeholders is hence a sensitive task that influences or even pre-determines the potential outcome. For example, if only environmental NGOs were included the evaluation process would be considerably different from a situation in which representatives from industry dominated the process.

The question of how roles and responsibilities amongst the stakeholders should be assigned for international and transboundary risks was given much attention in the comments. Commentators regarded as crucial an active dialogue between the countries affected by a particular risk. In particular, an international consensus on the rules and principles of inclusive governance structures – at least on governmental level – was seen as a necessary step toward initiating meaningful and effective transboundary deliberations between a broader spectrum of civil society actors. This kind of international dialogue was specifically recommended between industrialised nations, emerging economies and developing countries, all of whom may be affected in different ways by the same risks. However, emerging economies and developing countries have specific needs and it was pointed out that their particular interests and capacities should be taken into account in designing their roles in a risk governance process.

Risk Communication

Whenever risk communication was addressed by commentators, the message was unequivocal: risk communication is a necessary complement to both risk assessment and risk management. IRGC's placement of risk communication in a central role within and linking all four risk governance phases (see Figure 5, Chapter 1) found unanimous support among all commentators who dealt with this issue. Several comments even encouraged IRGC to place more emphasis on the importance of communication and elaborate more on the means and instruments for facilitating effective and mutually rewarding dialogues.

The importance of clear communication of knowledge and values between stakeholder and risk professionals was also emphasised. Several commentators saw a specific need for politicians, consumers and journalists to receive more accurate and balanced background information about risks because these groups are often believed to lack reliable knowledge. Scientists and experts from industry were seen under the moral obligation to serve as reliable sources and to communicate their knowledge and experience to others, as long as intellectual property rights or confidential information were honoured.

Numerous comments suggested that IRGC should provide a platform for risk communication between and among the various stakeholders. Scientists and experts can often be poor communicators. The commentators encouraged IRGC to act as a platform for organising training in risk communication for such professionals since competence in effective communication is in urgent demand. Mechanisms for forging liaisons between industry and government were also mentioned as being needed.

Revisiting and Testing

Finally, a number of comments were made regarding on the further development of the framework. Several comments suggested that the IRGC model needs to be revised periodically to account for the dynamic changes in knowledge and perceptions about risks. For this purpose, feedback systems should be in place to update the framework at regular intervals. Updating should include different stakeholder groups in order to ensure a broad consensus on the continuous applicability of the framework for guiding risk governance.

As part of the process of testing and revising the framework, actual risk governance issues should be monitored and the lessons learned from them incorporated into the conceptual model. The 'real test' of the framework would be a prospective study of the actual application of the framework to a risk issue facing society today.

Conclusions

The broad scope of the comments about the IRGC White Paper on Risk Governance demonstrates the diversity of opinions and illuminates the many challenges that the topic of risk governance embraces. Given the broad representation of different stakeholders and actors involved in risk assessment and management amongst the commentators, divergent opinions and advice were inevitable. For example, many critics wanted the authors to make the framework simpler, more practical and less sophisticated while others felt that the framework would need more differentiation, sophistication and scientific foundations. There is no easy way to meet both demands.

Nonetheless, several common messages emerged:

- Make the framework as simple as possible but more clearly specified.
- Void academic jargon and review the key terms used in the document for clarity and relationships with other taxonomies.
- Make sure that benefits are given equal treatment in the assessment and evaluation phases.
- Include malicious acts (e.g. intentional destruction), vulnerability assessment in the scope of the framework.
- Integrate combinations of risk agents (common mode failures or synchronous risks occurrences) and risk absorbing systems (multiple interacting vulnerabilities) into the framework (in particular relating to infrastructure risks).
- Avoid the high degree of rigidity with respect to the categories 'complex, uncertain, and ambiguous' and the 'boxing in' of activities (from pre-assessment to management).
- Focus more attention on the dynamic aspects of governance, in particular the change of knowledge and public attitudes over time.
- Let the framework become a flexible, adaptive and responsive instrument for all actors in the risk governance arena.
- Be more specific about stakeholder involvement and be more aware of the limits and problems posed by a more inclusive governance model.
- Test the framework in real applications, both retrospectively and prospectively.

Many of the comments were received not long after the IRGC framework was first presented and published. Since that time, IRGC commissioned several case studies as retrospective tests or evaluations of the applicability, efficacy and practicability of the framework. Authors used the IRGC framework to evaluate the risk governance approaches to diverse issues ranging from acrylamide in food products and Baltic energy security to nature tourism. The results of these case studies are presented in the next section of the book.

In the final part of the book, we examine the common lessons learned both from the critical comments presented in this part of the book and from applications of the framework in the next. Together, these results will serve as valuable input to future revisions and applications of the framework.

Part 3. A Framework for Risk Governance: Case Study Applications

Chapter 7
Risk Governance of Genetically Modified Crops – European and American Perspectives

Joyce Tait
Innogen Center, Institute for the Study of Science, Technology, and Innovation, University of Edinburgh, Scotland

Introduction and Background

Genetically Modified (GM) crops occupy a unique place in the evolution of risk governance approaches to dealing with modern, path-breaking technologies. They were the first such technology to be regulated on a precautionary basis, in a generic sense, from the earliest stages of a technology development process that began in the 1980s and is still evolving.

Today, distinctively different risk governance processes are in place in the European Union (EU) and the USA and the roots of these differences can also be traced back to the 1980s. The European regulatory process is more complex and demanding than that for any other technology; as a result, few GM crops are grown in or imported into Europe. And yet, although GM crops are grown on millions of hectares in the rest of the world, and GM foods are consumed on a daily basis by millions of people, under much less demanding regulatory regimes, there is so far no evidence of environmental or health risks associated with approved products based on this technology, and considerable evidence of their benefits.

The history of the risk governance of GM crops in Europe has been played out over the past 20 years without the benefit of the IRGC risk governance framework (hereafter, the IRGC framework). This case study examines that history in the light of the IRGC framework, considers whether and how it might have made a difference if it had been applied, and suggests where modifications to the framework could improve its applicability to such cases.

A range of interesting sub-texts is relevant to the governance of GM crops:

- The GM crop example has demonstrated the ability of internationally organised coalitions of advocacy of groups to counter successfully the power of multinational corporations, creating a new societal balance in power structures (Tait and Bruce 2004).

O. Renn and K. Walker (eds.), Global Risk Governance: Concept and Practice Using the IRGC Framework, 133–153.

- In Europe, these coalitions have led the way toward development of new processes of stakeholder engagement as part of a new *governance*, as opposed to *government*, policy agenda.[1]
- In Europe, which has experienced delays and difficulties in bringing GM crops to the market, this new risk governance process has led to major challenges to the evidence base for risk-related decision making, partly because the adoption of the precautionary principle in European legislation has enabled advocacy groups to invoke 'risk' as an issue to attain leverage in political debates which have very little to do with risk.
- The European approach to risk governance of GM crops, with heavy reliance on a precautionary approach, has been widely acclaimed as more democratic than that of the US, but its outcomes in practice have mainly been undemocratic.
- The US approach, on the other hand, is simpler and faster and has been more successful in enabling companies to bring GM crops into wide scale agricultural production. However, it has been showing some strains as the complexity of the technology and the product types to be regulated increases.
- Finally, the GM crops experience has illustrated the role of regulation in increasing the development time for new products and hence in increasing the number and variety of opportunities for stakeholder engagement including the integrated and co-ordinated framing of the technology as either negative or positive, depending on their perspectives, by a wider range of stakeholders.

Analysis of Risk Governance of GM Crops in Accordance with the IRGC Framework

This section is structured according to the different stages of the IRGC framework. It comments, where relevant, on the separate and distinctive risk governance approaches that have evolved in the USA and the EU from the late 1980s. In the USA, GM crops made a relatively rapid and straightforward passage through the existing risk governance process for comparable products and went subsequently into commercialisation, first in the USA itself and then in many other countries. In the EU on the other hand, the first European Commission (EC) Directive 90/220, which was developed after lengthy consultation, was abandoned and replaced by a temporary moratorium on GM crops. This step allowed the entire regulatory system and its basis to be re-assessed, leading to a much more restrictive set of regulatory regimes co-ordinated under a revised Deliberate Release Directive 2001/18/EC (von Homeyer 2002; Jaffe 2004). Compared to that of the US, the European risk governance approach has appeared to be less evidence-based and more driven by

[1] The '*Governance*' approach attempts to set the parameters of the system within which people and institutions behave so that self-regulation achieves the desired outcomes, implying a move away from the previous '*Government*' approach (a top-down legislative approach which attempts to regulate the behaviour of people and institutions in detailed and compartmentalised ways) (Pierre and Peters 2000; Lyall and Tait 2005).

political and advocacy group influences, rather than by formal approaches to risk governance.

Risk Governance Context

Three distinct periods in the governance of GM crops can be identified (referred to as 'periods' here to avoid confusion with the various 'stages' of the IRGC framework).

Period 1

In the early to mid 1980s, most scientists, industry managers and regulators in Europe, and many in North America, supported the adoption of a precautionary approach to the early development of GM technology. They mainly regarded this approach as an exercise in public reassurance, rather than a measure justified by expected risks.

Period 2

In the mid to late 1980s and early 1990s, some GM crop products were in the development pipelines and relatively close to market; and companies became frustrated by delays caused by the European precautionary approach. Monsanto was in a potentially leading position in bringing GM crops to market, moving ahead faster than other multinational corporations. The company was very influential in setting up an organisation, the Senior Advisory Group for Biotechnology (SAGB) to lobby the European Parliament for relaxation of the precautionary approach to GM crops. In the US, the Vice President's Committee on Competitiveness was equally active, and much more effective, in promoting a product-based approach to GM crop regulation. From this point on, the divergence between the US and EU approaches became increasingly marked. The role of the OECD in international co-ordination of regulatory systems was also very prominent in this phase, with the OECD favouring the industry/US position and opposing that of the EU (OECD 1993).

Period 3

From the mid 1990s onwards, relations between US and EU regulators became increasingly strained. A similar, but unprecedented, rift emerged between US-based and European-based multinational companies; and the concerted, co-ordinated opposition of European advocacy groups to GM crops became increasingly strident and influential in shaping public opinion.

Risk Pre-assessment – Framing New Technology

The IRGC framework, had it been applied to GM crop governance in Period 1, would have promoted a comprehensive pre-assessment of the technology, scientific, and regulatory contexts for the GM issue. Key aspects of these contexts should have included the overall framing of the technology, issues related to early testing and monitoring arrangements, linkages to existing regulatory systems (or alternatively judgements about the inadequacy of existing regulatory systems), and the scientific conventions and assumptions in use. All of these aspects were indeed discussed at one point in time or another. However, they were not part of a formal, overall, internationally co-ordinated approach. The work of the OECD came closest to such an approach, but it lacked several of the important features present in the IRGC framework. Thus, the roots of the eventual conflict can be found at this point but no nationally or internationally implementable mechanism for reconciliation was then available.

With any very new technology where there is no previous experience of either its benefits or its potential risks, it is the process of framing the technology *as a whole* that is important rather than the framing of any individual product. The IRGC approach to '*problem framing*' includes risk scope, risk perception and public awareness.

In Period 1, the early development period of this path-breaking technology,[2] framing of both benefits and risks was based more on conjecture than on evidence, as no products were yet available for testing. Those developing the technology were very active in trying to ensure that it was framed in terms of its benefits, rather than its risks. In general, for any new technology for which there is no obvious precedent, its framing by regulators is as important as its framing by those developing the technology in contributing to its subsequent framing by citizens.

In the competition to influence the public framing of GM crops, companies emphasised their potential contribution to the development of more sustainable farming systems, whereas advocacy groups emphasised their role in supporting intensive farming systems which they claimed were inherently unsustainable. Companies were at a disadvantage in this debate because they were unwilling or unable to use one of the strongest arguments supporting their case – the ability of GM crops to reduce the use of pesticides in intensive farming systems without reducing crop yields. As pesticide producers, they felt that they could not claim that it would be 'a good thing' to reduce the use of pesticides and most of them were not at that time prepared to discuss publicly the realities of developing an alternative product range that would undermine their existing product base in pesticide development (Tait and Chataway, 2007). Interestingly, the expectations and voices of farmers (other than organic farmers) were almost entirely absent from the debates throughout all periods of the development of GM crops.

[2] Path-breaking technologies have been defined as involving discontinuities in science and technology developments, in the nature of markets and in relationships among firms in a sector (Spinardi and Williams 2005).

The language used to describe the technology was also part of this framing process, with scientists initially referring to it as 'genetic engineering', then seeing 'genetic manipulation' as a less pejorative term, and finally settling on 'genetic modification' (Kornberg 1988). Likewise, in referring to the use of GM crops in an open farming environment, there was an unsuccessful attempt to move from the term, 'deliberate release' of GMOs (the term used in the European Directive) toward 'intentional introduction' instead, which was seen (by scientists and industry managers) as less pejorative.

Industry framing of GM crop technology in Periods 1 and 2 was, however, inconsistent. Presentations and publications from scientists and company managers seeking financial support to develop the technology emphasised its novelty as a radical break with previously available products (i.e. a path-breaking technology). At the same time, their papers and reports written in a regulatory context emphasised the continuity with previous generations of technology such as conventional plant breeding, baking bread and brewing beer, (i.e. its path-dependent nature), as a justification for avoiding additional regulatory constraints.

The differences between EU and US approaches to the regulation of GM crops can also be traced to a very early difference in the framing of the technology for regulatory purposes. In the EU, because GM crops were framed as a radical departure from any products that had previously been on the market, with potentially unpredictable properties, they were seen to require a *de novo* consideration of the risks they might present and the regulatory systems that could be put in place to control them, i.e. they were seen as requiring path-breaking regulatory approaches. The analogy most frequently used for GM crops by European regulators was the introduction of alien species with the attendant risks of uncontrollable spread in the natural environment (RCEP 1989).

Most companies and US regulators on the other hand, in line with the OECD approach, framed them as inherently similar to existing products developed through conventional plant breeding programmes and therefore not requiring any additional scrutiny beyond existing regulatory systems, for example for pesticides, food for human consumption or animal feeds (i.e. they were seen as requiring path-dependent and evolutionary regulation).

The regulatory language in which this debate was framed was thus that of 'product vs. process' (Tait and Levidow 1992) with the US looking for analogous *product* categories subject to existing regulatory systems and assigning GM crops to them according to their properties, while the EU viewed the *process* of genetic modification as potentially leading to novel properties requiring a new approach to regulation. This distinction has been a major contributor to World Trade Organisation disputes over GM crop regulation between the US and EU.

Some interesting parallels exist between IRGC's risk 'pre-assessment' phase, particularly the framing issues discussed here, and company innovation strategies. GM crops were 'path-breaking' for agrochemical company innovation strategies (see footnote 2) in that they required new approaches to research and development. The crops could not be marketed by the same routes as chemical pesticides, requiring them to be distributed through seed marketing routes. They also challenged the product base of other powerful industry sectors, namely food producers and supermarkets (Tait 2007). However, there was also considerable 'path dependency' in the strategies companies chose to develop GM crops. For example Monsanto's choice of herbicide tolerance as an early application of the technology fitted well with its earlier development of the very successful herbicide glyphosate, and had strong synergies with its existing product development and market strategies (Chataway et al. 2004).

Many of the framing debates surrounding the governance of GM crops can similarly be seen in terms of demands for either path-breaking or path-dependent regulatory systems. US regulators and multinational companies advocated path-dependence in the form of a product-based regulatory system for GM crops, while EU regulators saw a need for at least considering a path-breaking approach, if only until preliminary, precautionary risk assessments had been completed.

Risk Appraisal

The risk appraisal stage of the IRGC framework juxtaposes the results of risk assessment with the concerns of stakeholders and public groups.

Risk Assessment

As part of a formal precautionary approach, a wide range of risks has been evaluated for GM crops in Europe and elsewhere. An increasingly sophisticated array of experiments has been conducted throughout all three periods identified above, so far with no clear evidence of harm. For example:

- One early concern was that the use of antibiotic resistance markers[3] in crops used as food or animal feed could lead to the emergence of antibiotic resistant strains of micro-organisms in the intestines of humans and animals. Although it was demonstrated that there was a very small chance of this happening, the risk was calculated to be several orders of magnitude less than the risk of emergent strains arising from human or animal treatment with antibiotics. Given the negative publicity around this issue, however, companies agreed to phase out the use of antibiotic resistance markers.
- Laboratory experiments in the US demonstrated that that pollen from maize rendered insect resistant through incorporation of a gene coding for a toxin from *Bacillus thuringiensis*, was toxic to larvae of the monarch butterfly (Losey et al. 1999). These results were widely reported in the press, and were particularly

[3] Since not all attempts to insert genes into cells are successful, scientists use genetic 'markers' as a tool for recognising when they have been successful. An 'antibiotic resistance marker' is a gene that, when inserted into plant cells, conveys resistance to a particular antibiotic. Plant cells that survive exposure to that antibiotic are thus 'marked'.

promoted by environmental groups. However, the subsequent failure to demonstrate such effects outside the laboratory was not so widely publicised (Council for Biotechnology Information 2001).

- An influential experiment on food-related risks of GM crops carried out by Ewen and Pusztai (1999) purported to show that feeding GM potatoes to rats had damaging effects on their intestines. These results were widely reported in the press and are seen as one of the most important stimuli for the public backlash against GM crops in the UK, even although their experimental design was widely criticised by scientists expert in this field.
- Cross-pollination and uncontrollable spread of novel genetic material in the environment has been a long term concern for members of the public and also for some scientists. Several studies have shown that these events are possible and indeed that, under some circumstances, transgenic plants can be detected at considerable distances from the source crop. However, there is as yet no evidence for long term viability or spread of transgenic plants arising from such events. A major UK research initiative on *Gene Flow in Plants and Micro-organisms* by the Biotechnology and Biological Sciences Research Council (BBSRC) and the Natural Environment Research Council (see www.bbsrc.ac.uk, accessed on 7 May 2006), summarised in a BBSRC press release issued on 23rd June 2005, claimed that gene flow from GMOs to soil bacteria is vanishingly small and that introduced traits by GM methods can have less impact on overall gene expression than conventional breeding methods.
- Concerns have also been expressed about the impact of the adoption of GM crops on farming practices and consequently on farm wildlife biodiversity. Another series of experiments (GM Crop Trials) carried out in the UK examined such effects relevant to herbicide resistant oilseed rape, sugar beet and maize.[4] The differences attributable to genetic modification were small, but statistically significant, with GM oilseed rape (canola) and sugar beet showing a reduction of biodiversity and maize showing an increase. However, Les Firbank, who led the scientific team, has commented that the results reflect the effects of overall crop management practices rather than of genetic modification *per se*, and that similar evaluations of non-GM crop introductions in the past would have found similar impacts.[5] Thus, despite the challenges and uncertainties to which the regulatory system for GM crops has been subject, at least in the EU, before any significant exposure to GM crops, a wide range of potential hazards has been identified and their risks estimated with no evidence of harm.

As part of the risk assessment phase, the IRGC framework calls for the categorisation of risk with regard to the degree and cause of 'complexity, uncertainty, and/or ambiguity'. Categorisation of risks should be based on judgement by risk analysts, taking account of the nature and quality of evidence available including: hazard identification and estimation; exposure and vulnerability assessment; and risk es-

---

[4] The results of this experiment can be found in a special issue of *Philosophical Transactions of the Royal Society of London, B* (Biological Science), 29th November 2003.

[5] See http://www.innogen.ac.uk/Events/, accessed on 7 May 2006.

timation. However, in the case of GM crop regulation, the categorisation of risk has been primarily dependent on 'Concern Assessment' (see below) which drew largely from the 'Pre-assessment' stage during which the particular framing of the technology was established (as discussed above). Essentially, GM crops fell into the 'ambiguous' category, where agreement does not exist on the fundamental values driving evaluation of the risk. Under the precautionary regime that was established, the extent of public concern was determined more by the success of various stakeholder groups, mediated via the press, in raising public concerns for political purposes than by a formal, more balanced, risk assessment process.

When risk assessments are strongly influenced by advocacy groups that have a principled, ideological opposition to a particular technology, no amount of evidence, regardless of its scientific quality, will lead to a change of opinion or of risk-related behavioural responses (Tait 2001). For example, the precautionary approach adopted for the development of the regulatory system in the EU, required very careful control and monitoring of trial releases of GM crops. The failure of these early experiments to demonstrate any potential hazards was the trigger in Period 2 for industry to lobby for relaxation of the precautionary regime. On the other hand, it also triggered demands from activist groups for additional, more stringent testing. These positions related back to the original framing of the technology; on the one hand, many members of the UK public saw the adoption of the precautionary principle as reassuring (Martin and Tait 1992) while on the other, scientists and industry managers believed it was leading to unnecessary alarm, with members of the public questioning 'If this technology is as safe as you claim, why do we need to be precautionary?'

Concern Assessment

Public attitudes to GM crops are one of the most intensively surveyed technology-based issues, at least in the EU. Those surveys with a valid statistical base generally show that 30% or less of the population would avoid purchasing or eating GM foods, and yet the overall impression in the press is that most Europeans 'reject GM crops' (Bauer and Gaskell 2002). As the issue has faded from intense public debate in Europe, the proportion of the population expressing negative opinions on GM crops has also declined (Gaskell 2005), emphasising the labile nature of public attitudes as gauged by opinion polls, and as driven by a press that sells newspapers by generating controversy.

In the US, although some citizens express concern about GM crops, opponents of the technology have not been able to dominate its public framing as they have in the EU. Although there have been some risk-related incidents following the marketing of GM crops, they have not led to long term, sustained public opposition to the technology as a whole. To give just two examples:

- The monarch butterfly is an important icon for American conservationists, and there was a flurry of public concern about GM crops in the American press when the initial research about the toxicity of pollen from GM maize to butterfly larvae

was published, but it subsequently faded and did not prevent the adoption of insect resistant GM maize on farms.
- In another case which had widespread press coverage, GM Starlink corn which was approved for animal feed but not for human consumption, was found to be present in taco shells on sale to the public (Oliva et al. 2006). Because of fears of allergenicity arising from an introduced protein in the corn (Cry9c), the tacos were withdrawn from supermarket shelves and the crop itself was withdrawn from sale to farmers, at considerable cost to the companies concerned, and to US corn farmers through loss of export markets. However, this incident has not led to a generalised rejection of GM crops in the USA.

For the European public, the economic benefits from GM crops were perceived to be in terms of increased profits for farmers and for multinational companies, which was seen as unacceptable (Martin and Tait 1992). A constant refrain in the European press has been that there are no public benefits from GM crops. As noted above, benefits in terms of reduced use of pesticides were not emphasised by industry in the early stages of the development of GM crops. However, the press also largely passed over the fact that one of the first products to be available in Europe that was produced from GM crops, Zeneca's GM tomato paste, was cheaper than alternative products and very popular.

Both Europe and the US have a minority of the population that is fundamentally opposed to the introduction of GM crops, but the two regions have taken very different regulatory paths. The difference between the two regulatory contexts is related to the extent to which the concerned minority in each region has been able to influence wider public opinion and ultimately policy, thus determining the GM food consumption options available to citizens who are uncommitted and unconcerned.

Balancing Risk Assessment and Concern Assessment

The IRGC framework proposes an early stakeholder engagement that is well integrated into the overall governance process. Although stakeholder engagement was ongoing from the early period of GM crop development, the various initiatives were *ad hoc*, disconnected from one another, and not well integrated into the subsequent stages of the risk analysis. In such circumstances, risk categorisation becomes dominated by a political process, rather than being part of conventional risk assessment. In other words, risk categorisation carries with it the power to influence which technologies are developed and which are rejected, often on a basis of ideological preferences (for example, in the case of GM crops, opposition to globalisation or preferences for particular agricultural landscapes or types of farming system), rather than actual risks.

In such circumstances, there will be multiple risk categorisations which will evolve over time as part of an unstable and turbulent process. As outlined in Table 1, the balance of these multiple categorisations of GM crops shifted during the three

Table 1 Risk categorisation of GM crops.

Period of Development	Perspective	Dominant Categorisation	
		US	**EU**
Period 1	Regulators	Uncertain	Uncertain
	Industry	Uncertain	Uncertain
	Public	No opinion, uncertain	No opinion, uncertain
Period 2	Regulators	Complex	Uncertain, ambiguous
	Industry	Complex	Complex
	Public	Uncertain	Uncertain, ambiguous
Period 3	Regulators	Simple	Ambiguous
	Industry	Simple	Simple
	Public	Simple	Ambiguous

periods outlined earlier as more information about the products became available and as new players/stakeholders came onto the stage.

In the US, the dominant categorisation among industry, regulatory and public actors moved from 'uncertain' through 'complex' to 'simple'. In the EU on the other hand, although industry categorisations followed the US pattern, public and regulatory perspectives moved from 'uncertain' to 'ambiguous' with no sign as yet of a resolution of the ambiguities. In the EU, 'risk categorisation' itself, rather than risk appraisal or actual risk became the battle ground on which the political process was played out, with consequences which were largely negative for overall risk governance of this technology.

Risk Characterisation and Evaluation

Scientific, Evidence-Based Risk Profile

In the US, decisions to anchor GM crop regulation as a whole on the existing *product-based* system were taken on the basis of scientific extrapolation, rather than on new scientific evidence. This system was thus reactive, rather than precautionary, in the sense that it reacted to evidence of any hazard found to be arising from a GM crop following its introduction and put in place measures to prevent such hazards in future (Tait and Levidow 1992).

Another example of such a reactive process is the internationally applied principle of 'substantial equivalence' whereby a GM food is scrutinised to ensure that it is not significantly different from other foodstuffs available in the market place, and hence not in need of any additional regulation or restriction. For example, any foodstuff that had been genetically manipulated to incorporate nut proteins would be regarded as *not* equivalent because it might also contain nut allergens.

The European *process-based* approach, in contrast to that of the US, is much more precautionary. It is not based on evidence of harm but on societal concerns about potential risks that may arise at some future date. The principle of substantial equivalence has been strongly criticised in the EU as being insufficiently precautionary about currently unforeseen hazards in GM foodstuffs. Some authors have suggested that we should test GM foods in a similar manner to current drug testing regimes (Millstone et al. 1999), although how these might be operationalised for a foodstuff has not been explained.

The UK decision in 2003 on whether to approve cultivation of GM herbicide tolerant (HT) maize, oilseed rape and sugar beet was directly related to the scientific evidence from the GM crop trials. Genetically modified HT maize supported a higher level of biodiversity than non-GM and so was approved. In the trials for oilseed rape and sugar beet, the balance of evidence suggested that biodiversity had declined in the GM crops, so these were not approved. However, as noted in the next section, societal values played an important part in the initial design and subsequent interpretation of these experiments.

Societal, Value-Based Balancing of Benefits and Risks

The example of the UK GM crop trials illustrates the difficulty of maintaining a clear separation between scientific, evidence-based risk characterisation and societal, value-based characterisation. The assumption underlying the criteria chosen for evaluating the risk of GM crops was the societal value judgement that it was desirable to encourage weeds to grow in agricultural crops in order, in turn, to support a higher diversity of insect species as part of a wider food web. This value judgement is unlikely to be shared by many in the farming community. Yet assessment of crop yields or other management benefits were specifically ruled out of the comparisons made in these experiments – the participating farmers and companies were prohibited from collecting these data. Thus, one of the potential benefits, improved efficiency of crop production, was treated as irrelevant to the decision – there was no balancing of benefits and risks. Rather than collecting a range of evidence which would enable the balancing of an array of benefits and risks of interest to different societal sectors, the UK GM crop trials focused on only one environment-related aspect of GM crop production.

Where GM crops have been introduced with less public opposition, their widespread use by farmers implies that they do have benefits in the management and efficiency of crop production. The adoption of insect resistant GM crops, particularly cotton, has also led to major reductions in the exposure of farm workers to dangerous insecticides which is both a benefit to farming communities and a public benefit (James 2002; Bennett et al. 2006). The societal consensus in parts of the world where the technology has been adopted seems to be that, so long as there are no risks (or unacceptable levels of risk) and no public dis-benefits, the provision of agricultural benefits is sufficient justification for the adoption of the technology –

there is not a perceived need for the provision of additional public benefits before new technology can be introduced.

European demands for public benefits from a new technology that cannot be incorporated into the price which can be charged for that technology are probably not financially supportable in a globally competitive environment. In the case of GM crops, it seems likely that the eventual outcome will be the demise of a European-based GM crop industry sector, to the disadvantage of European farmers in a global trading environment.

Conclusions on Risk Acceptability or Tolerability

In the case of GM crops it is more appropriate to consider *product*, rather than *risk*, acceptability or tolerability, given the comments above on the relative lack of risk-related evidence underlying public opposition to GM crop development and use. In both the US and EU, there is a fairly large minority of the population for whom these products are intrinsically unacceptable, regardless of risks or benefits. This opposition relates to societal concerns about globalisation and the industrialisation of the human food chain, rather than to evidence of tangible risks associated with GM crops themselves. The difference between the US and the EU regulatory positions lies in the extent of the influence this minority has had on largely uncommitted and un-engaged members of the population and on politicians involved in developing risk governance processes. Nevertheless, it is still the case in both the US and EU that most members of the public do not particularly care whether their food is produced from GM or conventional crop varieties.

Risk Management

For those members of the public who are fundamentally opposed to the growing of GM crops, there are no acceptable risk management options.

Most others, including companies developing the technology, would recognise that, although no major hazards have yet been demonstrated for the GM crops in use, there may still be unexpected side effects. Several senior managers in multinational companies have suggested that there should be post marketing surveillance of GM crops to ensure rapid detection of any such effects (Chataway and Tait 2000). In the case of the Starlink corn incident noted above, the US regulatory system was changed to require that any GM crop approved for animal feed must also be approved for human consumption. The EU regulators have adopted a similar requirement. This rapid action on the part of the regulatory authorities seemed to pre-empt any further public opposition to the technology in the US.

However, this example also illustrates the complexity and inter-connectedness of risk governance and innovation systems. One potentially environmentally beneficial outcome of GM crop technology would be the development of animal feeds that are tailored to the nutritional requirements of particular species, pigs, cattle, chickens, etc. Such feeds could avoid the need to feed protein supplements to these animals and could also reduce the levels of phosphate pollution from farm effluent. Although such a product has yet to be tested in the regulatory process, it seems unlikely that a crop tailored to the nutritional requirements of, say, pigs, would be accepted also for human consumption. The choice of this particular risk management option thus seems likely to halt further development of such potentially environmentally beneficial products.

Although not yet included in the IRGC framework, there are cases where it would be useful to encourage technological innovation as a potential alternative or contributing factor to risk management, alongside risk regulation. Such options are often referred to disparagingly as 'technical fixes' but they can nevertheless be very effective. Indeed, generally speaking we are better at technical fixes than we are at 'social or regulatory fixes'. One such example would be the often-raised possibility that the genes engineered into GM crops might 'escape' to contaminate wild species, generating 'super-weeds' or other undesirable and uncontrollable new species. A potentially useful technology-based approach here would be for policy makers to require the incorporation of one of several genetic use restriction technologies (GURTs) into plants, restricting their ability to propagate through viable pollen or seeds (Daniell 2002), obviating the need for the complex societal and regulatory restrictions to maintain separation distances between organic and GM crops in Europe, which are likely to be difficult and expensive to monitor and enforce.

Risk Communication and Stakeholder Participation

Communication and stakeholder issues were intimately linked throughout the various stages of the development of GM crops and they have been brought together in one section here.

The GM crops case study provides numerous examples of communication failures:

- Linking the dialogue between industry and regulators in Period 1 to public groups and interested citizens.
- Communication among multinational companies involved in GM crop development, particularly in Period 3 and the later part of Period 2.
- Communication between the agro-biotechnology industry sector and the food processing and distribution sectors.
- Communication with 'wider society' – the largely un-engaged and un-interested public.
- Communication between policy makers and the public, particularly the failure by policy makers to explain the potential public benefits of the new technology, given the agro-biotechnology industry's reluctance to promote these benefits.

On the other hand, the advocacy coalition, involving groups with environmental, third world and consumer-related agendas, that came together in Period 3 and dominated the media presentation and the framing of the GM crops debate, provided an example of a very successful, integrated communication strategy that enabled this coalition to dominate the agenda in Europe and to have significant impacts internationally.

IRGC's framework suggests that there is a specific type of discourse that can be identified and used as appropriate to different risk categorisations and stakeholder groups. In this case study, the recommendation proved not to be feasible in practice. The type of discourse is intimately linked to the framing of the technology. For a situation like GM crops which involved numerous, actively engaged stakeholders competing to frame the technology for different audiences, the process could not be controlled by risk managers and regulators. This problem is particularly difficult where conflicting values and ideologies are involved. However, there are likely to be competing perspectives on any risk issue of sufficient complexity to warrant the application of the IRGC framework.

Failures of communication have been identified by activist groups as the main reason for the emergence of European public opposition to GM crops. However, policy makers made numerous attempts to encourage public engagement, and both policy makers and companies in the UK and Europe regularly took part in meetings with advocacy groups representing public opinion, needs, and desires. None of these efforts seemed to reduce the level of opposition or conflict that eventually emerged. As this case study has attempted to show, the evolution of the European response to GM crops was multi-dimensional and highly complex.

Nevertheless, building on this presumed communication deficit, more 'upstream' engagement with public representatives is now being advocated for emerging technologies like nano-technology as the route to avoiding future conflicts of the type experienced in Europe with GM crops (Willis and Wilsdon 2004). This simplistic analysis of the cause of the problem and its solution is unlikely to lead to improvements in risk governance and may indeed perpetuate and exacerbate problems of the type experienced by GM crops.

Conclusions and Recommendations

The value of a democratic governance process lies in its ability to prevent powerful vested interests from dominating decision making. There are many who regard the European GM crops regulatory outcome as an example of this process in action, the triumph of advocacy groups, acting in the public interest, over the power of multinational companies.

However, the European outcome could equally be seen as the replacement of one vested interest (the agro-biotechnology industry) by another more recently influential group, at least in the EU (public interest advocacy groups), with equally negative outcomes for democratic decision making on risk issues.

The following conclusions and recommendations mainly address the problems that emerged in the European risk governance of GM crops, although many of the recommendations could equally be applied to future developments of GM crops in a global context and also to new innovative technologies more generally.

GM crops and their risk governance provide a particularly complex example for a case study to test the IRGC framework. On the one hand, it would be legitimate to claim that the framework cannot be expected to deal with such very general cases. On the other hand, most of the complexity, turbulence and conflict arising from this case was related to public and stakeholder perspectives on the technology and the responses of governments and industry to these perspectives. The IRGC framework could usefully be developed further to improve its future applicability to such situations – the development of highly innovative technologies in a globally competitive environment which challenges the capacities of existing regulatory systems.

Experience in Applying the IRGC Framework to the Development of GM Crops

The most important deficits in the risk governance of GM crops in Europe, which the application of the IRGC framework might have prevented, related to the societal context and the categorisation of risk-related knowledge. Because these were both early stages in overall risk governance, difficulties experienced then, had serious implications for later stages of risk governance.

If the earliest pre-assessment and framing of the technology and its associated risks in Period 1 had been undertaken in a more formal manner, as a conscious component of a risk governance process, rather than the open competition to frame the technology, then greater control of subsequent stages of the analysis by risk policy makers and regulators might have been possible. However, the multiple framings of the technology that emerged among different stakeholders, leading to multiple risk categorisations that in turn evolved over time (see Table 1), led to a highly politicised debate within Europe and internationally that was beyond the control of any risk governance process. European policy makers, rather than having overall control of the risk governance process, were in the unenviable position of having to respond to increasingly vehement waves of public protest, amplified by the press and political lobbying, partially and temporarily countered by pressures from industry.

A window of opportunity for a less contentious process existed around 1990. If policy makers could have resisted industry pressures to relax the European regulatory system for GM crops and at the same time explained to the public the potential sustainability benefits of the technology, they might have been able to take a lead in the framing and subsequent governance of the technology (Tait 1993). Even so, the complexity of the interactions they faced may have defeated this purpose.

As it was, there was little input from the formal 'Risk Assessment' stage into the Risk Categorisation step for GM crops, the latter being influenced mainly by the risk perceptions and concerns of a vocal minority in European society. These factors also dominated the 'Risk Evaluation' and 'Risk Management' phases. The decision that the technology itself was not tolerable or acceptable to the European population was unrelated to any formal risk assessment, and risk reduction measures, particularly those demanded by the organic farming lobby, seemed more designed to make it impossible to develop the technology than to counter any demonstrable risks to health or the environment. The Risk Management options now being implemented in keeping with the EC Directive 2001/18 and subsequent regulations likewise bear little relationship to any evidence-based assessment of risks and are unlikely to be compatible with a profitable European agro-biotechnology industry sector.

One might argue that there has in fact been a gradual erosion of the evidence base for risk-related decision making about GM crops due to confusion generated by inputs from vested interests on all sides of this debate. The evidence produced by companies to support product registration is regarded as suspect by the public and is scrutinised carefully by regulators. In addition, any mistakes or deliberate biases in this evidence can have serious implications for the company concerned, so there are disincentives for a company to introduce such biases. The same does not apply to some public interest advocacy groups who quote selectively from evidence that supports their case, without suffering any loss of public confidence in their impartiality.

Unless we can develop standards and procedures to help decision makers to reach conclusions on the best available evidence from both social and natural sciences we risk retreating into a series of interlocking enclaves of indecision, challenge and counter-challenge. We should be building stakeholder engagement into the risk governance of new technology in a manner that reduces the so-called democratic deficit in such decision making (Tait 2004).

Further Development of the IRGC Framework

If we are to extend use of the IRGC framework beyond first generation GM crops to later developments of GM technology or other innovative technologies (e.g., stem cells or nanotechnology), a range of additional modifications could usefully be built into its operation. One of the most distinctive aspects of the IRGC framework is its careful consideration of public and stakeholder engagement processes, and this is perhaps where there is most need of further refinement if it is to prove of real value to risk regulators and policy makers and also to industry.

Timescales of Development of Innovative Technology

For the risk governance of many technologically innovative products, where the pace of development is very rapid, products appear on the market before there is time to begin to explore, far less prevent, any negative societal impacts. It then becomes a matter of consumer choice whether the product succeeds or fails. The speed of development in such cases, for example in information and communication technology (ICT), is one reason why innovations such as the world wide web, with major societal impacts, receive very little regulatory attention prior to their being publicly available. Where potential risks are discussed after a product is widely available (e.g. mobile phones and the associated transmission towers), risk management is evidence-based rather than precautionary.

In such cases, the speed of innovation is driven by intense competition. There are likely to be difficulties in convincing industry of the value of the careful and thorough engagement procedures, particularly on socio-economic implications and public concerns, which are part of the IRGC approach. It is therefore unlikely that there will be pressure for application of the IRGC framework to ICTs and other technologies with short development times.

The life science industries offer a dramatic contrast to innovation in ICTs. Innovative developments (e.g. pharmaceuticals and pesticides) arising from these industries are already subject to very demanding and lengthy regulatory processes. Risk regulation is the primary driver of innovation 'pipelines' in these industries. The process imposes major constraints on the dominant multinational companies, although by acting as a barrier to entry for small companies it helps maintain their dominance in the market (Tait 2007). Thus, for a company engaged in the economically risky development of new technology, the existence of a familiar regulatory system which supports its 'first mover' advantage is a considerable asset.

On the other hand, this lengthy development process (up to 15 years) also creates opportunities for extensive public and stakeholder engagement, as advocated by the IRGC framework. The twin circumstances in the USA, of more rapid passage of GM crops through the regulatory process and lower levels of effective public opposition to the technology, are probably related.

Framing Innovative Technology and Control of Engagement Processes

Ideally, public policy makers and regulators should take the lead in managing the framing of the risks and benefits of new technology to minimise the biases likely to be introduced by both industry and public advocacy groups.

Effective engagement processes require responsible behaviour by all stakeholders. As noted above, although industry managers do not always behave responsibly in such situations, there are major risks to the company, for example if biased or invalid evidence is used in support of risk regulatory processes. Most companies now accept the need for what has become known as a 'license to operate' – a general recognition of publicly responsible behaviour. Similarly, public advocacy groups should also exhibit responsible behaviour if they are to contribute properly to risk governance. Although some NGOs behaved very responsibly in representing the views of citizens, others adopted a strongly adversarial, uncompromising approach and were less careful about the validity of the evidence used to support their views. It would thus be helpful to build into the IRGC approach a set of standards for engagement covering responsible and unbiased use of evidence and willingness to compromise to accommodate the views of other groups.

Even with such safeguards, and given an effective application of the IRGC framework, there can be no reassurance that the kind of anomaly experienced for GM crop regulation in Europe will not be repeated for other technologies. Pressures for more 'upstream engagement', moving engagement processes to earlier stages in research and development (Willis and Wilsdon 2004), an approach that has considerable support from political and scientific communities in Europe, is likely to encounter several problems:

- The evidence base for decision making will be even weaker than it has been for GM crops.
- There will be even greater uncertainty about the validity of the science base and the eventual nature of products available on markets than there has been for GM crops.
- In framing the technology, public stakeholder groups are more likely to focus on potential risks than on benefits while industry stakeholders will focus more on benefits, exacerbating the potential for acrimonious conflict.
- Given the long timescale and uncertain nature of future risks and benefits, only those with a vested interest in the issues and outcomes will be prepared to engage in discussions and decision making.
- Public opinion is likely to change dramatically over a lengthy development period so early engagement cannot be a valid base for decisions taken later in the development process.

Choice of Regulatory Approach for Innovative Technology – Path-Breaking or Path-Dependent?

Although companies in highly regulated industry sectors can cope very well with existing, even if onerous, regulatory systems, they find it very difficult to operate in a climate of uncertainty over the eventual nature of the risk regulatory regime to which they will be subject. This is another issue which IRGC could usefully address – suggesting criteria for the development and choice of regulatory systems for innovative technologies which relate to the properties of the products and the nature of stakeholder views and requirements. Such criteria might remove some of the uncertainty from the development of regulatory processes and reduce the time required to develop such regulatory systems.

For similar reasons, a path-dependent regulatory approach, such as the product-based approach to the regulation of GM crops, is likely to encourage faster, and hence more profitable, development of new technology. A path-dependent approach should be desirable provided it can ensure effective and acceptable regulation.

Guidelines for policy makers for the governance of innovative technology should address the following questions:

- What are the relevant regulatory precedents?
- What are strengths and defects of various approaches?
- What kinds of technology will emerge from new scientific knowledge, how long will it take, who are the relevant stakeholders?
- What degree of influence should be given to conflicting stakeholder groups or to powerful advocacy coalitions?

What kinds of decision should they have the power to influence, e.g. should value-based or ideologically committed stakeholder perspectives be allowed to dictate the choices available to society as a whole, in the absence of evidence of risks to people or the environment, as has been the case for GM crops in Europe, particularly when labelling legislation allows consumers to avoid GM foods should they wish to do so.

Path-breaking technologies present particular challenges for policy makers and risk regulators. For such radical innovations there may be no obvious match between the properties of the new technology and an existing regulatory system. In the early phases of the technology development, the properties, benefits and risks of the new products may be difficult to judge. However, path-breaking technology does not necessarily imply the need for a path-breaking regulatory system.

A technology can be path-breaking for one group of companies in an industry sector and path-dependent for another. For example, stem cells would be a path-breaking technology for a multinational pharmaceutical company whose current innovation strategies are built around small molecule drugs. On the other hand, the technology would be path-dependent for a small company that has specialised in bone marrow transplants or tissue engineering products. Choosing a regulatory system for stem-cell-based therapies that follows that in place for the pharmaceutical industry will favour large companies and disadvantage small companies, with major implications for the scope and direction of innovation arising from this new set of technologies.

In other examples, path-dependent regulatory systems may be appropriate at one stage in the development of a technology but not at another. In the case of nanotechnology, path-dependent regulatory systems may be appropriate for early stage developments, but not for later more complex developments. Likewise, although our conclusion here is that the path-dependent, product-based approach was appropriate to the regulation of GM crops in the early stages of the technology's development, a more path-breaking regulatory approach may be required for later developments (for example the production of drugs in plants grown outdoors).

Risk Governance of Innovative Technologies

The development of internationally effective approaches to the risk governance of innovative technology, particularly in the life sciences and nanotechnology, is likely to remain a challenge for IRGC and could usefully become an important future activity. This analysis has identified some of the most important issues that should be part of an extension of the IRGC framework in this area:

- Development of effective and impartial systems of stakeholder engagement.
- Support, where possible, for individual choice.
- Maintaining and improving the integrity of the evidence base for risk governance-related decision making.
- Developing robust criteria for the development and choice of risk regulatory systems and instruments, including both existing and new regulatory frameworks.

References

Bauer, M.W. and Gaskell, G. (eds.), 2002, *Biotechnology: The Making of a Global Controversy* Cambridge University Press, Cambridge.

Bennett, R., Morse, S. and Ismail, Y., 2006, The economic impact of genetically modified cotton on South African small holders: Yield, profit, and health effects, *Journal of Development Studies* **42**(4), 662–677.

Chataway, J. and Tait, J., 2000, *Policy Influences on Technology for Agriculture: Chemicals, Biotechnology and Seeds – Novartis Agribusiness Monogaph*, Policy Influences on Technology for Agriculture (PITA): Report to the European Commission Targeted Socio-Economic Research Programme (TSER), Project No. SOE1/CT97/1068. Available at: http://www.technology.open.ac.uk/cts/pita/ and http://www.supra.ed.ac.uk/NewWeb/Reports.htm

Chataway, J., Tait, J. and Wield, D., 2004, Understanding company R&D strategies in agro-biotechnology: Trajectories and blindspots, *Research Policy* **33**(6–7), 1041–1057.

Council for Biotechnology Information, 2001, *Bt Corn and the Monarch Butterfly*, March 4.

Daniell, H., 2002, Molecular strategies for gene containment in transgenic crops, *Nature Biotechnology* **20**, 581–586.

Ewen, S.W.B. and Pusztai, A., 1999, Effect of diets containing genetically modified potatoes expressing Galanthus nivalis lectin on rat small intestine, *The Lancet* **354**, October 16, 1353–1354.

Gaskell, G., 2005, Double Eurobarometer survey: Citizens, science and technology, *RTD Info – Special Eurobarometer Issue*, November, 3–32.

Jaffe, G.D., 2004, Regulating GM crops: A comparative analysis, *Transgenic Research* **13**(1), 5–19.

James, C., 2002, *Global Review of Commercialised Transgenic Crops Featuring Bt Cotton*, ISAAA Brief No. 26, International Service for the Acquisition of Agri-Biotech Applications, Ithaca, USA.

Kornberg, H., 1988, Opening remarks, in: M. Sussman, C.H. Collins, F.A. Skinner and D.E. Stewart-Tull (eds.), *The Release of Genetically Engineered Micro-Organisms*, Academic Press, San Diego, CA, pp. 1–5.

Losey, J.E., Raynor, L.S. and Carter, M.E., 1999, Transgenic pollen harms monarch larvae, *Nature Magazine*, May 20.

Lyall, C. and Tait, J., 2005, *New Modes of Governance: Developing an Integrated Policy Approach to Science, Technology, Risk and the Environment*, Ashgate Publishing Ltd., Aldershot, Hampshire.

Martin, S. and Tait, J., 1992, Attitudes of selected public groups in the UK to biotechnology, in: J. Durrant (ed.), *Biotechnology in Public: A Review of Recent Research*, Science Museum for the European Federation of Biotechnology, pp. 28–41.

Millstone, E., Brunner, E. and Meyer, S., 1999, Beyond 'substantial equivalence', *Nature* **401**, 7 October, 525–526.

OECD, 1993, *Safety Evaluation of Foods Derived by Modern Biotechnology: Concepts and Principles*, OECD, Paris.

Oliva, M.J., Baumuller, H. and Mohan, S., 2006, *Guide to Trade, Biotechnology and Sustainability*, International Centre for Trade and Sustainable Development (ICTSD), p. 32.

Pierre, J. and Peters, B.G., 2000, *Governance, Politics and the State*, Macmillan, Basingstoke.

RCEP, Royal Commission on Environmental Pollution, 1989, *Thirteenth Report: The Release of Genetically Engineered Organisms to the Environment*, HMSO, London.

Spinardi, G. and Williams, R., 2005, The governance challenges of breakthrough science and technology, in: C. Lyall and J. Tait (eds.), *New Modes of Governance: Developing an Integrated Policy Approach to Science, Technology, Risk and the Environment*, Ashgate Publishing Ltd., Aldershot, Hampshire, pp. 45–66.

Tait, J., 1993, Written evidence on behalf of ESRC to Report of House of Lords Select Committee on Science and Technology on *Regulation of the United Kingdom Biotechnology Industry and Global Competitiveness*, 7th Report, Session 1992/93, HMSO, London, HL Paper 80-I, pp. 187–196.

Tait, J., 2001, More Faust than Frankenstein: The European debate about risk regulation for genetically modified crops, *Journal of Risk Research* **4**(2), 175–189.

Tait, J., 2004, *Science and Bias*, Paper presented at the BA Festival of Science, Exeter, 6 September, 2004 (www.innogen.ac.uk).

Tait, J., 2007, Systemic interactions in life science innovation, *Technology Analysis and Strategic Management* **19**(3), May, 257–277.

Tait, J. and Bruce, A., 2004, Global change and transboundary risks, in: T. McDaniels and M. Small (eds.), *Risk Analysis and Society: An Interdisciplinary Characterisation of the Field*, Cambridge University Press, Cambridge, pp. 367–419. (Commissioned by Society for Risk Analysis for the International Symposium on *Risk and Governance*, Warrenton, VA, USA, June 2000.)

Tait, J. and Chataway, J., 2007, The governance of corporations, technological change, and risk: Examining industrial perspectives on the development of genetically modified crops, *Environment and Planning C: Government and Policy* **25**(1), 21–37.

Tait, J. and Levidow, L., 1992, Proactive and reactive approaches to risk regulation: The case of biotechnology, *Futures*, April, 219–231.

Von Homeyer, I., 2002, *Deliberate Release Directive: Precautionary Interactions*, Project Deliverable No. D25, Final Draft, December 2002.

Willis, R. and Wilsdon, J., 2004, *See-Through Science – Why Public Engagement Needs to Move Upstream*, Demos, London, 71 pp.

Chapter 8
Nature-Based Tourism

Caroline Kuenzi[1] and Jeff McNeely[2]
[1]*IRGC, Geneva, Switzerland*
[2]*IUCN, the World Conservation Union, Gland, Switzerland*

Introduction and Background

Tourism is the largest industry in the world according to figures on employment and expenditures. Over 800 million people, the equivalent of roughly 12% of the world's population, travel internationally each year – with many more travelling within their countries of residence. Airlines, the hotel and restaurant business, and outdoor equipment manufacturers and vendors are among the supplier industries that are highly dependent on a successful tourism sector. Tourism furthermore relies on and provides income to national infrastructures such as airports, rail systems, road networks, electric power systems, agricultural production, and water supply systems. On a more personal dimension, travelling provides individuals with the opportunity to escape temporarily from the humdrum of every-day life, to experience the novel or unknown and, possibly, to make a long-time 'dream come true'. 'Discovering the world' arguably figures amongst the top life goals for many people in developed countries. Consequently, the tourism sector as a whole seems to be fairly resilient to disruptions from economic downturns, political crises, extreme weather events, or even natural disasters (UNWTO press release of 24 January 2006).

The fastest growing element of tourism is 'nature-based' tourism, often involving excursions to national parks and wilderness areas, to developing countries where a large portion of the world's biodiversity is concentrated (Olson et al. 2001: 936; WWF 2001; Christ et al. 2003: 5). It may also include an 'adventure tourism' element that may carry physical risks. More and more people are living an urban life and the amenities and conveniences that come with globalisation increasingly lead to a near complete disconnect from nature: the living creatures behind the neat slices of fish and meat that we consume or behind the clothes that keep us warm are no longer visible to us; the seasonality of fruit and vegetables virtually has ceased to exist; and the furniture in our homes is impossible to picture as the trees from which it came. Yet at the same time we may have more intimate insight into the mystery of a giant sequoia, the hibernating habits of a grizzly bear or the hatching behaviour

O. Renn and K. Walker (eds.), Global Risk Governance: Concept and Practice Using the IRGC Framework, 155–178.

of a hummingbird than our rural ancestors could ever hope to have had – insights that are brought to us in breathtaking close-up pictures via the many media channels that cater daily to our information needs. For many people, 'getting back in touch with nature' thus provides the ultimately different holiday experience. Indeed, from snow-covered mountains to earth-coloured savannas teeming with exotic wildlife, lush rain forests, vast desert landscapes and pristine coastal strips offering spectacular bird and marine life, the opportunities for immersing oneself in nature seem countless.

Tourism activities which focus on the natural environment exert a number of pressures on the very resources on which they rely. They create risks for ecosystems and the 'services' they provide (such as freshwater supply, soil regeneration and pollination). At the same time, the on-going protection of many of the world's protected areas and their natural resources depend on the well-being of the very tourism industry that also threatens them.

Nature-based tourism also creates a number of different kinds of risk for the larger socio-economic system, both within and across national boundaries. Tourism at many of the world's nature 'hotspots' contributes significantly to gross domestic product (GDP) in the host countries. Serious disruption of the tourism industry can therefore have negative impacts on local livelihoods for which it is often the only source of cash income; the impact of the Maoist insurgency in Nepal's tourism industry is a dramatic example. At the same time, a flourishing tourism industry also carries the risk of altering the social fabric of local communities as rural people exchange subsistence farming for cash economies such as inn-keeping, for example. Nature-based tourism also entails risks to the personal health and safety of individual tourists. Even in countries where tourism is mature and well-managed like Switzerland, accidents can still happen; mountain climbing, skiing, or climbing waterfalls can cause loss of life, with attendant legal and financial liabilities. Finally, the tourism industry itself is subject to external risks, including increases in the price of petroleum (which affects the cost of travel) and potential disruptions due to civil strife or disease epidemics (the recent example of SARS is a good indication of how a relatively minor outbreak can have major ramifications in specific tourism regions and sectors).

Box 1: Facts and Figures

- Tourism accounts for as much as 11% of global GDP, offers jobs to 200 million people, and includes more than 800 million international travellers per year (figure projected to nearly double by 2020).
- Tourism makes up 3–10% of GDP in developed countries and up to 40% in some developing countries.
- For 83% of countries tourism is amongst the top five shares of exported goods and services; for 38% of countries it is the primary source of foreign currency.

- While, over the past 40 years, the world's population has doubled, legally protected areas have tripled, now totalling well over 100,000 sites and covering about 12% of the Earth's land surface.
- Over the past two decades, both nature and adventure tourism have developed to be part of the high-growth segments within the tourism industry. With an annual growth rate of 10–30%, nature-based tourism seems to be the fastest growing tourism sector. Its share in the world travel market is currently about 20%.
- Within the same territory, employment related to recreational activities can exceed employment related to resource exploitation by more than five times.
- 1.4% of the Earth's land surface accommodates 44% of all endemic plants as well as 35% of all endemic mammals, birds, reptiles and amphibians ('endemic' species are those found only within the defined geographic area).
- 24% (1130 species) of mammals and 12% (1183 species) of birds are thought to be endangered with global extinction.
- 40% of the global economy is estimated to be rooted in biological products and processes, demonstrating that biodiversity matters.

Sources: Christ et al. (2003), Higginbottom (2004), UNWTO (2005).

The focus of this case study is on the risks associated with nature-based tourism, i.e. those tourism activities that draw on the natural environment as the primary attraction. Visitation to national parks, trekking, scuba diving and snorkelling, and wildlife watching and birding, if combined with travelling, are amongst the most prominent examples of such activities. However, from a terminological point of view, such a focus presents some challenges. The tourism literature has a profusion of terms conveying similar and partly overlapping meanings, all of which in some way relate to nature-based tourism as defined for this case study while differing in terms of emphasis or underlying philosophy. A relevant selection of these terms is presented in Table 1.

While the terms 'nature-based tourism' and 'nature tourism' are all-embracing in comparison to the other terms, the terms 'ecotourism' (at least in its second, more narrow definition above) and 'sustainable tourism' anticipate certain outcomes of tourism activities by attaching quality criteria to them. Clearly, not every form of nature-based tourism qualifies as 'eco' or 'sustainable'. In addressing nature-based tourism, this case study thus deliberately chooses a relatively broad focus which allows for discussion of a wide range of risks. Where appropriate, and in order to illustrate specific issues, the case study will use examples from the other types of tourism defined in Table 1.

Table 1 Definitions of 'nature-based tourism' and related terms.

Term	Definition
Tourism	'the sum of government and private sector activities that shape and serve the needs and manage the consequences of holiday, business and other travel' (Pierce et al. 1998, cited in Higginbottom 2004: 2)
Nature-based tourism	'the segment in the tourism market in which people travel with the primary purpose of visiting a natural destination' (March 2003 Symposium 'Tiger in the Forest: Sustainable Nature-Based Tourism in Southeast Asia')
Nature tourism	'travel to unspoiled places to experience and enjoy nature' (Honey 2002, cited in Christ et al. 2003)
Ecotourism	• 'traveling to relatively undisturbed or uncontaminated natural areas with the specific objective of studying, admiring, and enjoying the scenery and its wild plants and animals, as well as any existing cultural manifestations (both past and present) found in these areas' (Ceballos-Lascurain 1987, cited in Blamey 2003) • 'responsible travel to natural areas that conserves the environment and sustains the well-being of local people' (Honey 2002, cited in Christ et al. 2003)
Wildlife tourism	'based on encounters with non-domesticated (non-human) animals ... in either the animals' natural environment or in captivity. It includes activities historically classified as "non-consumptive" ... as well as those that involve killing or capturing animals ... ' (Higginbottom 2004: 2)
Adventure tourism	'nature tourism with a kick – nature tourism with a degree of risk taking and physical endurance' (Honey 2002, cited in Christ et al. 2003)
Sustainable tourism	'seeks to minimize the negative footprint of tourism developments and at the same time contribute to conservation and community development in the areas being developed' (Christ et al. 2003)
Tourists	people who 'travel to and stay in places outside their usual environment for not more than one consecutive year for leisure, business and other purposes not related to the exercise of an activity remunerated from within the place visited' (UNWTO 1995)

Analysis of Risk Governance for Nature-Based Tourism

Risk governance is a comprehensive concept requiring a broad analytic focus. Such analysis includes, and clearly goes beyond, the technicalities of pure risk assessment and management, gathering knowledge about a risk as well as its potential pathways and taking informed action in connection with its potential effects. Analysing risk governance also involves a thorough investigation of the environmental, economic and societal context which largely influences how a risk would actually materialise and be addressed. The following sections thus try to map out the existing risk governance systems covering the risks from and related to nature-based tourism and to provide some insights into what could be done to improve them.

Risk Governance Context

Tourism has numerous actors and stakeholders,[1] all inter-dependent and all having their own sets of rules, procedures, principles, responsibilities and objectives. The following major stakeholders and their decisions have an influence in identifying, assessing, managing, monitoring and communicating risks related to nature-based tourism: tourism industry players such as developers, tour agencies, guides, hotels, facilities and transport operators; governments (including relevant international organisations such as the World Conservation Union (IUCN) in which many of them participate and collaborate with non-state actors, and intergovernmental organisations (IGOs)); transnational organisations (mostly non-governmental organisations (NGOs) with international outreach such as World Wildlife Fund (WWF)); local authorities including planners and managers for public protected areas; local communities; and of course the tourists themselves (see Table 2 for more details; see also Tapper 2006, for a classification of stakeholder groups).

The sheer complexity of the networks of these various interest groups, their perspectives and responsibilities, makes governance of the risks related to nature-based tourism an enormous challenge. Considering the manifold repercussions tourism as a whole can have on the natural environment, the economy and the society of a host region and how these repercussions can differ across destinations, countries and the various forms of nature-based tourism, it is not surprising that the existing governance system does not comprise a single and overarching control and coordination mechanism. Instead governance rests on a mosaic of mechanisms which cover the spectrum from voluntary to legally binding and which may vary by stakeholder group. This mosaic is composed of multilateral and regional conventions (e.g. in the areas of conservation and environmental protection) and international agendas that touch upon aspects of nature-based tourism, national laws and policy (regulating e.g. hunting and fishing and outlining a strategy for e.g. tourism development), industry codes of conduct as well as standard setting and certification schemes, liability and insurance schemes and, eventually, world-views, values, traditions and taboos (see Table 3 for a brief overview).

The most important governance mechanism operating on an international level is the Convention on Biological Diversity. Its Decision VII/14, on Biological Diversity and Tourism, provides guidelines on biodiversity and tourism development which are fairly detailed and identify the environmental, socio-economic and cultural impacts that may be related to tourism. The guidelines also provide approaches to impact management and mitigation, call for the development of national strategies

[1] Actor is a term used to describe anyone or any organisation that may have a position on a subject whereas stakeholder implies having a more direct stake in or being more directly affected by the outcome of a risky activity. All are in some sense 'interested parties'.

Table 2 Major actors and/or stakeholders in the governance system and their roles.

Actors and/or Stakeholders	Roles
International Organisations (IGOs/NGOs) • World Tourism Organisation (UNWTO) • World Commission on Protected Areas (IUCN) • World Wildlife Fund (WWF) • United Nations Environment Programme (UNEP) • United Nations Conference on Trade and Development (UNCTAD) • International Institute for Environment and Development (IIED) • Conservation International (CI) • The Nature Conservancy of the USA (TNC) • Flora and Fauna International (FFI) • Wildlife Conservation Society (WCS) • Wetlands International • BirdLife International • National Audubon Society (in the USA) • The Royal Society for the Protection of Birds (RSPB) in the UK • Earthwatch Institute • International Council on Monuments and Sites (ICOMOS)	From agenda-setting activities to development of best practice guidance and guidelines, research, advocacy and day-to-day management of nature reserves
National governments	Develop legislation and policies/development plans setting the boundary conditions as well incentives for tourism development and the conservation of biodiversity. Governments sometimes also act as tour operators (e.g. in China and Vietnam)
Regional governments and local communities	Promoters, beneficiaries, and enforcers of conservation
Tourism industry • Major tourism companies (Club Med, TUI, Carson Wagonlit, Thomas Cook, etc.). • Trade and professional associations such as: – World Travel and Tourism Council (WTTC) – International Air Transport Association (IATA) – The International Ecotourism Society (TIES) – International Federation of Tour Operators (IFTO) – International Hotel and Restaurant Association (IHRA) – International Council of Cruise Lines (ICCL) • Guides' associations such as: – Swiss Mountain Guide Association – Professional Association of Diving Instructors (PADI) (USA) • Hotel and restaurant chains • Expedition outfitters, recreational equipment manufacturers • Financial services companies catering to tourists needs • Local handicraft makers producing tourist souvenirs	Tourism developers, operators, guides, etc.
Tourists	'Liability' and/or 'asset', atomised market demands

Table 3 Overview of component mechanisms of the governance system.

Multilateral conventions:	• United Nations Convention on Biological Diversity (CBD 1992), including – decision VII/14: Guidelines on Biodiversity and Tourism Development – decision VII/28: Programme of Work on Protected Areas • Convention of European Wildlife and Natural Habitats (Bern Convention 1982) • Convention on Migratory Species (CMS, 1979) and follow-up agreements • African Convention on the Conservation of Nature and Natural Resources (2003) • EC Directive on the Conservation of Natural Habitats and of Wild Flora and Fauna (Directive 92/42, Habitats Directive 1992) • Protocol on Environmental Protection to the Antarctic Treaty (1991) • Convention on Nature Protection and Wild Life Preservation in the Western Hemisphere (1940) • Convention Concerning the Protection of the World Cultural and Natural Heritage (1972) • Convention on Wetlands of International Importance Especially as a Waterfowl Habitat (1971) • Convention on International Trade in Endangered Species of Wild Fauna and Flora (CITES, 1973)
International agendas:	• Global system of protected areas, including e.g. World Heritage Areas, Biosphere Reserves, national parks • IUCN protected area guidelines and best practice • UNESCO's 'Man and the Biosphere Program' (biosphere reserves) • UN Millenium Development Goals • Agenda 21 (Earth summit) • Implementation plan from the 2002 World Summit on Sustainable Development
National/state legislation/ regulations:	• Regulation on hunting and fishing (licence system, district system, combined licence and district system, community-based system), using a number of strategies: fish/game reserve systems, open and closed seasons, bag limits, size and sex restrictions, firearms incl. calibres, type of bait, equipment • National wildlife protection regulation • Customs and quarantine controls at the port of arrival
Industry declarations as well as standard setting and certification schemes:	• Quebec Declaration on Ecotourism • Cairns Charter on Partnerships in Ecotourism • The European Charter for Sustainable Tourism in Protected Areas • Green Globe 21

and plans for tourism and biodiversity, and call for the training of tourism professionals in conservation, biodiversity issues, and impact assessment.

Risk Pre-Assessment

The benefits, both financial and personal, first and foremost accrue to those most directly involved in nature-based tourism – namely the different actors who comprise the tourism industry and the tourists themselves. For these actors, nature-based tourism is not primarily associated with risk but with opportunity. Depending on the actor's or stakeholder's perspective, it may be seen as a business opportunity, as an opportunity to gain new experience and learn (e.g. study Serengeti wildlife), to achieve a personal goal (e.g. reach the top of Mount Everest), to overcome a challenge (e.g. hunt a bear), or, simply as an opportunity for a break from everyday-life. Critical voices exist, but compared to those of the vast majority of unfettered proponents, they largely go unheard. These voices may include disgruntled and alienated local communities at tourism destinations who suffer from tourism-related restrictions in the traditional use of 'their' natural resources, as well as the many organisations active in conservation and environmental protection that are concerned about irreversible impacts on both the natural environment and the prevailing socio-economic system.

Based on these various viewpoints, at least four possible ways of framing nature-based tourism exist: as an expression of personal freedom and choice (potential tourists); as a major factor in local and regional development (governments, development agencies); as a legitimate and necessary source of income (industry participants); and as a critical success factor for both the conservation of biodiversity and local livelihoods (parts of the environmental community).

Relatively little work has been done specifically on the risks of nature-based tourism, although a substantial literature exists on both ecotourism and other components of the tourism industry, and on outdoor recreation and recreation ecology (Buckley 2004: 1). The pathways that a specific risk in relation to nature-based tourism actually might take are furthermore determined on the basis of a multitude of local conditions and are thus extremely context-specific. For instance, whether or not hiking causes damage to soil and vegetation depends on factors such as soil conditions, the sensitivity of vegetation to trampling, the frequency of hiking and its spatial distribution, the season, the weather, and the behaviour of the hikers, etc. (Cole 2004: 52–54). Although the methodologies (e.g. ecological risk assessment, natural resource damage assessment, cost benefit analysis, etc.) exist to evaluate the varied impacts of nature-based tourism, no substantial body of knowledge currently exists. Given the variability of site specific conditions, establishing globally applicable procedures – operational formulations that go beyond the status of mere principles and general requirements – will be challenging.

The situation may be slightly different with regard to the system of protected areas that now covers about 12% of the surface of the globe. These sites are usually managed by government agencies established for the purpose, though some countries have private protected areas or protected areas owned or managed by NGOs. Global standards for managing these areas are agreed under the auspices of IUCN and its World Commission on Protected Areas (WCPA), a 'network of protected area expertise' which brings together a large number of protected area managers, subject matter experts, scientists, and NGO representatives. WCPA has issued a series of best-practice guidelines that address pressing management issues and one of them, No. 8, is dedicated to sustainable tourism (Eagles et al. 2002). The guidance provided in this document deals with the planning and management of sustainable tourism to protected areas and, inter alia, suggests a basic risk management process and outlines requirements for a monitoring system for tourism-related impacts which, once it is in place, can also support early warning systems.

Risk Appraisal

Nature-based tourism poses risks to the natural environment, to local communities (especially indigenous peoples, who may be especially vulnerable in the more remote areas), to the tourism industry itself which suffers from economic fluctuations and, consequently, to the regional and national economy. As discussed earlier, tourists themselves may face health and safety risks. The risks to these various groups vary considerably, but all are amenable to identification, an assessment of exposure and vulnerability and, eventually, estimation. Whereas the assessment of risks to tourists' health and safety can be quite straightforward, the data required for estimating the risks to the natural environment and the social and economic setting present more of a challenge because these risks depend on a range of destination-specific observations. Because this study focuses on risks related to nature-based tourism in general (as opposed to the risks in a specific location) the ensuing discussion is limited to an outline of what some of these risks are with regard to ecosystems (i.e. comprising the living organisms in a defined area and the environment in which they are placed), to the socio-economic systems within which tourism takes place and finally, to tourists' health and safety.

Risks to Ecosystems

- Unplanned development of tourism infrastructure and facilities in an area – that is, development without management standards and guidelines ensuring participation of local communities in both conservation and the revenue from tourism – often results in significant alteration of ecosystems. The changes can include deforestation, drainage of wetlands, soil erosion or compaction, desiccation through excessive groundwater extraction, fragmentation and disruption of habitat, potential encroachment on protected areas, littering, air and water pollution, eutrophication (i.e. a process in which water bodies receive excess nutrients, leading to excessive vegetation growth and ultimately to a reduced concentration of oxygen), increased risk from fires, and, ultimately, loss of biodiversity.

- Tourists and their means of transportation can facilitate the introduction of invasive alien species (weeds, pests and possibly animals).[2]
- Wildlife tourism can have adverse affects on wildlife in three main ways (for a detailed discussion of potential ecological impacts of wildlife tourism, see, for instance, Newsome et al. 2005): by causing changes in the behaviour of birds and animals (e.g. deserting nests and eggs, stopping foraging due to restricted patterns of movement, coming close to campgrounds and roads in search of food, etc.), by leading to physiological changes (e.g. in heart rate, body temperature, etc.) or by causing damage to habitats (see Tapper 2006: 51, particularly for case studies outlining the impact of tourism on the Galápagos Islands and of whale watching on Valdés Peninsula, Argentina). The nature and magnitude of responses depend on a range of factors such as the species, the characteristics of an animal itself (age, sex, breeding status), its habitat, the frequency and intensity of tourists' interactions with wildlife, the approach distance, and the types of stimuli created by the interactions. In the longer term, wildlife tourism can affect the mortality rate of some species (e.g. through increased vulnerability to exotic infectious diseases introduced by tourists and/or lethal collisions with road vehicles and boats). For instance, road death is believed to be responsible for the local extinction of eastern quoll populations in Victoria and parts of Tasmania in Australia (Australian Museum 2003).

  - Wildlife observation often happens at critical stages in an animal's life history: e.g. for marine wildlife, during 'migration, breeding, feeding, resting and socialising' (Valentine and Birtles 2004: 28). Humans compound the problem by desiring to get too close. The tourism industry's need for predictability of viewing wildlife creates strong incentives to manipulate habitat, or, in particular, to provide wildlife with food. These actions can create a range of negative impacts (wild animals no longer behaving naturally, death/disease from inadequate diet, over-population, introduction of alien species, change in an ecosystem's natural 'balance of species') which need to be compared to perceived benefits (conservation of threatened populations, creation of positive attitudes towards wildlife). Tourists, out of misplaced concern, may misinterpret wild animals' needs (e.g. the needs of predators vs. prey) or the need for management policies such as culling programmes. Their well-intentioned interference may have adverse consequences for wildlife.
  - Hunting and fishing can affect the 'genetic fitness' of a species by changing the sex ratio and/or age distribution or through secondary effects such as poisoning following the ingestion of lead shot. They can also impact 'non-target species' via 'by-catch'. However, overexploitation and overharvesting of wildlife, leading to the local extinction of wildlife, has most often been associated with illegal hunting (poaching), subsistence hunting, and wildlife

[2] The main driving factor in the introduction of invasive species is probably international trade and in particular the use of ballast water in cargo ships.

trade (Green and Giese 2004: 89). A further risk for some species (in particular shells, corals and butterflies) derives from their value as popular tourist souvenirs, which leads to unsustainable specimen collection in some areas. Wildlife tourism also routinely practices the 'preventive' removal or killing of species such as snakes and insects in order to make tourists' wildlife experiences both safe and comfortable.

  - Ultimately, wildlife tourism can endanger the reproductive success of a breeding group, population or animal community and consequently alter the species structures of an ecosystem. These impacts can be brought about by 'even relatively minor stresses [including those that do not provoke a measurable change in an animal's behaviour], if experienced often enough, or experienced at key times of an animal's life cycle . . . ' (Green and Giese 2004: 92).

- The impacts of long-haul travel to ecotourism destinations, transportation between sites, and vehicle travel within a specific site may also be considerable. Substantial energy use, in particular burning of fossil fuels for traveling, results in the emission of greenhouse gases, thus contributing to climate change. In this respect, ecotourism is no less harmful to the environment than mass tourism. In an analysis of 'between-site travel' for different forms of tourism, an 'ecotourist's' daily energy consumption is estimated to be nearly three times bigger than that of a 'mass tourist' staying at an all inclusive resort holiday (Simmons and Becken 2004: 18–19).
- Climate change poses a major risk to tourism in multiple ways: bleaching and disappearance of corals; destruction of mangroves and coastal wetlands; general coastal erosion and degradation; loss of alpine glaciers, snow cover and meadows; shift of animal and plant ranges; extinction of species; and rising sea-levels threatening low-lying islands and other areas. Ultimately, these effects of climate change may contribute to the loss of the unique features that characterise both scenic landscapes and refuges for specific animals and plants.

Risks to Socio-Economic Systems

- Due to all kinds of tourism activities, local communities may find that access to or use of important resources may become more difficult or altogether restricted. For instance, in protected areas, traditional wood gathering or spiritual practices may no longer be permitted (Eagles et al. 2002: 32). Tourists' consumption of fresh water supplies, food, electricity, etc. in areas where such resources are scarce further competes with the needs of local population. For example, use of water for showers, swimming pools, and golf courses can conflict with local domestic and agricultural water uses.
- Displacement of local populations is another risk. An increased tourism-related demand for infrastructure, facilities and goods and services can put a financial burden on local communities both in the form of taxes as well as rising property values. These and other factors can eventually cause locals to move to places that

are more affordable, but ones that may be less suitable for earning a living from the land.

- A further concern is the potential disruption of indigenous culture and ways of life. Such disruption can be caused by an influx or the immigration of new residents in search of jobs and business opportunities. Increasing inequalities in local communities as tourism generates winners and losers can also strain the social fabric. Intergenerational and gender conflicts can result.
- Economic speculation about a rising tourism market can lead to overdevelopment and the creation of over-capacity, thus decreasing economic viability and leading to additional environmental damage. This phenomenon has been observed in some parts of Spain, Thailand, and Indonesia.
- Tourists can also be a significant vector for disease, as in the case of SARS. Travel entails multiple interactions, often taking place in crowded spaces with poor air circulation, and within short time frames before travelers know they are contagious or become overtly ill (transit times of 24–48 hours are shorter than the incubation periods for many infectious diseases). Even tourists traveling to and from remote locations typically rely on major transport hubs, often located in major cities, from which they disperse to home and work environments where the spread of disease can continue (Wilson 2002). By serving as a link between multiple locations, tourists thus have the potential to greatly fuel the dispersal of an infectious disease. Tourists can also facilitate the transport and dissemination of invasive alien species which can cause economic damage.

Tourism is a major source of national income for many countries, a fact clearly reflected in the share tourism contributes to their gross domestic product (GDP). And from a global perspective, tourism provides, on average, a large and steady contribution to global GDP. However, dependence on tourism presents a risk to the economy both on a local and national level; for tourism has proven to be a volatile industry in specific countries, regions and destinations (e.g. in the Caribbean, the Maldives). Extreme fluctuations in tourist volumes are often dependent on 'external factors' such as terrorism, civil unrest, natural disasters, outbreaks of infectious diseases, the price of petroleum, exchange rate fluctuations and changing tourist preferences, all of which are hard to foresee and/or control. For instance, the hotel bombings in Bali greatly reduced tourism for at least several months, and it took well over a year for the tourism economy to recover. Over the past decade, trekking tourism in Nepal has suffered because of civil unrest, undermining one of the few sources of income for many of the rural people. The tsunami of December 2004 dealt a devastating blow to tourism in southern Thailand although the country was able to recover relatively rapidly because it had the infrastructure to deal effectively with the disaster. Nonetheless, the disaster was extremely costly, especially to small operators who were essentially wiped out and whose risks were uninsured. Thus, it is often the local communities who are most vulnerable to of the 'boom or bust' nature of the industry.

Risks to Tourists' Health and Safety

Risks to tourists' physical well-being are inherent to travel, with traffic accidents being a particular – indeed the prevalent – hazard in many parts of the world. Accidental injuries or deaths are also not uncommon among hikers, skiers, and other mountain sports, or among enthusiasts of kayaking, rafting, scuba diving and other water-related sports. Some tourists are killed or injured every year by the very wildlife that they have come to observe, particularly by large carnivores, rhinos and elephants.

Fatalities and injury are features of adventure or extreme tourism as the recent increases in high profile deaths on Mount Everest illustrate (Krakauer 1997). Over 150 climbing deaths have occurred on Mount Everest since its summit was first reached in 1953. Climbing in the European Alps can also be deadly: a study of mountaineering accidents in the European Alps for the period of 1987 to 1997 reveals an average of 414 fatalities per year, with the number of deaths climbing to 470 for 1988, the most deadly year within the period investigated (Lischke et al. 2001). [3]

Characterisation of Risks as Simple, Complex, Uncertain, or Ambiguous

As part of structuring an approach to evaluating and discussing risks and identifying key stakeholders, the IRGC framework suggests that the risks be categorised as 'simple', 'complex', 'uncertain', and/or 'ambiguous'. For nature-tourism, the risks posed to ecosystems can be characterised as 'complex' and/or 'uncertain'; ecosystems rely on intricate interactions between species as well as with 'exogenous' elements such as temperature, atmospheric gases/nutrients and light, the total displaying the stunning dynamics of complex systems (Cooney 2004). Socio-economic systems, in addition to interacting with and transforming the natural environment, hinge on values and goals. Risks linked to the socio-economic systems should therefore be viewed as mostly 'ambiguous'. Health and safety risks to tourists, in contrast, would probably fall in the category of 'simple' risks, at least in comparison to the other risks affecting ecological and socio-economic systems.

Tolerability and Acceptability Judgement

Judging whether any of these risks are acceptable, tolerable or intolerable requires balancing potential negative impacts with likely benefits for each potential travel destination and form of nature-based tourism and across the stakeholder community.

[3] For lack of a full set of data for France, Italy and Switzerland, the authors had to resort to data for the Chamonix region, Southern Tirol and the Zermatt area, respectively. The totals given in this study therefore do not convey the full picture.

Overarching values and issues such as personal and economic freedom, right to development, right to subsistence and autonomy will come into play and need to be considered. Any judgement about acceptability or tolerability of risks will also have to take into account the potential impact that options to alleviate a certain risk might introduce: risk reduction measures themselves can introduce new risks, creating risk-risk tradeoffs and requiring compromises. These issues should be identified and put on the negotiation table early on.

For adventure tourism, judgements about acceptability or tolerability need to take into account that inherent risk is often part of the attraction of the activity. While the tourism operators may take reasonable measures to prevent a risk from occurring or at least mitigate its effects (for example, providing life vests for running rapids), those participating in the activity are theoretically, at least, aware of the risks. Adventure tourists may be assumed to have assessed the risks and have determined that they are acceptable, although it is not clear that they have been fully informed.

Risk Management

Successful management of the risks related to nature-based tourism is dependent on at least four factors: a favourable regulatory framework and far-sighted government policies; adequate funding for conservation and nature protection; education of all of the stakeholders; and intelligent and adaptive management at the site level. Some options that target each of these factors are outlined below (for a comprehensive discussion of the 'instruments' that governments can use in order to enhance the sustainability of tourism, see UNEP 2005).

Regulatory and Political Frameworks

- Regulatory frameworks (including prohibition/preclusion of tourism for particularly sensitive and endangered areas) in combination with permits/licenses for operators allows for governments to define the conditions under which tourism is acceptable or tolerable and provides for environmental standards as well as standards of safety and competence.
- Industry self-regulation including certification (e.g. Australia's Nature and Ecotourism Accreditation Program, Ecotourism Association of Australia's 'accreditation system for nature guides', Green Globe 21) and eco-labelling, in combination with dynamic and adaptive best-practices and codes of conduct, are also valuable and increasingly popular tools for ameliorating the ecological and social impacts of tourism. However, industry self-regulation – and in particular eco-certification schemes created by commercial ventures – may lead to adoption of the 'lowest common denominator' measures and thus may not actually do much to enhance industry standards (Buckley 2004: 10–11). This tendency

needs to be carefully watched and counteracted by those industry participants who adhere to the highest operational standards.

- Establishing incentive schemes which induce desirable behaviour patterns can be another way to manage risks. Private and/or communal landowners could, for instance, be encouraged to set up wildlife observation as well as hunting and fishing tourism on their properties as alternatives to large scale cattle, agricultural, logging and mining operations. In addition to restoring wildlife and increasing the habitat area for both fauna and flora, such practices are also likely to pay off economically: wildlife tourism activities that facilitate conservation can in fact lead to higher yields per hectare than alternative forms of land-use (Eagles et al. 2002: 26). Where necessary, indigenous and local communities could also be provided with economic incentives (e.g. employment or other compensation for managing nature-tourism sites) to change potentially detrimental resource management practices (e.g. poaching, capturing for live trade, extensive logging). Another option is rewarding tour operators and tourism guides by adding, for example, a 'conservation component' to the salaries of those who actively engage in conservation of both wildlife and other natural resources.
- Institutionalising collaboration between the public and private sectors and the affected local communities is a goal that both governmental and industry actors should work toward. Ideally, such collaboration will take the form of multi-stakeholder planning and decision-making with regard to tourism strategies and master plans, in which all parties are given equal opportunity to participate. Master plans should include a broader sustainable development rationale and address both biodiversity and socio-economic considerations. Collaborative schemes are also more likely to be successful at helping with surveillance and enforcement of tourism management strategies, activities which might otherwise be extremely difficult to carry out due to the remoteness of many nature-based tourism destinations. Collaboration amongst key stakeholders on political action and advocacy in support of conservation issues should also be encouraged.
- Environmental impact assessments, carrying capacity assessments, and other assessments relevant to specific regions should be made mandatory in tourism planning. Socio-economic impact assessments with a view to better integrate livelihood needs of local population with conservation goals are also an important component (McNeely 2005: 109). Setting indicators for monitoring the success of the goals established in any plans is also a key step. For nature-based tourism, specific site management plans should require the inclusion of physically measurable biodiversity and conservation indicators. In the case of wildlife tourism, the indicators might, for example, include 'species presence/absence, abundance, diversity, breeding success, behaviour or health; or attributes of the habitat' as well as 'amount of noise made by tourists, or degree of soil compaction' (Higginbottom 2004: 215–216). For monitoring to be effective it must include tourist data such as 'numbers, activities, distribution' (Higginbottom 2004: 216).

Importance of Funding

- Relevant government agencies need to be convinced that tourism should be seen as a major means to generate funding for the management of sensitive ecosystems – conservation and protection initiatives for endangered animals, patrolling for poachers and other illegitimate users of resources, control of visitor numbers and support of neighbouring communities. With regard to protected areas, for instance, income from tourism is seldom returned to the protected area but rather goes into countries' central budget. Ideally, the economic value generated from tourism would encourage the public and politicians to embrace the conservation of nature, wildlife and cultural heritage, which in turn relies on adequate government funding. This is particularly important in the context of protected areas, where funding for their management does not keep pace with the expansion of protected areas (McNeely 2005: 12).
- Tourists' willingness to support conservation should be better utilised to underwrite conservation programmes. Increasing entrance fees for protected areas, the active soliciting of donations, or offering conservation-related work for a fee to volunteers ('conservation holidays') can all raise revenues. However, in the case of protected areas it is crucial, though very difficult, to strike the right balance between income from visitor fees and the potential negative impacts of increasing the number of visitors.
- Reducing the share of 'revenue leakage', in which the economic benefits accrue primarily to individuals or industries outside the tourist destination, is imperative. This can be most effectively achieved if those in charge of planning and managing nature-based tourism operations employ local people and use local products.

Education

- Awareness about the impacts of tourism, the importance of biodiversity, and the need for conservation efforts needs to be raised within the tourism industry. It should be integrated all along the tourism supply chain, from tour developers to the indigenous communities. Tour guides who are knowledgeable about the environment and act responsibly can play an important educational role and can effectively influence tourists' behaviour (Littlefair 2004: 305–306). They can also influence tourists' continued engagement in conservation, both as donors and political campaigners. In areas where enforcement of regulations or guidelines is infeasible or impossible, education is often the only instrument to prevent, or at least mitigate, the harmful impacts of tourism activities.
- Increased research and monitoring efforts are necessary to improve knowledge about both the potential and actual impacts associated with nature-based tourism, particularly at the local destinations. Research efforts need to include evaluation of the effectiveness of available management approaches and monitoring methods so that adjustments can be made in accordance with the findings.

- Realistic marketing is a prerequisite for managing tourists' expectations (e.g. about hand-feeding and handling wildlife) and should include clear statements as to what experiences tourists can (and cannot) reasonably be expected to have. Publicising conservation efforts might help tour operators to attract environmentally-conscious tourists.

Site Level Management

- In wildlife tourism, specific measures to 'harden' the environment are now commonly taken to reduce the impact on animals. In these cases, viewing areas are strictly defined, for example by building physical structures like platforms, bridges, boardwalks, barriers or blinds. Although highly controversial, another measure involves the conditioning of wildlife such that 'being watched' does not cause them excessive stress; this has been done quite successfully with mountain gorillas in Uganda (Higginbottom 2004: 222) and in Rwanda. In many protected areas, tourists are encouraged to stay on marked paths, and in many African protected areas, tourists are forbidden to leave their vehicles without an armed guard.
- Requiring a minimum level of expertise among tourists undertaking certain risky activities can be used to reduce risks to tourists. For example, most countries require scuba divers to be certified before they are allowed to dive. This approach might be extended to the licensing of tourists to operate equipment such as mountain bikes or motorcycles.
- For some types of adventure tourism, improved equipment is also an option, but this would need to be traded off against costs.
- Finally, as adventure tourism continues to expand, insurance against liability will be likely to be more in demand and may play a role in reducing risks.

Internalising the environmental costs of tourism impacts in the cost of travel could be an important way to couple travel with conservation efforts. For instance, many international organisations are responding to the problem of climate change by making their travel 'carbon neutral'. The IUCN, for example, charges all staff air travel a 'carbon tax' which is assessed at the current European Union rate and paid into a special fund that is allocated to approved carbon sequestration projects. This tax is relatively modest, and many ecotourists may be willing to offset their carbon dioxide production against an appropriate carbon sequestration project, such as preservation of mature, old-growth forest (for a both fascinating and extensive discussion of the market approach to capturing carbon and conserving biodiversity, see e.g. Swingland 2002). Relying on a more altruistic approach, there are increasing calls for tourists to chain trips together in order to minimise air travel, to keep travel limited to a specific region, and to use less energy-intense travel alternatives (e.g. walking) at their destinations. Though such behavioural changes are desirable, implementing them faces numerous challenges including increases in the time and costs of travel and the need to persuade individual tourists to abandon the compelling logic that their marginal impacts are small – 'my own actions don't make a difference if you consider how many tourists are out there infesting almost any place on the world.' Improving transport efficiency through better logistics and planning, and more ample use of renewable energy sources would also reduce impacts.

Risk Communication

Risk communication is the key to raising awareness of a risk issue and consequently, to finding a remedy. In the context of nature-based tourism, communication may be described using several terms such as 'interpretation', 'information' or 'education' though each may be slightly different in application (McNeely 2005: 184). Unfortunately, awareness of the risks to ecological and socio-economic systems cannot simply be considered a given. While this conclusion applies to operators and guides, it ultimately applies to those whose choices and behaviour lie at the origin of these risks – the tourists.

While many potential tourists may have some general understanding of environmental and socio-economic problems in the places they visit, they may not link these issues to the footprint that they themselves leave behind when travelling. The existing governance system, with its multilateral conventions, principles of law, policies as well as spawning industry declarations and standards, is not an effective means of communication. It is not something the average tourist can realistically understand or connect with – assuming the issue makes it onto the tourist's 'radar screen' at all.

Raising awareness of risks related to nature tourism and of the existing governance system must be linked to the individual's travel experience and interests. Many protected areas have designed sophisticated communication strategies to this effect, sometimes even relying on them as their primary means of managing impacts (Eagles et al. 2002: 108–111; Littlefair 2004: 297). Sophisticated communication strategies may be more difficult to achieve for nature-tourism destinations outside the protected area system since responsibilities are often scattered amongst stakeholders with conflicting goals and interests.

Risk communication is however also crucial amongst the international community, governments, the tourism industry, NGOs – those stakeholders or actors who by virtue of their function and/or purpose should be aware of the risks and what they can do about them. Communication enables sharing of insights into aspects of risks that might not have been considered, other concerns, and strategies for risk management. Risk communication ultimately is a prerequisite for any form of coordinated action.

Risk communication directed at tourists' health and safety needs to be broken down between the different elements of the industry. In the travel sector, the airlines, for example, have a very comprehensive perspective of risk assessment, as do those involved in road transport, at least in the relatively advanced countries. Road transport standards are likely to deteriorate in the remote, and hence more risky, areas that are likely to be of greatest interest to adventure tourists.

In the case of adventure tourism, various standards have been developed for different kinds of activities. The mountaineering associations often have developed training and standards at a fairly sophisticated level. Some of the newer forms of adventure tourism, such as bungee jumping, are far less regulated, and the risks are poorly communicated. In areas where wildlife poses a danger, risk communication is often very mixed, with indigenous peoples being more aware of the risks than the tourists.[4]

Stakeholder Participation

In the long run, although opinions may diverge widely about the direction and methods that management should take, it is in the interest of all those involved in the tourism industry to manage the risks associated with nature-based tourism. The earlier all those generating or affected by these risks are involved in the management process, the more probable it is for any divergence to be resolved or addressed through a suitable compromise. Rarely does the responsibility and capability for risk management lie with one single actor.

The conservation and the development communities advocate broad stakeholder involvement from the planning stage of a tourism venture and through to the setting up and management of the operation and its impacts. In the case of protected areas, stakeholders are increasingly seen as 'constituencies', the most vulnerable of which need to be able to draw on tangible benefits (e.g. via employment and the recognition of 'customary' resource use and access rights) in return for their support for protection efforts. Those most vulnerable are local and indigenous communities who, if not meaningfully included in the governance of a protected area, often feel disenfranchised or are driven into depleting the very resources that are to be protected in order to secure a livelihood (McNeely 2005: 101–128). A broad stakeholder approach requires the forging of new alliances with partners who are not traditionally associated with supporting conservation such as the extractive industries (in particular the mining and energy sectors), urban dwellers, policy-makers and the security community[5] (see McNeely 2005).

Recent developments with wildlife tourism involving hunting illustrate the advantages of such alliances. Although hunting is highly controversial for some, it can alleviate some of the socio-economic risks related to nature-based tourism while limiting risks to the ecosystem. In fact, big game hunting in southern Africa has been both economically lucrative and a positive conservation force, enabling rural people to actually earn money from their wildlife. Photo safari tourism or wildlife observation tourism have had similar benefits. South Africa especially has developed a thriving industry around these types of tourism, often on private lands. It

[4] That said, local people continue to suffer mortality from animal attacks, for example, from attacks from rhinoceros and tigers in Nepal, from hippos and lions in Africa.

[5] Security community includes those involved in border control, including the army, navy, a coast guard, and border patrol police.

has changed local perceptions of the value of wildlife, creating widespread support for the conservation of numerous species. Similarly, Ducks Unlimited, a non-profit organisation devoted to the conservation of waterfowl and wetlands in Canada and the US primarily to benefit duck hunters, has earned substantial income over the years from selling duck stamps and other fundraising activities. A result has been considerable expansion in waterfowl populations and reductions in the loss of wetland habitats.

While the need for such broad stakeholder participation may be less obvious and more difficult to organise in 'unregulated' nature-based tourism destinations, it is nonetheless crucial. After all, the income from tourism activities that accrues to stakeholders will only continue to flow if the environmental and socio-economic features of a destination remain largely intact. Good governance of the risks that come with tourism, however, can only be achieved if all those interested, affected, or able to help manage the risks work toward the same goal.

Conclusions

Tourism has experienced rapid growth over the past 50 years and is expected to continue to develop, particularly in biodiversity 'hotspots'. Ecosystems constitute the main capital not only for nature-based tourism, but for other critical resource 'services' – watershed management and local climate regulation – on which our societies and other economic activities rely. They need to be protected and conserved in order to allow both ecological and socio-economic systems to thrive. In addition to tourism, challenges to conservation of these ecosystems come from short-term high-yield alternatives in land-use (e.g. oil-drilling in the Arctic National Wildlife Refuge in Alaska, deforestation in order to enable industrial agriculture) or, in the case of poor countries, from the pressure of a growing population and the needs of local communities to earn a living.

Nature-based tourism, when well managed, can contribute both to biodiversity conservation and to alleviating poverty. It is one means by which local people can derive economic benefit from protected areas, their habitats and wildlife, creating incentives for protection of those resources. Tourism and conservation can then coexist or may even be seen as symbiotic. The challenge for developing nature-based tourism is to make it symbiotic with conservation (Lynam 2006).

Such a symbiotic relationship is of course not only desirable for protected areas: those in charge of nature tourism destinations outside the protected area system should aspire to achieve the same. It will require a very delicate balancing act in terms of sharing accrued benefits as well as 'governing' the risks in relation to tourism activities. Such a balance can probably be achieved best if based on a polycentric, nested system of governance (McNeely 2005: 19–20) in line with the characteristics of a tourism location. Effective risk governance however requires that the risks be understood and evaluated in the wider system within which nature-based tourism operates, including 'multi-origin' risks such as those arising from transportation and motorised travelling as well as climate change. Such a system furthermore has to 'have teeth' in that it must be able to regulate those whose behaviour creates risks, that is, the industry and ultimately, the tourists themselves. If done successfully, nature-based tourism could become a role model for sustainable development, thereby clearly outperforming other forms of land-use.

Lessons Learned and Recommendations

Risks in relation to nature-based tourism, irrespective of whether they potentially affect the ecosystem, the overarching socio-economic fabric or the well-being of individual tourists, are of predominant concern only to a few players with special interests such as the conservation and development movements, parts of the tourism industry and, possibly, local communities. Unlike the risks related to genetically modified organisms (GMOs) or nuclear energy, these risks have yet to develop into a burning issue for the general public. Within the public realm, they seem largely overshadowed by the obvious benefits that potential tourists associate with travelling and nature-based tourism.

Consequently, it is doubtful that the systematic framework for risk governance that this case study advocates would have changed much in the way risks from nature tourism are generally perceived and dealt with, even if had it been available earlier on. It seems more likely that they would still have been the primary preoccupation of a relative few for whom the benefits and economic opportunities of tourism do not outweigh the risks. While the framework might have helped individuals or particular sectors to better understand and manage the risks within their control, a fragmented approach to managing risks is unlikely to have been successful.

Where the framework might have made a difference, however, is with regard to stakeholder involvement in general. Had all those concerned had their say and been part of decision-making, some of the more egregious examples of the development of tourism resorts could perhaps have been prevented (for example, the beach resort of Cancun, Mexico, where mangroves, swamps, dunes and inland forests, home to a vast number of animal species, had to give way to a town of 300,000 inhabitants that attracts 2.6 million visitors per year and has a major waste water problem; see Christ et al. 2003: 21). Similarly, the establishment of protected areas, might have been less conflict-laden had local communities and their ancestral rights been acknowledged as part of the process. In the past, these areas have typically been created by government in a top-down approach despite the fact that an estimated 50% of the main eco-regions of the world are located on the ancestral territories of indigenous peoples (McNeely 2005: 179 and 116). While it is clear that stakeholder involvement in itself does not automatically lead to good risk governance, it can nonetheless define the boundaries for both making and implementing risk-related decisions.

This case study demonstrates that the major advantage of using the IRGC framework is that it encourages the establishment of the 'big picture', the appropriate framing of a risk issue with which any detailed assessment begins. Investigating the relevant stakeholders' networks along with the rules, interests and values that affect their actions (i.e. the governance context) as well as existing bodies of knowledge regarding risks, concerns and management alternatives (i.e. pre-assessment) provides opportunities to identify where major gaps in risk governance might lie that can prevent a technically adequate solution from being embraced and successfully implemented.

Applying the framework to the risks related to nature-based tourism also met with a number of difficulties, pointing to areas where the framework could benefit from more clarity:

- The analysis required under the pre-assessment component proved difficult because the terms used to outline the structure of this analysis seem insufficiently defined and delineated (namely 'early warning', 'risk assessment policy' or 'screening' and 'scientific conventions for risk assessment and concern assessment'). It would be very helpful if this part of the framework were complemented with specific examples providing illustrations of all four of these elements.
- The framework's development of four distinct risk classes (simple, complex, uncertain, and ambiguous) which are then juxtaposed with specific risk management strategies (see Table 6 in the Risk Governance Framework in Chapter 1), is very helpful in setting priorities for risk management. As usual with summary tables, however, a user might mistake the distinctions made in this table for distinct and mutually exclusive choices. For many risk fields, such clear-cut compartmentalisation hardly reflects reality. Instead these risk classes may simply describe different aspects of the same risk. Risk management would therefore have to consist of a mix of the offered strategies and instruments as well as possibly others. A word of caution to this effect might prevent the user of the framework from misinterpretation.
- Even if the framework is being applied to well-defined, very specific risk topics, the requirements it poses to its users are substantial. The framework components presuppose a substantial background and in-depth knowledge, including on how the issue has evolved within the wider socio-economic context. It therefore seems that any user who is not a seasoned expert with regard to the topic in question might struggle to make best use of the framework. In its current format, the framework is not particularly 'user-friendly'. Despite the many summary tables, the framework requires the first-time user to basically 'juggle' the content of up to 50 pages. Framework users who have to work with tight deadlines might therefore benefit from a condensed and easy-to-handle version of the framework which comes in an accessible format (such as a 2–3 page template or blueprint). That said, such a format would need to ensure that the analytic structure it provides is not mistaken for a rigid frame – or all-embracing form – which merely has to be filled in, thus replacing genuine thinking. Instead the format should provide the impetus for the right questions to be asked and the relevant issues to be considered in both a systematic and creative manner with regard to the governance of a particular risk.

References

Australian Museum, 2003, Sydney Mammals Database, Easter Quoll, Retrieved on February 12, 2007.

Blamey, R.K., 2003, Principles of ecotourism, in: D.B. Weaver (ed.), *The Encyclopedia of Ecotourism*, Cabi Publishing, Wollingford, UK.

Buckley, R. (ed.), 2004, *Environmental Impacts of Ecotourism*, Cabi Publishing, Wollingford, UK.

Ceballos-Lascurain, H., 1996, *Tourism, Ecotourism and Protected Areas*, IUCN, Gland, Switzerland.

Christ, C. et al., 2003, *Tourism and Biodiversity – Mapping Tourism's Global Footprint*, Conservation International, Washington (available at http://www.unep.org/PDF/Tourism_and_biodiversity_report.pdf).

Cole, D.N., 2004, Impacts of hiking and camping on soils and vegetation: A review, in: R. Buckley (ed.), *Environmental Impacts of Ecotourism*, Cabi Publishing, Wollingford, UK.

Cooney, R., 2004, *The Precautionary Principle in Biodiversity Conservation and Natural Resource Management: An Issues Paper for Policy-Makers, Researchers and Practitioners*, IUCN, Gland, Switzerland.

Eagles, P., McCool, S. and Hains, C.D., 2002, *Sustainable Tourism in Protected Areas; Guidelines for Planning and Management*, IUCN, Gland, Switzerland.

Green, R. and Giese, M., 2004, Negative effects of wildlife tourism on wildlife, in: K. Higginbottom (ed.), *Wildlife Tourism: Impact, Management and Planning*, Common Ground Publishing, Altona, Victoria, Australia.

Higginbottom, K. (ed.), 2004, *Wildlife Tourism: Impact, Management and Planning*, Common Ground Publishing, Altona, Victoria, Australia.

Higginbottom, K. and Tribe, A., 2004, Contributions of wildlife tourism to conservation, in: K. Higginbottom (ed.), *Wildlife Tourism: Impact, Management and Planning*, Common Ground Publishing, Altona, Victoria, Australia.

Huey, R.B. and Eguskitza, X., 2000, Supplemental oxygen and mountaineer death rates on Everest and K2, *Journal of the American Medical Association* **284**(2), 181.

Krakauer, J., 1997, *Into Thin Air: A Personal Account of the Mount Everest Disaster*, Villard, New York.

Lischke, V. et al., 2001, Mountaineering accidents in the European Alps: Have the numbers increased in recent years?, *Wilderness and Environmental Medicine* **12**(2), 74–80

Littlefair, C.J., 2004, Reducing impacts through interpretation, Lamington National Park, in R. Buckley (ed.), *Environmental Impacts of Ecotourism*, Cabi Publishing, Wollingford, UK.

Lynam, A., 2006, *Sustainable Nature-Based Tourism*, published on the website of the Tourism Authority of Thailand at http://www.tatnews.org/conservation/1927.asp, accessed on 30 June 2006.

McNeely, J. (ed.), 2005, *Friends for Life. New Partners in Support of Protected Areas*, IUCN, Gland, Switzerland.

Newsome, D., Dowling, R.K., and Moore, S.A., 2005, *Wildlife Tourism*, Aspects of Tourism: 24, Channel View Publications, Clevedon, Buffalo, Toronto.

Olson, D. et al., 2001, Terrestrial ecoregions of the world: A new map of life on earth, *BioScience* **51**(11), 933–938 (available at http://www.worldwildlife.org/sciencc/pubs/bioscience.pdf).

Renn, O., 2005, *Risk Governance – Towards an Integrative Approach*, IRGC White Paper No. 1, IRGC, Geneva, Switzerland.

Simmons, D.G. and Becken, S., 2004, The cost of getting there: Impacts of travel to ecotourism destinations, in: R. Buckley (ed.), *Environmental Impacts of Ecotourism*, Cabi Publishing, Wollingford, UK.

Swingland, I.R. (ed.), 2002, *Capturing Carbon and Conserving Biodiversity*, The Royal Society, London.

Tapper, R., 2006, *Wildlife Watching and Tourism*, Study on the benefits and risks of a fast growing tourism activity and its impacts on species, published by the United Nations Environment Programme (UNEP) and the Secretariat of the Convention on the Conservation of Migratory Species of Wild Animals (CMS), Bonn, Germany.

Tiger in the Forest: Sustainable Nature-Based Tourism in Southeast Asia, March 2003 Symposium, organised by the American Museum of Natural History's Center for Biodiversity and Conservation in collaboration with the Wildlife Conservation Society and World Wildlife Fund – online at http://cbc.amnh.org/symposia/archives/tigerintheforest/ (site accessed on 30 May 2005).

UNEP, 2005, *Making Tourism more Sustainable – A Guide for Policy Makers*, UNEP, Paris.

UNWTO, 1995, *Concepts, Definitions, and Classifications for Tourism Statistics*, Technical Manual No. 1, UNWTO, Madrid.

UNWTO, 2005, *Tourism Highlights – 2005 Edition*, Summary Information Brochure, UNWTO, Madrid.

UNWTO, 2006, *International Tourism up by 5.5% to 808 Million Arrivals in 2005*, Press Release of 26 April 2006 (available at http://www.world-tourism.org/facts/menu.html).

Valentine, P. and Birtles, A., 2004, Wildlife watching, in K. Higginbottom (ed.), *Wildlife Tourism: Impact, Management and Planning*, Common Ground Publishing, Altona, Victoria, Australia.

Wilson, M.E., 2002, Ecotourism – Unforeseen effects on health, in: A. Aguirre et al. (eds.), *Conservation Medicine: Ecological Health in Practice*, Oxford University Press, New York.

WWF, 2001, *Global 200 Ecoregions*, Map, WWF, Washington
(online accessible at http://www.nationalgeographic.com/wildworld/global.html).

Chapter 9
Listeria in Raw Milk Soft Cheese: A Case Study of Risk Governance in the United States Using the IRGC Framework

Andrew J. Knight[1,5], Michelle R. Worosz[1], Ewen C.D. Todd[1,2],
Leslie D. Bourquin[2,3] and Craig K. Harris[2,4]
[1] *Food Safety Policy Center, Michigan State University, USA*
[2] *National Food Safety and Toxicology Center, Michigan State University, USA*
[3] *Department of Food Science and Human Nutrition, Michigan State University, USA*
[4] *Department of Sociology and Michigan Agricultural Experiment Station, Michigan State University, USA*
[5] *Department of Sociology and Anthropology, Susquehanna University, USA*

Introduction and Background

Between 1980 and 1996 there were 30 known and reported outbreaks of foodborne illness associated with cheese consumption in the United States, Canada, Europe, and Scandinavia (Cody et al. 1999), and 16 of these outbreaks were associated with cheese produced using unpasteurised milk contaminated with one or more of the following pathogens – *Brucella sp.*, *Escherichia coli*, *Listeria monocytogenes*, *Salmonella spp.*, and *Yersinia enterocolitica* (Teuber 2000). In this chapter, we will focus on only one of these pathogens – *Listeria monocytogenes* (*Lm*). Of the above outbreaks, three were caused by *Lm*, which resulted in 284 reported illnesses and 86 deaths (Teuber 2000). Periodic outbreaks of listeriosis from cheese have continued to occur; at least another six *Lm* outbreaks in the US, four in Europe, and two in Canada have been associated with cheese consumption since 1996 (de Valk et al. 2005; Food Safety Network 2005; Pagotto et al. 2006; US Food and Drug Administration Center for Food Safety and Applied Nutrition [FDA CFSAN] et al. 2003).

Listeria is a genus of bacteria that includes six separate species that can be found throughout the natural environment, for example, in the feces of mammals, on vegetation, and in silage. The *Lm* strain was first identified in 1926 following an outbreak in rabbits (CFSAN 1992), but has only gained significant interest by the US federal regulatory bodies in the past 20 years (Woteki and Kineman 2003). *Lm* is commonly found in the gastro-intestinal tract of several animal species and humans. It has been found in at least 37 mammal species, 17 species of birds, and both fish and shellfish; and is believed to be present in up to 10% of humans (CFSAN 1992).

O. Renn and K. Walker (eds.), Global Risk Governance: Concept and Practice Using the IRGC Framework, 179–220.

Lm is the primary causative agent of listeriosis. Listeriosis can be distinguished as two types: invasive and non-invasive. Invasive listeriosis is the severe form of the disease which typically has a two to three week incubation time, but can extend up to three months. Adverse outcomes can include septicemia, meningitis, encephalitis, abortion or stillbirth, endocarditis, cutaneous infections, and, though rare, it may cause focal infections, such as endophthalmitis, septic arthritis, osteomyelitis, pleural infection, and peritonitis (FDA CFSAN et al. 2003). Non-invasive listeriosis causes gastrointestinal illness, which may result in chills, diarrhea, headache, abdominal pain and cramps, nausea, vomiting, fatigue, and myalgia. The frequency of contracting non-invasive *Lm* is unknown because most of the cases are not reported to public health officials (FDA CFSAN et al. 2003).

Although contracting listeriosis is relatively rare compared to other foodborne pathogens, it is of concern because of its high fatality rate, which has been estimated as high as 30%. Certain vulnerable populations (e.g., neonates, pregnant women, the elderly, and those with compromised immune systems) are particularly at risk of contracting severe cases of listeriosis. Potential food sources for *Lm* include dairy products (e.g., soft cheese), ready-to-eat meat (e.g., hotdogs, delicatessen meats), liquid whole eggs, fish and shellfish, vegetables (predominantly raw vegetables), and salads made with mayonnaise (Economic Research Service [ERS] 2000). According to the International Life Sciences Institute [ILSI] (2004), high-risk foods have five common properties: they have the potential for contamination with *Lm*; they are capable of supporting the growth of high numbers of *Lm*; they are ready-to-eat; they require refrigeration; and they are stored for an extended period of time.

While there are several styles of cheese including hard (e.g., parmesan), semi-hard (e.g., cheddar), Frischkäse (e.g., cottage cheese), and soft with red smear (e.g., Münster) (Teuber 2000), we will limit this analysis to those classified as soft, meaning cheeses that have a high moisture content, that are aged for less than 60 days (e.g., Camembert, Brie), and that are made from unpasteurised milk. These types of cheese are produced around the world including, but not limited to European countries (e.g., France and the United Kingdom), and Latin America countries (e.g., Mexico). In the US, the production and sale of raw milk soft cheese is illegal because it is considered to be a high food safety risk. Despite this ban, there is still a small market for these varieties; raw milk soft cheese is often produced in the home, purchased in local markets and restaurants, obtained from door-to-vendors, or illegally imported into the US.

This chapter serves as a case study of the International Risk Governance (IRGC) risk governance framework (see Chapter 1). The risk governance system for raw milk soft cheese warrants analysis because it raises a number of issues with respect to food safety standards, values, science, cultural sensitivity, and economic development, including international trade. The organisation of the chapter follows the framework outlined by IRGC.

Risk Governance Context

In the US, the food safety statute of most significance to our discussion is the Federal Food Drug and Cosmetic Act. It grants authority to the US Food and Drug Administration (FDA) to regulate most areas of food safety and foodborne pathogens, including milk and milk products, when those foods and products are considered to be in interstate commerce or in trade with the federal government. Several other agencies are involved in the monitoring and surveillance of foodborne illness, including listeriosis, with the Centers for Disease Control and Prevention (CDC) having the primary role. Each state also has its own laws and regulations governing the production of milk and cheese, and foodborne illness is monitored by state and county health departments. The FDA banned the interstate sale of raw milk in retail packages in 1987. However, at the state level, laws and regulations governing the sale of unpasteurised milk vary because the issue of mandatory pasteurisation of milk is controversial. At last count, 28 of the 50 states continue to permit the sale of unpasteurised milk, although some restrict the volume of sales (Headrick et al. 1998).

Current federal regulations governing the use of raw, heat-treated, and pasteurised milk for cheese-making were promulgated in 1949 (Donnelly 2005). Under these regulations, cheesemakers have two options to meet food safety requirements. First, the standard practice is that milk destined for cheese-making is pasteurised. Second, raw milk can be used for cheese manufacture as long as the resulting cheese is held at a temperature of not less than 35°F/1.7°C for a minimum of 60 days. Although raw milk is often heat treated at a slightly lower temperature than pasteurisation to prevent spoilage, it is still considered to be raw for legal purposes. Aging cheese under the temperature and time conditions above is believed to destroy foodborne pathogens (Teng et al. 2004:580). Since soft and fresh cheeses are aged less than 60 days, they are required under US federal regulations to be made from pasteurised milk.

At the international level, there is currently no agreement on a standard for *Lm*. According to Todd (2007), some countries have a zero tolerance policy for *Lm*, but most others believe 'zero' is not only unattainable, but serves as an international trade barrier. The Foods Standards Programme of the Codex Alimentarius Commission is currently charged with developing an international standard for food that protects human health and ensures fair trade (Todd 2007). Under present regulation, raw milk soft cheese cannot be imported into the US.

Table 1 Management framework for *Lm* and raw milk cheese.

Illness Prevention	Consumer Sovereignty
• Standardisation	• Choice
• Unsafe	• Safe
• Zero tolerance	• Tolerance
• General population	• Susceptible populations

Pre-Assessment

Problem Framing

The overall frame within which we will discuss the management of the hazard of *Lm* in raw milk soft cheese is defined by two alternative principles, illness prevention versus consumer sovereignty (shown in Table 1). This overarching frame is composed of four subframes: standardisation versus choice, unsafe versus safe, zero tolerance versus tolerance, and general population versus susceptible populations. The illness prevention principle argues that one of the proper functions of the state is to protect consumers from avoidable harms by insuring that food is as safe as possible. Under this principle, regulations requiring the pasteurisation of milk and banning raw milk soft cheese are necessary to protect consumers from high-risk foods that might contain potentially lethal foodborne pathogens. Consumer sovereignty is associated with autonomy or freedom of choice and implies that sufficiently well-informed consumers should have the opportunity to acquire, without excessive transaction costs, whatever goods or services they desire, including fresh raw milk soft cheese. This principle implies, in turn, that producers, distributors, and marketers must have the freedom to respond to this demand, and that consumer demand will determine the market for raw milk soft cheese. It should be noted that this overall framing of the *Lm* management issue is not intended to represent a dichotomy between two opposing extremes. Instead, it is designed to capture the arguments underlying the current regulatory structure and those advocating for change and highlight the underlying issues of governance, science, and values.

Standardisation versus Choice

The illness prevention principle promotes the standardisation of regulations and practices. To protect the public, raw milk soft cheese is banned. The consumer sovereignty principle argues that current regulations are too restrictive and limit choice. Issues like preservation of culture, quality, and economic development underlie this position. In some cases, consumption of raw milk soft cheese is not simply a preference for taste, but is part of a complex collection of culturally based preferences that include notions of family and tradition. It can also be viewed as a method to preserve 'traditional' cheese-making methods (Reed and Bruhn 2003).

Most artisan cheese producers in the US use pasteurised milk because they believe it increases safety by decreasing the bacteria and yeast that may spoil the flavour or produce an undesirable gas (Teng et al. 2004: 580). Some artisan cheese producers, however, prefer cheese made with unpasteurised milk because they believe pasteurisation would mean the extinction of the 'best' cheeses. They argue that pasteurising the milk prior to use decreases the flavour and lengthens the time for ripening (Teng et al. 2004). Moreover, it is believed that the pasteurisation of milk results in an increased homogeneity of cheese products.

Quality is an important marketing mechanism for many small and specialty cheese producers. Studies have found that consumers intentionally seek out raw milk soft cheeses because they are perceived to be fresher, more natural, as well as more interesting with respect to both the complexity of their flavour and their story – consumers are interested in the cheese-maker, the farm of origin, and cheese-making practices (Reed and Bruhn 2003; Teng et al. 2004). Some consumers also express socio-economic and/or political goals in their purchasing behaviour. Rather than supporting a transnational food manufacturer and/or a retail giant (e.g., Kraft and Wal-Mart), they are interested in the origin and process of production (e.g. local, organic, and family owned operations) (Reed and Bruhn 2003). Those favouring the consumer sovereignty principle point out that the pasteurisation requirement limits the production and sale of some cheeses; US cheese producers are not allowed to sell their products in other states nor access international markets. Cheese producers in other countries are also unable to sell their products in the US.

Unsafe versus Safe

Under the illness prevention principle, raw milk soft cheese is unsafe. From a regulatory perspective, the sale of raw milk soft cheese (e.g., Panela, Camembert, Feta, Brie, and blue-veined) is simply too risky (FDA CFSAN et al. 2003). Raw milk soft cheese is considered high-risk, particularly because of the higher risk of contamination from raw milk. Even though outbreaks may be rare, *Lm* in soft cheese made from unpasteurised milk was implicated in at least six outbreaks in the US in 1985, 2000–2001, 2001, 2003, and 2005. It is a public health issue, particularly for the Hispanic population which tends to consume specific types of these cheeses in much higher quantities than the rest of the US population (Shiferaw et al. 2000).

Under the consumer sovereignty principle, the regulation and control of raw soft milk cheese are seen as '*Listeria* Hysteria' (Shaw 2000); the argument is that foodborne illness from raw milk cheese is simply too infrequent to cause concern and to warrant regulation. Reviews of all outbreaks of foodborne illness in the US by Altekruse et al. (1998) identified only 32 cheese-associated outbreaks between 1973 and 1992, and Johnson et al. (1990) identified only six cheese-related outbreaks from 1948–1988. Critics of the current regulations that ban raw milk soft cheese tend to view the risk of foodborne illness from consuming raw milk cheese relative to risk associated with the ingestion of other foods. For example, the highest degree of risk of contracting listeriosis is from consumption of uncooked delicatessen meat and unheated frankfurters (FDA CFSAN et al. 2003). Some view the risk associated with raw milk cheese as similar to non-food risks like smoking and alcohol, which are not banned but require warning labels; Stein (2001) writes that "we can only hope that [FDA] will decide to put a warning label on all cheese – pasteurised and raw – as it does for alcohol and tobacco, instead of lowering the curtain on that luscious Fourme d'Ambert'.

Zero Tolerance versus Tolerance

Debates within this subframe centre around the issues of what are the appropriate levels of food safety and when should the regulatory triggers take effect?[1] The FDA has had a 'zero tolerance' policy for *Lm* since the mid 1980s. Thus, the FDA has defined the appropriate level of food safety to be zero tolerance and have set the regulatory trigger at this value. This policy was reaffirmed by a US District Court decision, United States v. Union Cheese Co., in 1995 (Todd 2007: 26). On the basis of that decision, *Lm* was defined as an 'adulterant', which gave the FDA authority to regulate it. The FDA has specified the zero tolerance standard to be the absence of *Lm* in 25 grams (<1 cfu in 25 g) in the food under consideration. Advocates for tolerance raise questions about whether this zero tolerance standard is realistic and based on science. For instance, Donnelly (1989) points out that that detection limits for *Lm* vary by the detection methodology and because of limitations in detection, these tests may result in a false negative. Thus, advocates of tolerance argue that the regulatory trigger is set too low for the appropriate level of risk.

General Population versus Susceptible Populations

Most people are not at risk of contracting severe listeriosis. However, particular subsets of the general population are more susceptible (e.g., pregnant women, neonates, the elderly, and those with compromised immune systems). Because FDA rules prohibit the manufacturing and sale of soft cheese made from raw milk, the questions posed in this subframe are: (1) 'Why should raw milk soft cheese be banned if the risk associated with severe listeriosis is high for only a segment of the population, especially if this standard is not applied to other high-risk products like alcohol and tobacco?', and (2) 'Should the regulatory trigger be enacted to prevent poten-

[1] An appropriate level of food safety refers to acceptability of the risk of human illness. The regulatory trigger is the point at which government sets the regulation and can take action based on this regulation. In this context, there is debate over what is an appropriate level of risk (e.g. could be 0, 100 cfu/g, or even higher depending who is at risk), but the government sets a particular trigger or value (in US 0 cfu/g, Canada and some European countries 100 cfu/g). Thus, the trigger can be below or above what is considered to be an appropriate level of risk. In the case of the US, critics of zero tolerance argue that the regulatory trigger is set too low for the appropriate level of risk, at least for majority of the population.

tial illness to minority vulnerable populations or the majority population who is not high-risk?' The response in the context of the illness prevention principle is that the entire population should be protected; under the consumer sovereignty principle, the response would likely be that the most vulnerable should be accorded the greatest protection and that the ideology of protecting everyone should not be consistently applied to all products.

Monitoring and Early Warning

Monitoring, surveillance, and enforcement activities are carried out by a combination of state and federal agencies and across several programme areas. Listeriosis is one of many infectious diseases that the CDC tracks. When a case is confirmed it is reported to the CDC via the National Notifiable Diseases Surveillance System (NNDSS) (Hopkins et al. 2005). The CDC publishes statistics on all reported and confirmed incidences of listeriosis in the Morbidity and Mortality Weekly Report. The CDC also uses the Foodborne Diseases Active Surveillance Network (FoodNet), which is part of CDC's Emerging Infections Program (EIP), to determine the burden of 10 pathogens (including *Lm*), to monitor trends, to attribute the burden to specific foods, and to assess interventions. Currently, microbiological data are collected from 650 laboratories in ten regions across the US, representing approximately 15.3% of the total population. The Pulsed-Field Gel Electrophoresis Network (PFGE) (PulseNet) is the CDC's programme for DNA 'fingerprinting' of certain pathogens, including *Lm*. The intent of this programme is to assist epidemiologists by subtyping the pathogen, identifying case clusters, and facilitating source identification.

A *Listeria* surveillance programme in the US dairy industry has been in place since the mid 1980s. In response to the 1985 *Lm* outbreak, the FDA established a Dairy Safety Initiative, which was 'designed to correlate sample collection of finished product with a physical plant inspection' (Kozak et al. 1996:217). Between April 1986 and September 1988, 1,370 milk plants were inspected and thousands of samples were collected. These inspections found less than 3% of the plants had a finished product that was contaminated with *Lm*, and the contamination level that was found was quite low (i.e., <10 colony forming units per millilitre [cfu/ml]). In 90% of the positive samples, the contamination was traced to the production environment (i.e., not to the raw milk itself). As a consequence, the FDA, the States, the National Conference on Interstate Milk Shipments (NCIMS), and the dairy industry created a joint *Listeria* Prevention Program for improving production and handling practices for milk.

Early warning indicators of the potential for listeriosis outbreaks come from both programmes that monitor the potential presence of prohibited milk products in the market place (e.g. confiscation of raw milk soft cheese, surveillance of homes, stores, and door-to-door vendors making or selling raw milk soft cheese) and from medical monitoring efforts (e.g., surveillance of illnesses and stool cultures to detect

Lm outbreaks). The California Department of Food and Agriculture, for example, conducts periodic seizures of illegal cheeses and some of these contain *Lm*. An early warning indicator of a listeriosis outbreak is when two or more persons are reported to a local health agency with symptoms consistent with foodborne diseases and with recent histories of consuming one or more common foods. After confirmation of diagnosis (i.e., microbiological laboratory test), these cases are reported to the CDC and included in the *Morbidity and Mortality Weekly Report*. In the case of an outbreak, the affected individuals could be reported through more than one health department.

Institutional Pre-Screening

Once a foodborne disease outbreak is suspected, attempts are made by local health officials to determine its cause (i.e., the pathogen or other hazard that is the etiologic agent) and its source (i.e., the food). Patients suspected of suffering from a foodborne illness are administered a questionnaire by a public health professional. The CDC produced a foodborne outbreak investigation toolkit for public health professionals to help diagnose the causative agent, to administer the questionnaire, and to collect samples. The Bacterial Analytical Manual outlines 'standard methodology, and permitted alternative rapid methodologies, to be used by FDA laboratories for detection and isolation of *Listeria monocytogenes*' (Hitchins 2003). If tests are positive for *Lm*, then further tests are conducted to quantify the level of contamination. However, in most instances, because the incubation time of listeriosis can vary from 1 to 90 days, the food item associated with the outbreak is not available for testing. Moreover, non-invasive listeriosis is unlikely to be reported or detected.

Scientific Conventions

The scientific conventions governing input to decisions about management of *Lm* comprise epidemiological testing and outbreak data, as well as risk communication that is geared toward educating both the general and high-risk populations about the risks associated with consuming raw milk soft cheese. Once an outbreak has been acknowledged, risk management strategies involve containment. Occasionally, microbial testing and risk assessments may be conducted to revisit current regulations. When microbial testing showed that pathogens in raw milk cheese could survive more than 60 days, for instance, the FDA conducted a review to consider whether the use of pasteurised milk should be required for all cheeses (Donnelly 2005).

Risk Appraisal

Under the IRGC risk governance framework, risk appraisal comprises two parts: (a) risk assessment which focuses on the 'generation of knowledge linking specific risk agents with uncertain but possible consequences' (IRGC 2005) and (b) concern assessment which is the scientific assessment of the concerns or perceptions of relevant stakeholders regarding a particular risk.

Risk Assessment

The following section presents an overview of risk assessment of *Lm* in soft cheeses. It is broken down into three sections recommended by the IRGC framework: hazard identification and estimation, exposure and vulnerability assessment, and risk estimation.

Hazard Identification and Estimation

Three risk assessments of *Lm* in soft cheeses – FDA CFSAN et al. (2003), Bemrah et al. (1998), and Sanaa et al. (2004) – were identified and will be referenced in this chapter. The FDA CFSAN et al. (2003) risk assessment comprised 23 ready-to-eat foods, including various soft cheeses, and predicted risks for the US population. Both Bemrah et al. (1998) and Sanaa et al. (2004) focused on risk assessments for France. Bemrah et al.'s (1998) study examined soft cheese made from raw milk, and Sanaa et al. (2004) looked at Camembert and Brie made from raw milk. It should be noted that the assumptions and methods employed in each of these three risk assessments are different and that none of the three addressed raw milk cheese made in the home.

Hazard identification describes the adverse effects of a particular substance, organism or other entity. In addition, the relationship between exposure level (dose) and frequency of illness is evaluated, often by some biological endpoint, like infection, morbidity, or fatality (FDA CFSAN et al. 2003). Three factors affect the dose-response relationship for *Lm* and its adverse outcomes: the environment (i.e., food matrix), pathogen virulence (i.e., the virulence of the particular strain of *Lm*), and the host (i.e., susceptibility of contracting listeriosis and the ability of the immune system to respond to a *Lm* infection). In the case of *Lm*, the relationship between dose and likelihood of severity of illness is not well understood. Because *Lm* has a high fatality rate, human trials have not been completed. Instead, dose levels have been calculated using animal, particularly mice, data and/or epidemiological data (Chen et al. 2003). A limitation of animal studies is that the relationship between *Lm* infection in mice and illness in humans is not well understood, particularly at lower doses, so the FDA CFSAN et al. (2003) risk assessment used mortality as the endpoint.

While contracting a severe case of listeriosis is relatively rare in humans, its consequences are often severe. For instance, US data shows that *Lm* led to higher rates of hospitalisation than any other foodborne pathogen and was responsible for over one-third of all reported deaths associated with foodborne pathogens in 2000. The infection rate of *Lm* in the US is estimated to be 3.4 infections per 1,000,000 population, with the overall number of listeriosis cases estimated to be 2,500 per annum (FDA CFSAN et al. 2003). The infection rate in the US is similar to those in other developed countries such as Canada and France (Bemrah et al. 1998).

There are several high-risk groups for contracting listeriosis. Susceptible populations often have immature or compromised immune systems. A precise number of susceptible individuals is difficult to calculate because they include diverse groups such as the elderly, cancer and transplant patients, and persons with immunosuppressive diseases (FDA CFSAN et al. 2003). A further limitation is that since an immunocompromised state is based on qualitative or circumstantial criteria, the criteria may apply to some, but not all members of a particular group. Despite these limitations, high-risk subpopulations can be separated into non-perinatal and perinatal groups. Non-perinatal groups include the elderly (over the age of 60); perinatal groups include pregnant women and neonates.

Most of what is known about the epidemiology of listeriosis has been derived from outbreak data. As mentioned in the introduction, several outbreaks in the US have been associated with raw milk soft cheese. According to FDA CFSAN et al. (2003: 23), 'Outbreaks due to dairy products were most often the result of raw milk being present in a product such as soft (fresh and mold-ripened) cheese, or from post-pasteurisation contamination'.

The first outbreak occurred in 1985 when consumption of California-made Jalisco-brand Hispanic soft cheese contaminated with *Lm* serotype 4b was linked to 142 listeriosis cases in Los Angeles County with a mortality rate of 34% (Linnan et al. 1988). The total number taken ill in the US was estimated at 300, mainly Hispanics. Factory records indicated that raw milk might have been illegally added to pasteurised milk used in cheese-making. The epidemic strain was also widespread in the factory environment, suggesting ample opportunity for post-pasteurisation contamination. It was also found that the cheese was kept in cold storage for a period of weeks which, as discussed later, can increase risk of listeriosis by allowing any *Lm* present to multiply (ECHCP 1999).

In a second outbreak in 2001, consumption of a homemade Hispanic soft cheese was directly linked to 12 cases of listeriosis in North Carolina (CDC 2001). All 12 victims – 11 women (10 pregnant, one postpartum) and a 70-year-old immunocompromised man – were of Hispanic origin. This outbreak resulted in five stillbirths, three premature deliveries, two infected newborns and two cases of meningitis. Fourteen *Lm* isolates from patients, the implicated cheese, and the dairy supplying the milk all belonged to the same PFGE type (or 'fingerprint'), which confirmed that the source of the outbreak was homemade raw milk cheese sold illegally by street vendors or by several small Hispanic grocery stores.

A third outbreak in 2001 occurred when three pregnant Hispanic women in the state of Washington reported purchasing unlabelled *queso fresco*, two from a door-

Table 2 Predicted annual number of listeriosis cases in the susceptible population at different dose levels.

Level (cfu/g)	Max. Dose[a] (cfu/g)	% of Servings at Max. Level[b]	# Cases/ Year[c]
0.04	1	100	0.5
0.10	3	3.6	0.5
1.00	32	1.7	0.7
10.00	316	0.8	1.6
100.00	3,160	0.4	5.7
1,000.00	31,600	0.2	25.4

[a]Serving size of 31.6 g.
[b]Number of servings in the highest *L. monocytogenes* level assumed divided by 6.41×10^{10} times 100.
[c]Levels of *L. monocytogenes* per serving used to calculate predicted number of cases based on the overall distribution from the FDA et al. risk assessment (2001). A total of 6.41×10^{10} servings per year was assumed.
Source: WHO/FAO (2004)

to-door vendor and one through a friend (Stewart 2002). During 2003 and 2004, at least five additional listeriosis cases in pregnant Hispanic women were linked to the purchase of illegal Hispanic soft cheeses from door-to-door street vendors in California (Food Safety Network 2004a), North Carolina (Food Safety Network 2004b), and Texas (Food Safety Network 2003). Also in 2003, 13 cases of listeriosis were reported in the state of Washington with three deaths; risk factors included immunodeficiency, ingestion of soft homemade cheese and travelling in Mexico. In 2005, another outbreak occurred in Texas affecting three Hispanic pregnant women, two newborns, and one Hispanic elderly woman. While the source was not identified, it was suggested to be unpasteurised soft Hispanic cheese purchased at a flea market (Food Safety Network 2005).

A complication with using epidemiological data from outbreaks to assess dose-response relationships is that the incriminated food is rarely available because of *Lm*'s long incubation period (1 to 90 days) (ECHCP 1999). That is, the original dose received by the affected individuals is often unknown. In foods, *Lm* is usually present in relatively low numbers (<100 cfu/g), and the outbreak and sporadic cases data suggest that high doses are required for infections through food (>100 cfu/g). However, the ECHCP (1999: 7–8) report states that the 'possibility of infection from low numbers of *L. monocytogenes* especially among the immunocompromised cannot be discounted'. The World Health Organisation and the Food and Agriculture Organisation of the United Nations [WHO/FAO] (2004) technical report attempted to shed more light on this issue by estimating the annual number of listeriosis cases in the susceptible population in the US with varying dose levels. Table 2 shows that the number of estimated listeriosis cases increases rather substantially with levels greater than 100 cfu/g.

Table 3 Estimates of the total number of annual servings of cheese consumed in the US by population.

Cheese	Intermediate age	Perinatal	Elderly	Total
Fresh soft	6.9×10^7	4.08×10^5	1.3×10^6	7.1×10^7
Soft unripened	3.4×10^9	2.3×10^7	1.0×10^9	4.4×10^9
Soft ripened	1.7×10^9	1.2×10^7	1.8×10^8	1.9×10^9
Semi-soft	1.6×10^9	1.12×10^7	1.5×10^8	1.8×10^9

Source: FDA CFSAN et al. (2003)

Exposure and Vulnerability Assessment

The risk of listeriosis associated with consumption of raw milk products to any given population is a function not just of the virulence of the pathogen itself, but of exposure to the pathogen, where exposure is a function of the quantity of food consumed and the level of contamination in that food. The exposure assessment component of the risk assessment therefore typically contains data on food consumption, food contamination, pathogen growth, and prevalence of the pathogen.

Food consumption data is often used to determine the quantity of food consumed. In the case of raw milk soft cheese these estimates are rather problematic for various reasons. First, consumption surveys in the US do not collect information on unpasteurised milk; estimates are based on pasteurised milk. Second, cheese portions tend to have a small number of servings, and can be considered snack items rather than meal components, all of which may cause estimates to be less reliable statistically. Third, demographic information delineating consumers who are immunocompromised and the elderly in nursing homes or assisted living outside of the home are not collected in food consumption data, nor do the food consumption data contain a large sample of pregnant women. Fourth, the number of servings is generally reported over a one or two day period, requiring yearly estimates to assume that foods are consumed in the same frequency over the entire year.

The risk assessment estimates for the total number of annual servings of various cheeses consumed by the intermediate-age, perinatal, elderly, and total populations in the US are presented in Table 3. The FDA CFSAN et al. (2003) risk assessment is based on serving sizes of 31 grams (g) for fresh soft cheese, 29 g for soft unripened cheese, and 28 g for soft ripened and semi-soft cheeses. The fresh soft cheese category comprises high moisture (>50%) cheeses such as traditional Hispanic-style soft cheeses (i.e., panela, *queso de crema*, *queso fresco*, and *queso de puna*). Soft unripened cheeses include high moisture cheeses such as ricotta, cottage, cream, baker, and Neufchatel. Soft ripened cheeses include high moisture cheeses, such as Brie and Camembert, and pickled or white-brined cheeses, like feta and mozzarella. Semi-soft cheeses include cheeses with a moisture content of between 39% and 50% such as Blue, Brick, Edam, Gouda, Havarti, Limburger, Monterrey Jack, and Provolone.

Bemrah et al. (1998) also found that accurate data on individual consumption patterns of raw milk soft cheese were not available, so they used data from the

Centre Interprofessionel de Documentation et d'Information Laitières and estimated that the consumption of ripened soft cheeses of any type made from raw milk would be 50 servings of 31 grams per capita per year in France. In addition, they presented three scenarios based on different consumption estimates (10, 20, and 50 servings).

Contamination data can be calculated in two ways. First, qualitative data measure whether a pathogen is present or absent in a particular food. Second, quantitative data enumerate the presence of a pathogen in a particular food, usually expressed as the number of colony forming units (cfu) in a gram. A review of prevalence studies shows that 4.6% of soft, mold-ripened cheese samples tested positive for *Lm* (27 studies); 5.1% of raw milk soft, mold-ripened cheese samples tested positive (11 studies); and 25.6% of smear-surface cheese tested positive (10 studies). In 23 studies that investigated the prevalence of *Lm* in soft cheese, but type not identified, 2.5% tested positive (ILSI 2004). A review for the FDA CFSAN et al. (2003) risk assessment found that 1.4% of fresh soft cheese samples tested positive, compared to 3.9% of soft unripened cheeses, 3.8% of soft ripened cheeses, and 3.1% of semi-soft cheeses. It should be noted, however, that these studies may have utilised different levels of detection and that a positive test would not necessarily result in someone developing listeriosis. It simply means that *Lm* was detected at some level.

Several other prevalence studies have been completed that were not captured in the previous review. Genigeorgis et al. (1991b) recovered *Lm* from 2 of 100 Hispanic soft cheeses in California, both of which had a pH >6.2 and were positive for the phosphatase test, indicating that they were likely prepared from raw milk. More recently, Gombas et al. (2003) detected *Lm* in 5 of 2,931 (0.2%) retail Hispanic soft cheeses purchased in Maryland and California at levels as high as 100 cfu/g. Sagoo and Little (2004) tested 8 samples of unripened soft cheeses made from raw or thermised milk from production establishments in the United Kingdom and did not detect any *Listeria.* In addition, 8 samples of ripened soft cheeses tested negative, and of seven semi-hard samples, one tested positive at >199 cfu/g. Sagoo and Little (2004) also tested cheeses made from raw or thermised milk from retail premises. Only one sample of 62 tested positive for *Lm* in unripened soft cheese at <100 cfu/g. *Lm* was only detected in 1% of the ripened cheese samples (8 of 806) with all 8 being acceptable (<100 cfu/g). Similarly, *Lm* was only detected in less than 1% of the semi-hard cheese samples (8 of 943) with only one positive test resulting in an unsatisfactory rating (>100 cfu/g). An FDA study of 57 Mexican personal cheese importations confiscated during a blitz at the Mexican border showed that 68% of cheeses were made with raw milk and 8% contained *Lm* (FDA/CFSAN/CDCP 2005).

The FDA CFSAN et al. (2003) risk assessment assumed that contamination levels at consumption of a given food did not vary significantly from the contamination distributions observed in Western Europe and other developed countries. Similarly, it was assumed that all foods had a similar pattern of contamination and virulence with all *Lm* present having the potential to cause human illness. However, *Lm* is frequently consumed in small amounts by the general population without apparent ill effects. Because previous studies were conducted in the late 1980s and early 1990s, the effect of improved sanitation and other control measures implemented by the food industry since 1993 were not included. For this reason, recent data were given greater weight in the food contamination calculations than older data.

Bemrah et al. (1998) took a different approach in their exposure assessment by estimating the cumulative effect of contamination at various points in the cheese-making process (see Figure 1). Using data from a French study on the origin of raw bovine milk contamination by *Lm*, they estimated that the concentration of *Lm* in milk from infected cows and environmental sources before cheese processing to be 0 to 32.68 cfu/ml with a mean of 1.29 and a median of 0.32 cfu/ml. This model predicted that 67% of raw milk would be contaminated with any concentration of *Lm*. However, the contamination was predicted to generally be of low concentration, with only 2.5% having a concentration of at least 10 cfu/ml and less than 0.01% having a concentration of at least 100 cfu/ml. The simulations in their model predicted that the contamination in 250 g of cheese could range from 0 to 259.6 cfu/g. They estimated that 1.4% of 250 g raw milk cheese samples could have a contamination level greater than 100 cfu/g.

Sanaa et al. (2004) collected actual *Lm* contamination data for milk samples obtained from two regions in France. A 50-ml sample of milk was taken from each bulk tanker that arrived at the dairy and tested for the presence of *Lm* for an entire year. While the contamination level varied for each month, the percentage of positive tests ranged from about 2% to about 6% for milk tankers at Camembert cheese plants in Normandy, and from 0 to just over 3% for milk tankers at Brie cheese plants in Meaux.

Growth data of *Lm* is a function of storage time, storage condition, and rate of growth in specific foods. *Lm* is particularly problematic because it has the ability to survive the manufacturing and ripening of many types of cheeses, and is capable of surviving even after refrigeration, freezing, surface dehydration, and simulated spray-chilling (ECHCP 1999). Growth is highly dependent on temperature, pH, type of food, and presence of microflora (ECHCP 1999). The key factors in determining the exposure of *Lm* are the initial numbers of bacteria present, temperature, and storage time (ECHCP 1999). In the FDA CFSAN et al. (2003) risk assessment, storage times were multiplied by the rate of *Lm* growth to provide an estimate of the level of *Lm* growth that would be expected to occur between retail sale of the food and its consumption. As data did not exist on the storage of foods in the home, storage time estimates were calculated based on expert judgement of the risk assessment team and others such as the Food Marketing Institute and external experts. Storage times for fresh soft cheeses were calculated to be about 1 to 5 days (15 to 30 days maximum storage), and 6–10 days (15–45 days maximum) for other types of soft cheeses. Bemrah et al.'s (1998) risk assessment estimated that the growth of *Lm* in raw milk stored in farm bulk tanks, transported in tanker trucks, and stored again in cheese manufacturers' silos would follow the curves of Peeler and Bunning (1994). They assumed that there was no net growth of *Lm* during the first month after milk curdling, that contamination did not occur during processing and post-processing, and that temperature abuse did not occur at the distribution and consumption stages.

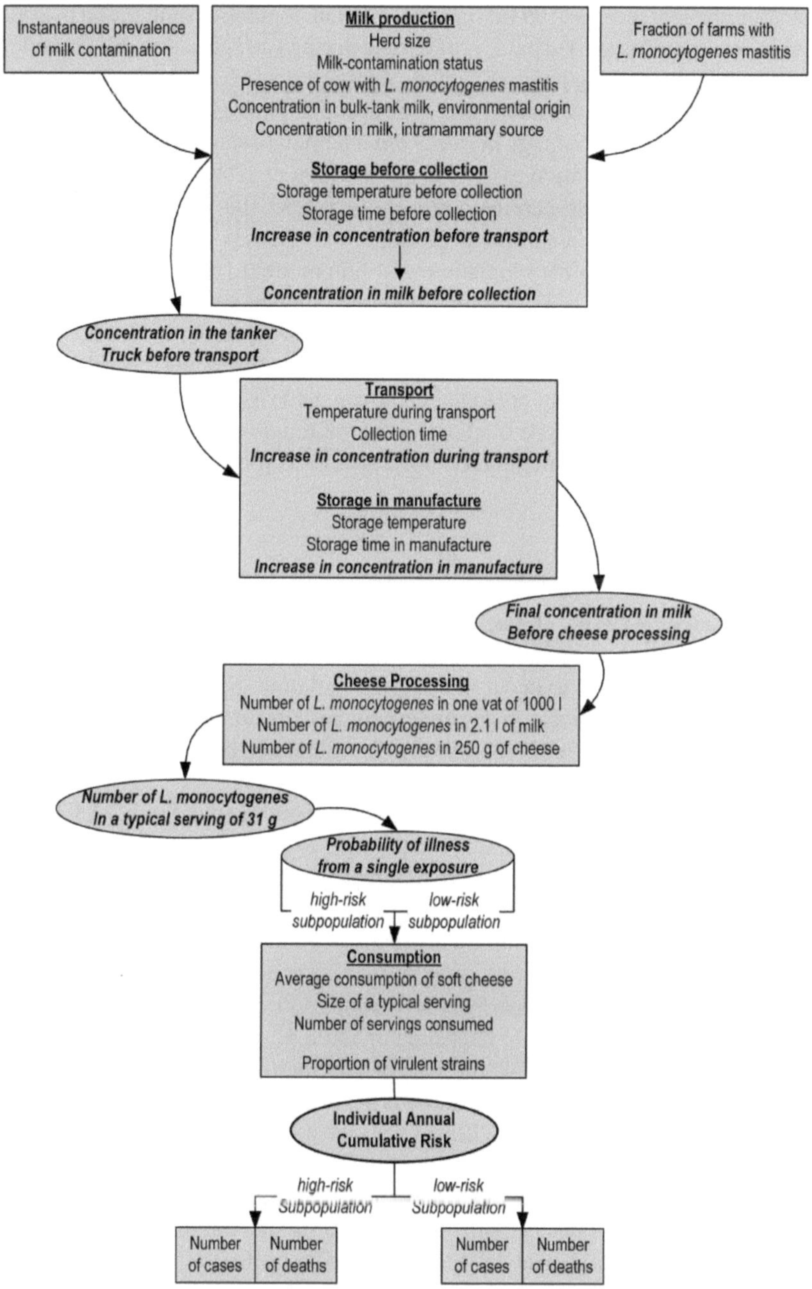

Fig. 1 Commodity chain risk assessment for raw milk cheese (Bemrah et al. 1998).

Although Genigeorgis (1991a) reported that *Lm* populations increased over three logs[2] in inoculated retail Hispanic soft cheese during storage at 4 to 30°C, we know little about food handling practices among consumers of Hispanic fresh, soft cheese that may allow temperature abuse and growth of the pathogen in these cheeses. *Lm* can reportedly increase 1.4 logs in *queso blanco* after 14 days of storage at 4°C and attain a maximum population of 7.9 log cfu/g (Glass et al. 1995). Bolton and Frank (1999) applied several probabilistic models to predict the effects of salt, pH, and moisture content on the fate of *Lm* in Hispanic soft cheeses stored for 21 or 42 days at 10°C. They tested a binary logistic regression model to predict the probabilities of growth or no growth (i.e., only two possible outcomes), and an ordinal logistic regression model to predict the probabilities of growth, stasis, or death (i.e., three possible outcomes). By validating their models with independent data, they showed that this approach accurately predicted pathogen behaviour in Hispanic soft cheese. Therefore, storage of this type of cheese either at retail or in the home will cause a gradual increase in the *Lm* population if the species is present.

Risk Estimation

The risk estimation or risk characterisation step of a risk assessment integrates data acquired during the hazard identification and exposure assessment steps to estimate the adverse effects likely to occur in a given population. For foodborne illness, it is often calculated in two ways: risk per serving and/or number of cases per annum. The FDA CFSAN et al. (2003) risk assessment calculated both using the approaches outlined in Figure 2.

Several limitations of *Lm* risk estimations are mentioned in the FDA CFSAN et al. (2003) risk assessment. First, it is not possible to separate the risk attributable to sporadic and outbreak cases.[3] Second, outbreaks typically result from a breakdown in food production, manufacturing, or distributing systems that result in contamination. The prediction of system failure is usually beyond the scope of risk assessment. The estimated median number of cases of listeriosis per serving and the estimated median number of cases of listeriosis for various soft cheeses in the FDA CFSAN et al. (2003) risk assessment are presented in Table 4.

As demonstrated in Table 4, the risk of contracting listeriosis from cheese is rather low. Out of the 23 ready-to-eat foods assessed by FDA CFSAN et al. (2003), fresh soft cheese ranked 8th on an illness per serving basis, and 5th in the total number of cases it could cause each year in the US. While this places fresh soft cheese in the moderate risk category, this standing is based on data derived from cheese purchased at retail stores made with pasteurised milk. In addition to these estimates, FDA CFSAN et al. (2003) also defined a 'what-if' scenario to estimate the risk from consuming fresh soft cheese made from unpasteurised milk. This scenario assumed

---

[2] Logs are based on 10, meaning the logs increased by 10, i.e. 1 log = 10, 2 = 100, 3 = 1000, etc., so 7.9 is close to 100,000,000.

[3] A sporadic case is defined as a single unrelated case whereas an outbreak is defined as a cluster of cases with a common source.

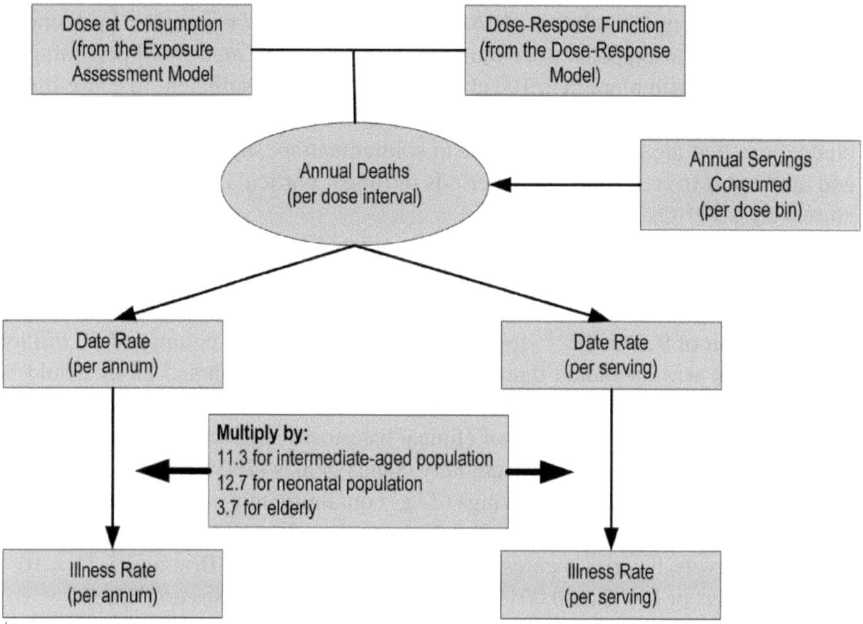

Fig. 2 Components of a risk characterisation model (FDA CFSAN et al. 2003).

Table 4 Estimated median number of cases of listeriosis per serving and per annum.

Type of cheese	Intermed. age	Elderly	Perinatal	Total
		Median Cases per Serving		
Fresh soft cheese	1.2×10^{-10}	1.0×10^{-9}	4.2×10^{-8}	1.7×10^{-10}
Soft unripened cheese	5.8×10^{-10}	4.9×10^{-9}	2.0×10^{-7}	1.8×10^{-9}
Soft ripened cheese	2.1×10^{-12}	2.2×10^{-11}	1.3×10^{-9}	5.1×10^{-12}
Semi-soft cheese	2.9×10^{-12}	3.0×10^{-11}	1.6×10^{-9}	6.5×10^{-12}
		Median Cases per Annum		
Fresh soft cheese	<0.1	<0.1	<0.1	<0.1
Soft unripened cheese	2.0	5.1	0.5	7.7
Soft ripened cheese	<0.1	<0.1	<0.1	<0.1
Semi-soft cheese	<0.1	<0.1	<0.1	<0.1

Source: FDA CFSAN et al. (2003)

that 50% of *queso fresco* cheeses would test positive for *Lm*. According to the assessment, '(t)he risk per serving was 43 times greater for the perinatal population and 36 times greater for the elderly population' than the healthy adult population or intermediate aged population (FDA CFSAN et al. 2003: 221). These risk estimates are considered high-risk. The percentage of contaminated cheeses would be expected to be substantially higher for soft cheese purchased from door-to-door vendors or from home producers because they would likely be made with raw milk under less sanitary conditions (Gombas et al. 2003). Although most *Lm*-positive fresh cheese servings would be expected to contain very low levels of *Lm*, a small percentage is predicted to contain more than 10^6 cfu/g, which would be sufficient to cause illness in susceptible populations. Ready-to-eat products of greatest risk to consumers include those that are particularly prone to contamination, support the growth of *Lm*, and are kept refrigerated for long periods of time – particularly at elevated temperatures (e.g., 5–9°C).

Bemrah et al.'s (1998) risk assessment predicted the probability of illness associated with one soft cheese serving. The risk ranged from 0 to 3.373×10^{-4} with a median of 1.86×10^{-8} for a high-risk sub-population, and from 0 to 1.96×10^{-8} with a median of 9.74×10^{-13} for a healthy population. In a country of 50 million inhabitants, it was estimated that the mean number of *Lm* annual cases would be 57 and the mean number of annual deaths would be 12. For the low-risk healthy population, the estimated number of clinical listeriosis cases would range from zero to four, with zero to three deaths annually. Sanaa et al.'s (2004) risk assessment estimated that the percentage of servings (27 g) containing more than 100 cell/g of *Lm* would be 0.03% for Camembert and 0.22% for Brie. For 100 million servings, the number of severe listeriosis cases would be 3.46×10^{-3} for Brie and 5.11×10^{-4} for Camembert or less than 0.001% of servings for both cheeses combined.

Concern Assessment

> It doesn't matter if food is safe if consumers think that it is not.
> Brewer et al. (1994: 64)

Regardless of regulations, practices, technology, and assurances that foods are safe to eat, if consumers ultimately do not believe that the foods they eat are safe, those foods are considered unsafe. Previous food safety scares, such as Alar[4] on apples and Mad Cow Disease in Europe, have demonstrated that even if the risk is low, the perception that foods are unsafe may lead to reduced consumption of certain foods, and demands for increased regulation, trade restrictions, or other actions.

Consumer risk perceptions are also important because consumers play a large role in food safety, particularly when dealing with how foods are cooked, handled, prepared, and stored. While contamination of foods may occur at various points in the food chain (e.g., farms, processors, restaurants, and retailers), in the case of *Lm* in cheese, most outbreaks are the result of homemade cheeses. If consumers' perceptions affect their food purchasing and handling practices, understanding these links can be important in both predicting risks and introducing appropriate measures to reduce risk.

Another reason for understanding risk perceptions is the fact that most people who become ill from foodborne diseases do not report their illness to a health care provider or a food safety agency (Mead et al. 1999). Thus, most foodborne illnesses

[4] Trade name for daminozide, a chemical designed to delay ripening of fruit.

remain undetected and thus not captured in official statistics. In other words, a misperception of the prevalence and risk of foodborne illness can result in under- reporting of cases thereby interrupting a potentially important feedback loop to food safety agencies and consumers. *Lm*, however, is the most likely foodborne pathogen to be reported because of its severe consequences (Mead et al. 1999).

Risk Perceptions

Previous studies have continually shown that consumers have high levels of concern about food safety. In particular, microbial contamination or disease consistently ranks as one of the top food safety concerns (Priest 2000). In comparison to other foodborne pathogens, *Listeria* is not very well known by consumers. For example, only 32% of respondents were aware of *Listeria* compared to 94% for *Salmonella* and 90% for *E. coli* (Lin et al. 2005). Altekruse et al. (1995) reported that only 9.6% of respondents were aware of *Listeria* and only 0.4% knew of a food vehicle that transmitted *Listeria*. The psychometric risk literature indicates that *Listeria* appears in the quadrant labelled familiar and dreaded (Knight and Warland 2005), where the effects of *Listeria* are somewhat known but its consequences are dreaded.

Concern about food safety risks vary by socio-demographics. A consistent finding in the food safety literature is that females have higher levels of concern about food safety issues than men (e.g. Dittus and Hillers 1996; Knight and Warland 2005; Lin 1995; Williams and Hammitt 2001). The significance of other socio-demographic variables have varied among studies with race, age, education, presence of children, and income sometimes being related to concern about food safety issues. Lin et al. (2005) found that risk perceptions, awareness of other food safety issues, contracting a foodborne illnesses, meal preparer, income, household size, presence of child, education, race, and age were related to awareness of *Listeria*. Respondents who believed that microbial contamination was a serious problem, were more likely to be aware of other food safety issues, to have someone in the household who had experienced a foodborne illness, to be the principle meal preparer, to have higher incomes, to have households with more than four persons, to have at least a college degree, and to be black or of a race other than white. Those between the ages of 30-49 had a higher awareness of *Listeria*. Frewer et al. (1994) state that optimistic bias occurs when individuals believe that negative events are more likely to happen to other people than to themselves. This is important considering that respondents typically state that food safety risks are greater for others than themselves.

Anecdotal evidence from US outbreak data and newspaper reports suggests that two groups in particular – Hispanics and 'yuppies' – may be at greater risk of contracting listeriosis because their consumption of raw milk and raw milk cheese is higher than that of the general US population. Hispanics, particularly immigrants from Latin America, are disproportionately represented in outbreak data. 'Yuppies', or young urban affluent professionals, are also more likely to be consumers of raw milk products, as are members of the organic and natural foods movement. While there are no studies on raw milk soft cheese consumers, studies have been conducted on consumer preferences for specialty cheeses. In a study on consumer preferences among farmer's market consumers, Teng et al. (2004) found that consumers purchased cheese at farmer's markets because of selection (60%), freshness (28%), flavour (28%), price (10%), and origin (2%). These consumers also purchased soft, semi-soft, and hard cheeses about equally. Reed and Bruhn (2003) found that consumers purchased specialty cheeses because of perceived health benefits (94%), they were direct from farm (79%), they were locally produced (76%), they were organically produced (73%), and they were produced using sustainable methods (65%).

In a California consumer survey of specialty food stores, 38% of respondents said that they purchased raw milk cheese and did not have any health related concerns, while 45% did not know if they had or had not purchased raw milk cheese (Reed and Bruhn 2003). Bell et al. (2000) found that 78% of Hispanic participants they surveyed believed consuming raw milk *queso fresco* cheese posed a health risk. Despite awareness of the risks, almost half of the participants reported that they made *queso fresco* cheese with raw milk. Even though pasteurised milk *queso fresco* cheese could be purchased in supermarkets, the making of raw milk *queso fresco* has continued at home. In this instance, the desire to continue the tradition of making *queso fresco* at home outweighed the possible risks of becoming ill. Survey data from the same collaborators also found that half of the respondents did not know if the *queso fresco* cheese was made from raw milk, and 40% did not believe this cheese could cause illness; an additional 25% were unsure (Hillers et al. 2002). Half of the respondents indicated that they purchased or received *queso fresco* from a family member, neighbour, or a door-to-door vendor.

Social Concerns

Obvious social concerns surrounding *Lm* and raw milk cheese are illness and outbreaks, and the high mortality rate associated with *Lm*, especially among vulnerable populations. While relatively large outbreaks garner some media attention, the concern demonstrated about raw milk soft cheese by the public and affected parties remains unclear because there is lack of data on this subject. As the sale of raw milk cheese is relatively rare and outbreaks have been largely localised, affecting only particular groups, these outbreaks do not appear to have resulted in widespread 'food scares'.

Debates surrounding raw milk and raw milk soft cheese have reflected the basic divide between principles of consumer sovereignty and illness prevention. Sovereign consumers believe that they have a right to make or purchase raw milk products regardless of governmental regulations and will probably continue to seek them or make them at home. The debate between consumer sovereignty and illness prevention proposes some interesting questions concerning food safety, the role of the state, values, and risk in general. For instance, should consumer preferences, such as taste and texture, be taken into consideration when determining food safety policy? Does the public want a zero-tolerance policy? Or is the public willing to assume some risk for particular products? Is unpasteurised cheese relatively safe? Is it safe when made under certain conditions?

Socio-Economic Impacts

The total estimated economic costs related to all *Lm* contamination problems is $2.3 billion (Frenzen and Buzby 2000). This estimate includes medical costs, productivity losses from missed work, and the estimated value of lives lost to premature death. The estimate does not include other factors such as travel costs in obtaining medical care, lost leisure time, loss of work to care for sick children, pain and suffering, and the costs of other chronic complications. The proportion of these economic costs that are due to raw milk soft cheese consumption is unknown.

Although current US regulations restricting the sale of raw milk cheese may reduce the economic costs due to illness, they also have potential economic impacts on farmers. According to the Oldways[5] website the demand for raw milk cheese is increasing. To meet this demand, specialty food stores and farmer markets have increased their selection of cheeses. The website argues that because raw milk cheeses are high-end products, farmers are able to receive a premium price, which aids the sustainability of small farms and cheese producers. As demonstrated in the survey results of Reed and Bruhn (2003), production methods and support for local farmers play a large role in consumer purchase intentions. Thus, the current practice of banning raw milk soft cheese prevents farmers and cheese manufacturers from producing a potentially profitable product.

US regulations requiring pasteurisation also have international trade implications. Currently, European nations and Canada allow the importation and sale of raw milk soft cheese if it is produced under acceptable manufacturing conditions. Cheese producers in these countries are unable to sell their products in the US, and US cheese producers are unable to access those international markets which allow the sale of raw milk soft cheese.

Because of its flavour, smell, and texture, raw milk soft cheese is highly valued in many cultures; it is often deeply rooted in cultural and community traditions. The Oldways website claims that cheese has been made with raw milk for 3,000 years. Raw milk soft cheese not only sustained American ancestors, but helped shape the Western diet. While raw milk soft cheese consumers may represent a tiny fraction of the population, they appear to be loyal to these products, and consumption of these cheeses is part of their quality of life. For instance, Hispanic soft cheeses, particularly *queso fresco*, are part of the Hispanic culture, and the making and sharing of raw milk cheese is often a communal event.

[5] Oldways is a food issues 'Think tank', website: http://www.oldwayspt.org/

Tolerability & Acceptability Judgement

A review of the anecdotal and academic literatures suggests that a small minority will seek out raw milk soft cheese regardless of restrictions imposed by the government. Survey data show that people may not be aware of the risks associated with the consumption of raw milk cheese, which makes it difficult to gauge public opinion. Whether the public finds the risk of contracting listeriosis from raw milk soft cheese unacceptable or acceptable remains largely unknown as data on this topic are very sparse. It is probably safe to assume that the most people remain unaware of the safety debate surrounding the consumption of raw milk soft cheese. However, previous research has shown that people dread the consequences of listeriosis and, because of that, there is likely to be support for its reduction and/or elimination in foods.

It appears that the public's definition of tolerability and acceptability might depend on risk communication strategies and whether the public believes there is much of a risk from consuming raw milk soft cheese. As the survey data indicate, there is a segment of the population that does not believe that raw milk poses much of a risk. These individuals are likely to find the risks associated with consuming raw milk soft cheese to be at an acceptable level because they believe the risks to be small or similar to those associated with other foods such as fresh fruits and vegetables. In addition, people who believe in consumer choice are also likely to have high levels of acceptability and tolerability. These people might support the consumer's right to purchase raw milk cheese regardless of the risk, even if they may themselves choose not to eat it.

Risk Characterisation

The risk characterisation step compiles scientific evidence based on the risk appraisal phase and contains three components: (a) risk profile, (b) judgement of the seriousness of the risk, and (c) identification of risk reduction options.

Risk Profile

Consumption of unpasteurised milk and milk products is a significant public health concern for Hispanics and 'yuppies' in the US. In particular, consumption of Mexican-style soft cheese made from unpasteurised milk has been responsible for multiple outbreaks, including outbreaks of multidrug-resistant *Salmonella* Typhimurium DT104 (Cody et al. 1999), and *Lm* (Linnan et al. 1988). An analysis of active surveillance data shows that the incidence of listeriosis among Hispanic females of 15–39 years of age is 11 times higher than non-Hispanic females in this age group (Lay et al. 2002). In addition to Hispanics, populations at high-risk for contracting invasive listeriosis from raw milk soft cheese consumption include preg-

Table 5 Relative susceptibility of contracting listeriosis for different populations.[a]

Condition	Relative susceptibility
Transplant	2,584.0
Cancer – Blood	1,364.0
AIDS	865.0
Dialysis	476.0
Cancer – Pulmonary	229.0
Cancer – Gastrointestinal and liver	211.0
Non-cancer liver disease	143.0
Cancer – Bladder and prostate	112.0
Cancer – Gynaecological	66.0
Diabetes, insulin dependent	30.0
Diabetes, non-insulin dependent	25.0
Alcoholism	18.0
Perinatal	14.0
Elderly (over 65 years old)	7.5
Less than 65 years, no other condition (reference population)	1.0

[a]Relative susceptibility values were calculated by 'taking the total number of listeriosis cases for a subpopulation and dividing it by the estimated number of people in the total population that have that condition. This value is then divided by a similar value for the general population' (WHO/FAO 2004: 141).
Source: WHO/FAO (2004)

nant women, neonates, elderly, and people with immunocompromised systems. The WHO/FAO (2004) technical report calculated the relative susceptibility to listeriosis for different populations. Table 5 shows that groups suffering from immune-compromising diseases or treatments have the greatest susceptibility to listeriosis.

Judgement of the Seriousness of Risk

Risk assessments on contracting invasive listeriosis from consuming raw milk soft cheese are variable. For example, the FDA CFSAN et al. (2003) risk assessment estimated that the risks are high, although this is based on a 'what-if' scenario, assuming a large *Lm* contamination prevalence rate. Bemrah et al. (1998) and Sanaa et al. (2004) found the risk to be much lower. However, there are differences in the seriousness of rates for different populations. For the general population, the risk is probably low. On the other hand, for pregnant women, neonates, elderly, and people with immunocompromised systems, the risk is rather high, although there are questions surrounding at what dose *Lm* would make people ill. Survey data collected by the Food Safety Policy Center in 2005 and 2006 illustrate that in the US, the public expects the foods they eat to be safe, and that they are dissatisfied with the current number of foodborne illnesses; these data suggest the public might favour more stringent food safety regulations. Hispanics and 'yuppies' (sovereign group), however, will likely continue to consume these cheeses regardless of regulations and may be at increased risk of consuming contaminated cheeses that tend to be produced under unsanitary conditions.

Several questions have been raised about what causes contamination and where does contamination occur. Dairy safety experts see the problem of cheese contamination as starting with the cow, as cows carry a number of bacteria including *Lm*, which they acquire 'naturally' from their environment – grazing and pasturing. Therefore, dairy safety experts believe the use of raw milk poses a high-risk of contamination in cheese-making, and needs to be prevented because the end-product has a high likelihood of *Lm* contamination solely because of the use of raw milk. Thus, pasteurisation is the only method assured of eliminating foodborne pathogens in the resulting soft cheeses. Soft cheeses are considered to be a particular food safety risk because they are favourable to pathogen growth due to their higher 'water activity' and pH (Teng et al. 2004). Furthermore, recent evidence suggests that *Lm* and other foodborne pathogens can even survive beyond the 60 days of aging incorporated into regulations (Donnelly 2001).

Published reviews of illness outbreaks associated with cheese consumption (Altekruse et al. 1998; Donnelly 2001; Johnson et al. 1990) show, on the other hand, that the presence of pathogens is most often due to environmental contamination, such as unsanitary practices at the farm or pasteurisation or post-pasteurisation errors during manufacturing. Donnelly (2001: 16) concluded that 'in the majority of instances, confounding parameters other than use of raw milk contributed to pathogens being present in the product at the time of consumption'. Some of these confounding parameters include use of pasteurised milk versus raw milk in cheese-making trials; inadequate development of acidity during cheese-making; low salt levels; contamination by sick employees during manufacturing; temperature abuse of milk designated for cheese production; or environmental contamination during cheese-making. Fletcher (2006), in an article analysing the myths surrounding raw milk and raw milk cheese, also points out that outbreaks are most commonly associated with improper pasteurisation, contamination after pasteurisation, or cheeses prepared under non-commercial conditions. She states that the risk of contracting *Lm* from raw cheeses aged at least 60 days at 35°F/1.7°C or above, as described by law, is minute even for pregnant women, and it would make more sense for pregnant women to avoid higher risk foods such as fruits, vegetables, and deli meats.

The risk reduction strategy that follows from this second argument is one that focuses on the safety of the processes in which raw milk is used, rather than on prevention of the use of raw milk. This view holds that the safety of raw milk cheese and cheese in general can be achieved by using proper milk screening procedures, following good manufacturing practices (GMP), and following Hazard Analysis and Critical Control Point (HACCP) procedures. France is an example of a country that allows raw milk soft cheese to be produced and sold, but has been able to reduce the incidence of invasive listeriosis through improved milking hygiene, rapid detection and elimination of cows excreting *Lm*, and requiring farms that produce milk to adopt the highest level of hygiene (Sanaa et al. 2004). Critics of pasteurisation also point out that there is no compelling data to indicate that mandatory pasteurisation will lead to a safer product (Donnelly 2001:24). Proponents of raw milk cheese also argue that these cheeses pose little risk of contamination in comparison to other foods. While recognising that soft and semi-soft cheeses made from unpasteurised milk do pose potentially greater risks than hard cheeses aged for 60 days, Donnelly (2005) states that environmental contamination poses a far greater risk to cheese safety.

Conclusions & Risk Reduction Options

Based on the information gathered for this analysis, five risk reduction options have been identified and are presented in Table 6. Option 1, banning the retail sale of soft cheese made from raw milk cheese that is aged less than 60 days, is the current US regulation. The advantage of this option is that it prevents the manufacture and sale of potentially harmful products and reduces the likelihood of *Lm* outbreaks from raw milk soft cheese. The risk and concern assessments, however, show disaffection with the current regulation. From a food safety perspective, the regulation is not entirely effective: outbreaks, though rare, still occur, especially with raw milk soft cheeses made in the home or bought from door-to-door vendors. From a socio-economic perspective, the regulation limits economic development; even though there is likely to be consumer demand for these products, they cannot be produced or sold legally and costs are incurred as border agents and public officials seize these products. From a cultural perspective, the regulation interferes with traditional social practices associated with the making and distribution of raw milk soft cheese.

Option 2, allowing the retail sale of specific raw milk soft cheese commodities, is similar to the first option except rather than banning all varieties of raw milk cheeses, it allows the retail sale of particular varieties that are deemed to be low risk through a review of existing manufacturing practices, regulations, and microbial testing. Under this option, specific varieties of raw milk soft cheese that are deemed to be low risk may be imported into the US for retail sale, and perhaps can be produced domestically under certain conditions. Risk assessments have shown that Camembert and Brie would both be low risk, with Camembert being a lower risk than Brie (Sanaa et al. 2004). However, Hispanic or Mexican-style soft cheese may be a higher risk (Gombas et al. 2003). Advantages of this option are that it would allow consumers to purchase some varieties of raw milk soft cheese; it might allow domestic production of particular varieties of raw milk soft cheeses under certain conditions; and it would be more consistent with regulations in other countries. This option is consistent, for example, with the Australian and New Zealand position on French Roquefort cheese, which is considered safe after storage for 90 days before sale, even though it is made using raw milk. At the same time, this option provides a scientific level of food safety and protects the public from the manufacture and sale of high-risk products. However, raw milk soft cheese varieties deemed to be low risk may not coincide with the preferences of particular groups, and may not eliminate the desire for particularly high-risk varieties.

Option 3 requires the use of warning labels. This option offers several potential advantages: it allows consumers to make informed choices; it could contain or reduce the number of outbreaks by making illegal cheese products less prominent; and it could allow contaminated cheeses to be identified through microbial testing and recalled. In an event of an outbreak, the source of contamination could be more easily traced back to its origin. This option may also result in increased economic opportunities for cheese producers and the preservation of social cultures engaged in the making of traditional cheeses. However, it may carry potential public health costs, since raw milk soft cheeses would be accessible to the general population, including high-risk groups. Producers would have to assume at least some of the costs associated with labelling, and increased competition from larger producers might drive smaller local cheesemakers out of business. Finally, additional public costs would be incurred by the enforcement of labelling laws and the inspection of imported cheese.

Option 4 involves changing the *Lm* tolerance standards. The most frequent alternative to a zero tolerance policy (defined as <1 cfu in 25 g of food (cfu/g)) is <100 cfu/g. Chen et al. (2003) argue that foods containing low levels of *Lm* (<100 cfu/g) pose very little risk, since prevention of higher concentrations would eliminate >99% of listeriosis cases. Thus, control efforts should focus on reduction of higher concentrations of *Lm* in ready-to-eat foods. This same argument can be derived from the FAO/WHO (2003) risk assessment, which showed that it is the increase in contaminated servings that drive the number of cases up, not the tolerance level. The question then becomes: at what point in time does raw milk soft cheese become a high-risk product? Since a tolerance level can only be set at production or retail, any actions by the consumer may increase the risk. In the case of soft cheese, it supports growth of *Lm* even when stored at refrigeration temperatures over time, and no practical method exists for decontaminating it before it is eaten. Thus, any concentration of *Lm* in soft cheese at purchase could result in *Lm* levels that could increase the risk of illness.

Countries such as Germany, the Netherlands, and France, have adopted the higher tolerance level for a number of years and they have about the same frequency of listeriosis per capita as the US. Canada and the European Union have a mixed tolerance standard where some foods are subjected to zero tolerance and others to a <100 cfu/g standard (Todd 2007). Since January 2006, the European Union has promulgated a new standard for all member states. This standard requires zero tolerance in ready-to-eat foods prepared for infants and for special medical purposes. For other foods, the zero tolerance standard applies before the food has left the immediate control of the food producer or manufacturer; the higher tolerance standard, <100 cfu/g, must be met at purchase (European Union 2005). The advantages of this option are that it allows consumers to purchase raw milk soft cheese, and it is consistent with the regulations of other countries, while at the same time providing a science based level of food safety.

Option 5 would require the implementation of HACCP programmes. This option has several possible advantages: it allows consumer choice, may increase public safety by reducing the potential for contamination, and might reduce the production, importation, and sale of illegally produced cheeses. Other benefits include an increased ability to identify where contamination occurred, facilitation of the recall of contaminated cheeses, increased economic development opportunities, and the preservation of traditional cheese-making cultures. The potential limitations of this risk reduction option are that cheeses produced under these procedures would be available to the entire population, including high-risk populations, substantial costs might be incurred by dairy producers and cheesemakers to comply with HACCP requirements, and the government would have to inspect and enforce these standards throughout the product chain. A sixth option in which raw milk soft cheese would not be regulated was not presented because it is not a risk reduction strategy and this option has not been advocated by either proponents or opponents of current regulations regarding raw milk soft cheese.

Risk Evaluation

The risk evaluation assesses broader value-based issues that also influence the judgement. It comprises two parts: (a) judging the tolerability and acceptability of risks, and (b) the need for risk reduction measures.

Judging the Tolerability and Acceptability

The US has adopted a zero tolerance policy, which requires the absence (<1 cfu in 25 g) of *Lm* in 25 grams of foods. Risk assessments suggest that this stringent tolerance level may be unnecessary and could be substituted with a limit of <100 cfu/g at consumption. Outbreak data suggest that outbreaks of *Lm* in raw milk cheese have occurred because the milk used in production derived from animals that were infected (e.g., had mastitis, or were asymptomatic carriers) or was stored under unsanitary or improper storage conditions, and not because of the use of raw milk *per se*. These suggest that proper practices surrounding the production of cheese, rather than prohibition of raw milk might reduce this risk.

While risk estimates predict a low risk of contracting invasive listeriosis for the general population, susceptible populations are at higher risk and precautions should be taken, preferably avoiding consumption of raw milk soft cheese. Because there is little data that can inform policymakers on what the public deems to be an acceptable risk, it is difficult to judge the tolerability and acceptability of raw milk soft cheese. One possibility is that the availability of commercially made raw milk soft cheeses under HACCP guidelines may act as a substitute for homemade or imported products. Bell et al. (2000), however, suggest that some Hispanic Americans will continue to make *queso fresco* cheese until an alternative is presented. What we do know is that consumer demand persists for raw milk soft cheese amongst a relatively small, but dedicated proportion of the population, and that current regulations have not prevented outbreaks affecting these people.

Table 6 Options for risk reduction.

Option	Possible Advantages	Possible Disadvantages
1. Ban the retail sale of all raw milk soft cheese (RMSC) (aged <60 days)	• Limits exposure to RMSC and consumption of potentially harmful product especially by vulnerable groups • Reduces the likelihood of outbreaks associated with raw milk and RMSC in the future • Lower costs associated with care and treatment of listeriosis	• Loss of traditional cheese-making culture • Does not affect levels of current 'illegal' production in homes or importation of RMSC • May result in traditional local sales (i.e. to friends and neighbours) being pushed further underground • Consumers of illegal RMSC at risk because milk and cheese are not subjected to microbiological sampling • Does not allow consumers of RMSC to make informed choice about risk • Costs of enforcement (inspections and seizures)
2. Allow the retail sale of commodity specific RMSC (aged <60 days)	• Limits exposure to some RMSC and consumption of potentially harmful product especially by vulnerable groups • Reduces the likelihood of outbreaks associated with some raw milk and RMSC in the future • May lower costs associated with care and treatment of listeriosis • May reduce illegal importation of RMSC • Allows some microbiological testing • Legal unsafe products can be recalled • Can trace legal contaminated products to source • Increases selection of cheeses available for purchase • May increase economic development as RMSC producers could sell their products across states and on international markets	• May result in loss of traditional cheese-making culture • May not affect levels of current 'illegal' production in homes or importation of RMSC • May result in traditional local sales (i.e. to friends and neighbours) being pushed further underground • Consumers of illegal RMSC at risk because milk and cheese are not subjected to microbiological sampling • Does not allow consumers of RMSC to make informed choice about risk • Costs of enforcement (inspections and seizures) • If outbreaks occur, FDA may be blamed for allowing unsafe products on the market

Table 6 Continued

Option	Possible Advantages	Possible Disadvantages
3. Warning labels required for RMSC	• Allows customers to make an informed choice about whether to consume RMSC or not • May reduce incidence of foodborne illness among high-risk populations • May reduce RMSC production in homes as cheese can be purchased in retail outlets • May reduce illegal importation of RMSC • Allows microbiological testing • Unsafe products can be recalled • Can trace contaminated products to source • Increases selection of cheeses available for purchase • May increase economic development as RMSC producers could sell their products across states and on international markets • Preservation of traditional culture	• Limited public health benefits – will not prevent anyone consuming RMSC, potentially increasing risk of foodborne illness • Some cost to producers in implementing new labelling requirement • May heighten competition leaving local producers unable to compete • Enforcement of labelling laws and inspection of imported RMSC • If outbreaks occur, FDA may be blamed for allowing unsafe products on the market • Costs of enforcement (inspections and seizures) • Costs of microbial testing • Consumers may not notice label or take action based on label
4. Change in tolerance levels	• Scientifically based tolerance level • Allows customers to make an informed choice about whether to consume RMSC or not • Consistent with some countries, enabling trade • May reduce RMSC production in homes as cheese can be purchased in retail outlets • May reduce illegal importation of RMSC • Allows microbiological testing • Unsafe products can be recalled • Can trace contaminated products to source • Increases selection of cheeses available for purchase • May increase economic development as RMSC producers could sell their products across states and on international markets • Preservation of traditional culture	• Limited public health benefits – will not prevent anyone from consuming RMSC, potentially increasing risk of foodborne illness • May heighten competition leaving local producers unable to compete • Costs of enforcement (inspections and seizures)

Table 6 Continued

Option	Possible Advantages	Possible Disadvantages
5. Implementation of HACCP and/or best management practices	• Allows consumer choice • May increase public health protection by reducing pre- and post-contamination • May reduce RMSC production in homes as cheese can be purchased in retail outlets • May reduce illegal importation of RMSC • Increases economic development of raw milk farmers and artisan cheesemakers by creating new markets • Allows microbiological sampling • Unsafe products can be recalled • Can trace contaminated products to source • Increase selection of cheeses available for purchase • May increase economic development as RMSC producers could sell their products across states and on international markets (open new markets) • Preservation of traditional culture	• Limited public health benefits – will not prevent anyone consuming RMSC, potentially increasing risk • Cost to producers in implementing HACCP • May heighten competition leaving local producers unable to compete • Inspection and enforcement of farms, transport vehicles, production facilities, and retail stores • Inspection of imported cheese

RMSC: Raw milk soft cheese
HACCP: Hazard Analysis and Critical Control Point

Need for Risk Reduction Measures

Ideally, a zero risk scenario would be preferred, but this goal is not realistic. The need for risk reduction measures is made evident by examining the outbreak data. The five options presented in Table 6 offer different benefits and risks. The current US regulation bans the sale of raw milk soft cheese, and has been in place for over 50 years. It is conceivable that with proper guidelines in place, this policy could be changed. One possible scenario would be to allow raw milk soft cheese production using HACCP, or other similar procedures, and to require a warning label be applied to the packaging. Another scenario would be to allow the importation of particular raw milk cheese varieties from specific countries that can insure that these varieties are low risk and contain a warning label on the packaging. The risks assumed under either of these options would be similar to those of other countries, like Canada, France, and Australia, and are congruent with the interests of the sovereign frame.

Risk Management

Risk management involves recommending and implementing actions and remedies to deal with risks with an aim to avoid, reduce, or transfer them. It involves two steps: decision-making and implementation.

Decision Making

Based on a reconsideration of the knowledge gained in the risk appraisal phase and the options evaluated in the tolerability and acceptability judgement phase, the decision-making process involves the selection of a risk management strategy. Option identification and generation

Table 6 outlined five risk management options. The current risk management strategy practiced in the US is Option 1 (banning the retail sale of raw milk soft cheese). The assumption of this strategy is that raw milk soft cheese represents a high food safety risk, because it is conducive to the presence and growth of *Lm* pathogens. Because it is considered a dangerous food, the risk management strategy is to make the production and sale of raw milk soft cheese illegal, thereby reducing the exposure to the population. The issue is complicated by rules and regulations surrounding the sale of raw milk. Federal law prohibits the interstate sale of raw milk, and most states either prohibit or limit its sale. However, there is a small but persistent consumer demand for raw milk soft cheese, particularly among the Hispanic population and 'yuppies'. Several outbreaks of listeriosis and other foodborne illnesses have occurred because these cheeses are made in the home, imported illegally for personal consumption, and purchased from retail stores and door-to-door vendors. An argument for maintaining the ban on the sale of raw milk soft cheese is that, in comparison to other foods, there have been few outbreaks suggesting that the status quo is working.

Should risk management options be based on a rather substantial high-risk susceptible population (an estimated 20% of the general population) or based on a larger low-risk population? If the goal is to protect the high-risk population, then the current risk management strategy is probably the best. If the goal is the latter, then other risk management strategies may be more desirable. Chen et al. (2003) state that the level of risk deemed acceptable is ultimately decided by society. In this case, a question is who constitutes society? Is it food safety experts, politicians, regulators, cheese artisan industry, farmers, and/or the public? In the US, the FDA is primarily responsible for food safety and would probably require substantial scientific data to overturn existing legislation, especially since it has been in effect for over 50 years.

Option Assessment

While education programmes are suggested for all five risk management options, they become more prominent if raw milk soft cheese production becomes legalised, as susceptible populations would have greater access to these products. Consumers, particularly susceptible groups, would have to be educated about the dangers associated with consuming these cheeses. However, as discussed previously, there are scientific and cultural arguments for legalisation of at least some varieties of raw milk cheese, including the fact that outbreaks have occurred primarily because of pre- or post-contamination practices; the source of outbreaks can be established and perhaps reduced through microbiological testing; economic development opportunities can be explored; consumers would have greater choice; and traditional cultures can be maintained. To reduce the risk of listeriosis, a HACCP programme would have to be introduced for raw milk soft cheese production. A risk management strategy banning the production and importation of high-risk raw milk soft cheese varieties, manufacturing low-risk varieties under HACCP, and placing warning labels on all raw milk soft cheese products might be an acceptable compromise to all parties, including both sovereign and illness prevention groups. A similar approach has been proposed by The Institute of Food Science & Technology [IFST] (1998) in the United Kingdom.

Option Evaluation & Selection

Evaluating and selecting the best option for the production of raw milk soft cheese is difficult because of the lack of data in risk and concern assessments. Risk assessments with realistic data need to be conducted on each type of raw milk soft cheese, and concern assessments need to be completed on how consumers and stakeholders determine trade-offs between preferences and food safety. As well, microbiological testing procedures need to be improved to determine the actual risk levels for different populations. Once this additional information is gathered, the information can be provided to the public, interest groups, and stakeholders to participate in the decision-making process.

Implementation

This section comprises the implementation of the risk management design with particular emphasis on option realisation, monitoring and control, and feedback from the risk management process.

How might a HACCP programme for the raw milk soft cheese industry operate? There should be an overall framework to help managers develop a consistent approach, to include all relevant inputs, and make consistent decisions. These components are also important for regulatory authorities to better balance risks and benefits across society in an acceptable way, and to make the decision-making process more explicit by being more open and consistent. The WHO/FAO approach to risk assessment is to have four steps: risk evaluation, option assessment, option implementation, monitoring and review. With *Lm* the complication is that different strains of the pathogen may enter at different steps in the process, and though it may be rare, contamination of the product in the processing plant may be the most critical. ILSI's expert panel (ILSI 2004) states that

> Control strategies are needed at all stages from pre-harvest to consumption to minimize the likelihood that food will become contaminated by *LM* and to prevent the growth of the organism to high numbers. The panel recommends five control strategies: 1. Good Manufacturing Practices, sanitation standard operating procedures and HACCP programmes, to minimize environmental LM contamination and to prevent cross-contamination in processing plants and at retail; 2. an intensive environmental sampling programme in processing plants along with an effective corrective action plan to reduce the likelihood of contamination of high-risk foods; 3. time and temperature controls throughout the entire distribution and storage period including establishing acceptable storage times of foods that support growth of LM to high numbers; 4. reformulating foods to prevent or retard the growth of LM, and 5. using post-packaging treatments to destroy LM on products.

In the US, there are already voluntary HACCP programmes in place for milk. However, the dairy industry in the US is not subject to HACCP regulations, but rather is subject to regulation by individual states according to procedures published in the 'Grade A Pasteurised Milk Ordinance'. The FDA NCIMS HACCP Pilot Program suggests that every milk producer, milk distributor, bulk milk hauler/sampler, milk tank truck, milk transportation company and each milk plant, receiving station, transfer station, milk tank truck cleaning facility operator must hold a valid permit to produce and transport milk. Also, each entity in the milk chain is subject to inspection and/or audit by the appropriate regulatory agency if the milk is to be used for consumption.

Dairy farms and milk tank cleaning facilities are inspected at least every six months, and milk plants are inspected at least every three months. Inspectors are particularly concerned that proper pasteurisation is taking place and that there are no violations with regards to cross-contamination. In addition to inspections, bulk milk haulers are to collect representative samples from each farm bulk tank so that the milk samples can undergo microbial testing (FDA CFSAN et al. 2003). However, tests do not appear to include testing for *Lm*. An additional requirement would be the testing of unpasteurised milk destined for cheese production for foodborne pathogens, including *Lm* at the farm level. As well, rapid detection methods would have to be implemented at the farm level so that cows excreting *Lm* can be identified and eliminated, and a selection and sorting system would have to be established to insure the production of milk to the highest level of hygiene (Sanaa et al. 2004).

In France, if milk is found to test positive for *Lm*, or other foodborne pathogens, like *Salmonella*, *E. coli*, and *Staphylococcus*, it is checked for hygienic practices. If, despite corrective hygienic procedures, the contamination persists, the milk of individual cows is tested and those with somatic cell counts higher than 300,000/mL are tested for mastitis caused by *Listeria*. A positive test results in the cow being slaughtered. Microbial testing is also completed in cheese samples before they are released. If the samples test positive, the lot is withdrawn (Sanaa et al. 2004).

Based on the regulations and practices in place in France, Food Standards Australia and New Zealand (FSANZ 2005) recently conducted a scientific evaluation of Roquefort raw milk cheese made in France to determine whether this variety of cheese can be legally imported for commercial sale in Australia and New Zealand. The scientific evaluation comprised three stages: (1) a scientific valuation of the safety of the cheese; (2) a review of the safety control measures implemented by producers and enforced by the French government; and (3) an on-site audit of control measures. This evaluation concluded that the risk of the prevalence and growth of seven pathogens that cause foodborne illnesses, including *Lm*, were low, and that regulatory control was demonstrated. Thus, it was recommended that Roquefort cheese from France be permitted for sale in Australia with the following requirements: compliance with French Ministerial Orders; test raw milk for *Lm*; monitoring of pH, salt concentration and moisture; minimum storage period of no less than 90 days; must meet microbiological limits in the Food Safety Code; and the packaging requires labelling. While Roquefort cheese cannot be considered soft because of the 90 days storage requirement, it does provide an option of how an importation process might be implemented for raw milk soft cheese.

While it may be impossible to insure that raw milk is free from potential pathogens, IFST (1998) suggested four measures in the production of raw milk cheese. First, raw milk should be collected and maintained in good hygienic conditions. Second, if the raw milk is not used immediately, it should be refrigerated to minimise pathogen growth. Third, risk assessments should determine high-risk products, and those products should undergo full pasteurisation. Fourth, good conditions of hygiene should be maintained throughout the cheese manufacture, ripening, distribution, sale, and storage until consumption to prevent contamination. Thermisation of raw milk, where raw milk is heated below pasteurisation temperatures, may also be utilised to reduce the growth of pathogens (Donnelly 2005). Pritchard (2005) advocates that pre-requisite programmes as well as HACCP procedures should be in place that detail cleaning and sanitation procedures. Furthermore, there should be an understanding of microbiology and factors that suppress or promote the growth of organisms, and good agricultural practices. While the HACCP process may be daunting and labor intensive for a small cheese-making facility, Pritchard (2005: 151) states that 'The time taken to develop a hazard analysis and to identify areas of special interest in a manufacturing process is time well spent, if for no other reason than gaining a better understanding of the principles behind the process of cheesemaking'.

Risk Communication

This risk governance framework has identified two problem frames and five risk reduction options. The two problem frames comprise scientific, economic, cultural, and value debates surrounding current regulations requiring the pasteurisation of milk and the production of raw milk soft cheese. The first step of risk communication may be to determine if risk professionals, regulators, and politicians are aware of these debates. If deemed necessary, a risk communication strategy may be to bring risk professionals, regulators, politicians, stakeholders, media, and the public together to understand the breadth of these debates to insure that credence is given to political, scientific, economic, and cultural considerations, and that the most effective risk management approach can be formulated and communicated. Once this is accomplished, the FDA must make a decision whether current regulations need to kept or changed.

ILSI's expert panel (ILSI 2004) recognised the need for a science-based approach to risk communication, and data from risk assessments show that some interventions work better than others. The challenge of risk communicators is to educate various target groups about the effects of *Lm* and practices to lessen the risk. Consumer practices are especially important as long-term storage in refrigerators can make a large impact on growth of the pathogen at home. ILSI's expert panel (ILSI 2004) stated that science-based education and risk communication strategies should be aimed at susceptible populations and focused on high-risk foods. Information should be delivered through health care providers or other credible sources of information. Campaigns are most successful if the audience is segmented, based on risk factors, demographics and other factors, and the messages are carefully targeted and tailored for audience members (Rimal and Adkins 2003). Because consumption patterns of raw milk soft cheese vary by ethnicity, socio-economic status, and consumer preferences, messages need to be tailored to these specific target groups. Risk communication strategies should also be tailored to specific vulnerable populations such as pregnant women, the elderly, and the immunocompromised.

Conventional wisdom suggests that many of the raw milk cheese consumers are simply unaware of the risks associated with its consumption and unaware of proper handling practices. Following a 2001 outbreak among Hispanic women in Washington, it was found that none of them was aware of the risks of consuming soft cheese during pregnancy, nor were they aware that unpasteurised milk is typically used to make *queso fresco* (Stewart 2002). However, lack of knowledge and/or concern about *Listeria* and risks associated with raw milk cheese are not limited to Hispanics. Educational strategies to increase awareness of the risks associated with raw milk soft cheese must take into account the possibility of optimistic bias or the perception that others are a greater risk of contracting listeriosis than themselves. A particular problem for risk communicators is that efforts to educate the US Hispanic population about the risks of *queso fresco* consumption have been hampered by their transient nature.

Passive education strategies have generally been adopted as the main approach for reducing the incidence of listeriosis among Hispanics in the US who consume raw milk soft cheese. After the first two listeriosis cases were identified in the 2001 Washington outbreak, for example, public health officials visited the neighbourhood where the cases resided and to near-by health care providers and distributed Spanish-language fact sheets describing the potential risks associated with consuming *queso fresco* (Public Health Seattle and King County 2002). The Texas Department of Health has included warnings about listeriosis in its Nutrition Fact Sheet: A Quick Consumer Guide To Safe Food Handling. It states that *Lm* has been found in unpasteurised milk, imported soft cheese, hot dogs, luncheon meats and spreads. It suggests a number of control strategies including recommendations to avoid soft cheeses such as Mexican style cheeses.

A more active approach to risk communication has been the Abuela ('Grandmother') project, which was designed and implemented in the state of Washington (Bell et al. 1999). After an outbreak of *Salmonella* Typhimurium associated with raw milk soft cheese in 1997, a multi-agency intervention was developed. An element that sets this approach apart from others is its inclusion of the local Hispanic population in the design of the intervention. The centre piece of the intervention focused on workshops, which introduced a pasteurised milk *queso fresco* recipe. In addition to the workshops, a mass media campaign about the risks of consuming raw milk cheese was implemented, and newsletter articles warning about the risks of selling or giving away raw milk were aimed at dairy producers. The goal of the intervention was to reduce the incidence of *Salmonella* while maintaining the traditional, nutritious food in the Hispanic diet.

Dairy scientists at Washington State University modified a local *queso fresco* pasteurised milk recipe to inhibit undesirable microbial growth, increase shelf life, and improve the ease of preparation. Members of the Hispanic community tested the recipe to insure that flavour and texture were satisfactory. Fifteen older Hispanic women, abuelas or grandmothers, from the Hispanic community were recruited to participate in workshops that provided training on safe production of traditional cheeses (Bell et al. 1999; Medeiros et al. 2001). Each abuela educator signed a contract to teach at least 15 additional members in the community on how to safely make *queso fresco* from pasteurised milk. Pamphlets containing instructions for making the cheese were also distributed at these workshops and made available to others. A mass media campaign followed the workshops with radio public service announcements, newspaper articles, posters, and medical alerts to physicians to raise awareness of the risks associated with raw milk and raw milk soft cheese consumption. Data from the Abuela project indicate that this intervention had a significant impact on behaviours related to making Hispanic soft cheese with a significant number of participants subsequently using pasteurised rather than unpasteurised milk to make their cheese. However, the impacts from this project were not sustained over time, probably due to the transient nature of the Hispanic population in this area, and the fact that the mechanisms were not in place for community sustainability of the intervention. Although this intervention was a success in many respects, additional strategies are required.

Conclusions

The current approach of the US government (illness prevention) to *Lm* in raw milk soft cheese is rather simple in that the management strategy is rather routine, the appropriate instruments for risk reduction are rather 'traditional', and there is no need for stakeholder participation or revision of current legislation. However, a more thorough analysis using the IRGC framework illustrates that looks can be deceiving. This review of risk governance of *Lm* in raw milk soft cheese shows that there is much uncertainty as well as ambiguity with the *status quo*. First, an analysis of the risk assessments reveals that a fair amount of uncertainty exists in risk assessment models. One particular issue of scientific debate is the establishment of an appropriate *Lm* tolerance level. What initially appears to be simple (tolerance defined as absence of *Listeria in 25 g*), actually becomes rather complicated (tolerance defined as <1 cfu in 25 g at manufacture or as <100 cfu/g at consumption). Similarly, other debates surround the dose response and the amount of cheese that needs to be eaten to contract listeriosis. In hindsight, risk assessments on cheese tend to have a relatively high level of uncertainty in their models. Second, while the current legislation suggests that raw milk soft cheese cannot be produced safely, it appears that other countries have achieved a similar level of food safety with less restrictive regulations. Third, debates surrounding values are ambiguous. The two frames – illness prevention and consumer sovereignty – identified in this research represent different values. For example, illness prevention focuses on consumer protection whereas consumer sovereignty focuses on consumer choice.

The IRGC framework was particularly useful in conceptualising the range of stakeholders throughout the commodity system, as well as the range of positions that any one group may or may not have with respect to raw milk soft cheese governance. It is possible that each stakeholder group may devise a very different range of governance options not otherwise considered. Moreover, it opens up the possibility to question the ethics of any one decision. The use of the model also draws our attention to how little is known about risk decision-making at the regulatory level! The IRGC framework sheds light on what we believe may be a set of entrenched organisational dynamics. At the federal level these dynamics appear to lack foresight and reflexivity, as well as institutional memory. As a consequence, it is unclear why such a strict reactionary standard was created for *Lm* in raw milk soft cheese (aside from reaction to raw milk outbreaks), and how this standard was justified (i.e., if *Lm* is ubiquitous, how it could be determined to be an adulterant?). Understanding how and why these decisions were made is one step in the process of developing efficient and effective statutory and regulatory standards. Another advantage of the IRGC framework is that it makes risk governance more transparent. It forced us to conceptualise risk management options not currently employed, identify inconsistencies across regulations and risk management procedures, identify uncertainty in risk assessment and risk perception data, incorporate social science and public perception into risk management, to identify contentious issues surrounding raw milk soft cheese, identify stakeholders, and provide an understanding of different stakeholder views.

In conducting this project, we also encountered several problems using the framework. In particular, we found that the definitions of the concepts are not well defined. Second, we chose a pathogen/commodity pair that is, relatively speaking, heavily regulated. Regardless of the issues we raised in our analysis, it is unlikely that there will be any major changes to the current US regulations. Third, we found the intent of the framework to be unclear (i.e., the deliverables). Is the purpose: (1) to develop a risk communication strategy?; (2) to recommend policy?; (3) to be a guideline during the policy formation process?; and/or (4) to develop criteria from which to evaluate policy? Because of these questions we were unsure if the risk judgement and management sections were to focus on what had already been done or if the intent was for us to provide our own analysis of what had been done and to raise alternative management options.

It seems intuitive that a framework incorporating social science would be an improvement to existing risk governance procedures and decision-making. However, examining concern assessment with anecdotal evidence may lead to a position that a particular frame has more public support than it actually does. In addition to public concern assessments, it would be beneficial to understand the concerns of risk professionals and regulators. In the case of the US, regulatory proposals or changes are usually reactive and involve placement in the federal register where the public and interest groups may comment if they are aware of the issues. The system does not actively engage public opinion. Science based regulation is the key phrase, so why should experts engage the opinion of the public when they are the experts who know best? In other words, should regulations be based on science or values or a combination of both? This seems to be a fundamental question underlying the IRGC framework.

References

Altekruse, S.F., Street, D.A., Fein, S.B. and Levy, A.S., 1995, Consumer knowledge of foodborne microbial hazards and food-handling practices, *Journal of Food Protection* **59**, 287–294.

Altekruse, S.F., Timbo, B.B., Mowbray, J.C., Bean, N.H. and Potter, M.E., 1998, Cheese-associated outbreaks of human illness in the United States, 1973 to 1992: Sanitary manufacturing practices protect consumers, *Journal of Food Protection* **61**, 1405–1407.

Bell, R.A., Hillers, V.N. and Thomas, T.A., 1999, The Abuela project: Safe cheese workshops to reduce the incidence of *Salmonella typhimurium* from consumption of raw-milk fresh cheese, *American Journal of Public Health* **89**, 1421–1424.

Bell, R., Hillers, V. and Thomas, T., 2000, *The Abuela Project: A Community-Based Food Safety Intervention.*

Bemrah, N., Sanaa, M., Cassin, M.H., Griffiths, M.W. and Cerf, O., 1998, Quantitative risk assessment of human listeriosis from consumption of soft cheese made from raw milk, *Preventive Veterinary Medicine* **37**, 129–145.

Bolton, L.F. and Frank, J.F., 1999, Defining the growth/no-growth interface for Listeria monocytogenes in Mexican-style cheese based on salt, pH, and moisture content, *Journal of Food Protection* **62**, 601–609.

Brewer, M.S., Sprouls, G.K. and Russon, C., 1994, Consumer attitudes towards food safety issues, *Journal of Food Safety* **14**, 63–76.

Center for Food Safety and Applied Nutrition, CFSAN, 1992, The Bad Bug Book: Food-borne Pathogenic Microorganisms and Natural Toxins Handbook, Retrieved March 2, 2006: http://vm.cfsan.fda.gov/~mow/intro.html.

Centers for Disease Control and Prevention, 2001, Outbreak of listeriosis associated with homemade Mexican-style cheese – North Carolina, October 2000–January 2001, *Morbidity and Mortality Weekly Report* **50**, 560–562.

Chen, Y., Ross, W.H., Scott, V.N. and Gombas,D.E., 2003, Listeria monocytogenes: Low levels equal low risk, *Journal of Food Protection* **66**, 570–577.

Cody, S.H., Abbott, S.L., Marfin, A.A., Schulz, B., Wagner, P., Robbins, K., Mohle-Boetani, J.C. and Vugia, D.J., 1999, Two outbreaks of multidrug-resistant Salmonella Serotype Typhimurium DT104 infections linked to raw-milk cheese in Northern California, *Journal of the American Medical Association* **28**, 1805–1810.

de Valk, H., Jacquet, C., Goulet, V., Vaillant, V. Perra, A., Simon, F., Desenclos, J.C. and Martin, P., 2005, Surveillance of listeria infections in Europe, *Eurosurveillance* **10**, 251–255.

Dittus, K.L. and Hillers, V.N., 1996, Attitudes about the nutritional benefits and pesticide-exposure risks from fruit and vegetable consumption, *Family and Consumer Sciences Research Journal* **24**, 406–421.

Donnelly, C., 1989, *Listeria Safety and Government Regulation of Cheese*.

Donnelly, C.W. 2001, Factors Associated with Hygienic Control and Quality of Cheeses Prepared from Raw Milk: A Review, IDF Bulletin 369:16-27.

Donnelly, C., 2005, The pasteurization dilemma, in: P.S. Kindstedt and the Vermont Cheese Council (eds.), *American Farmstand Cheese: The Complete Guide to Making and Selling Artisan Cheeses*, Chelsea Green Publishing, White River Junction, VT, pp. 173–196.

Economic Research Service, 2000, *Economics of Foodborne Disease: Listeria Monocytogenes*, Briefing Room, Retrieved March 1, 2006, http://www.ers.usda.gov/briefing/foodborneDisease/listeria/.

European Commission Health and Consumer Protection Directorate-General (ECHCP), 1999, *Opinion of the Scientific Committee on Veterinary Measures Relating to Public Health on Listeria Monocytogenes*, edited by M. o. S. C. I. Scientific Health Opinions: European Commission Health & Consumer Protection Directorate-General, 44 pp.

European Union, 2005, *Commission Regulation (EC) No. 2073/2005 of 15 November 2005 on Microbiological Criteria for Foodstuffs*, edited by The Commission of the European Communities, 26 pp.

Fletcher, J., 2006, The myths about raw-milk cheese, *Specialty Food Magazine*, Retrieved March 1, 2006, http://www.specialtyfood.com/do/news/ViewNewsArticle?id=1841.

Food Safety Network, 2003, *Health Warning Issued for Illegally Sold White Cheese*, Retrieved October 4, 2006, http://archives.foodsafetynetwork.ca/fsnet/2003/8-2003/fsnet_august_26.htm#CHEESE).

Food Safety Network, 2004a, *Listeriosis, Neonatal*, Retrieved October 4, 2006, http://archives.foodsafetynetwork.ca/fsnet/2004/3-2004/fsnet_march_8.htm#story0.

Food Safety Network, 2004b, *Listeriosis, North Carolina*, Retrieved October 4, 2006, http://archives.foodsafetynetwork.ca/fsnet/2004/6-2004/fsnet_june_4-2.htm#story3.

Food Safety Network, 2005, *City Fears Potential Outbreak of Bacterial Illness; Health Officials See 6 Listeriosis Cases in a Month*, Retrieved October 4, 2006, http://archives.foodsafetynetwork.ca/fsnet/2005/8-2005/fsnet_aug_23-2.htm#story0.

Food Standards Australia New Zealand (FSANZ), 2005. *Final Assessment Report Application A499 to Permit the Sale of Roquefort Cheese*, Retrieved October 24, 2006, http://www.foodstandards.gov.au/_srcfiles/A499_Roquefort_FAR_FINALv2.pdf#search=%22roquefort%22.

Frenzen, P. and Buzby, J., 2000, ERS updates foodborne illness costs, *Food Safety* **23**, 1.3, Retrieved March 1, 2006, (http://www.mindfully.org/Food/Foodborne-Illness-Costs-USDA.htm.

Frewer, L.J., Shepherd, R. and Sparks, P., 1994, The interrelationship between perceived knowledge, control and risk associated with a range of food-related hazards targeted at the individual, other people and society, *Journal of Food Safety* **14**, 19–40.

Genigeorgis, C., Carnicui, M., Dutulescu, D. and Farver, T.B., 1991a, Growth and survival of *Listeria monocytogenes* in market cheeses stored at 4 to 30°C, *Journal of Food Protection* **54**, 662–668.

Genigeorgis, C., Toledo, J.H. and Garayzabal, F.J., 1991b, Selected Microbiological and chemical characteristics of illegally produced and marketed soft Latino-style cheeses in California, *Journal of Food Protection* **54**, 598–601.

Glass, K.A., Prasad, B., Schlyter, J.H., Uljas, H.E., Farkye, N.Y. and Luchansky, J.B., 1995, Effects of acid type and ALTATM2341 on *Listeria monocytogenes* in a Queso Blanco type of cheese, *Journal of Food Protection* **58**, 737–741.

Gombas, D.E., Chen, Y., Clavero, R.S. and Scott, V.N., 2003, Survey of Listeria monocytogenes in ready-to-eat foods, *Journal of Food Protection* **66**, 559–569.

Headrick, M.L., Korangy, S., Bean, N.H., Angulo, F.J., Altekruse, S.F., Potter, M.E. and Klontz, K.C., 1998. The epidemiology of raw milk-associated foodborne disease outbreaks reported in the United States, 1973 through 1992, *American Journal of Public Health* **88**, 1219–1221.

Hillers, V., Bell, R. and Thomas, T., 2002, The Abuela Project: A community based food safety intervention involving Queso Fresco, a raw-milk cheese, in: *Thinking Globally – Working Locally: A Conference on Food Safety Education*, FSIS/USDA, Orlando, FL.

Hitchins, A.J., 2003, Detection and Enumeration of Listeria monocytogenes in Foods, in: US Department of Health and Human Services and US Food and Drug Administration Center for Food Safety & Applied Nutrition (eds.), *Bacteriological Analytical Manual*.

Hopkins, R.S., Jajosky, R.A., Hall, P.A., Adams, D.A., Connor, F.J., Sharp, P., Anderson, W.J., Fagan, R.F., Aponte, J.J., Jones, G.F., Nitschke, D.A. Worsham, C.A., Adekoya, N. and Chang, M., 2005. *Summary of Notifiable Diseases – United States, 2003*, Division of Public Health Surveillance and Informatics, Epidemiology Program Office, Coordinating Center for Health Information and Service, Centers for Disease Control and Prevention, Atlanta, GA.

Institute of Food Science and Technology (IFST), 1998, *IFST: Current Hot Topics. Food Safety and Cheese*, Retrieved March 14, 2006, http://www.ifst.org/hottop15.htm.

International Life Sciences Institute (ILSI), 2004, *Achieving Continuous Improvement in Reductions in Foodborne Listeriosis – A Risk Based Approach*, Report from the ILSI Institute Working Group on Listeria monocytogenes in Foods.

Johnson, E.A., Nelson, J.H. and Johnson, M., 1990, Microbiological safety of cheese made from heat-treated milk, Part II, *Journal of Food Protection* **53**, 519–540.

Knight, A.J. and Warland, R., 2005, Determinants of food safety risks: A multi-disciplinary approach, *Rural Sociology* **70**, 253–275.

Kozak, J., Balmer, T., Byrne, R. and Fisher, K., 1996, Prevalence of Listeria monocytogenes in foods: Incidence in dairy products, *Food Control* **7**, 215–221.

Lay, J., Varma, J., Marcus, R., Jones, T., Tong, S., Medus, C., Samuel, M., Cassidy, P., Hardnett, F., Barden, C. and EIP FoodNet Working Group, 2002, Higher incidence of Listeria infections among Hispanics: FoodNet, 1996–2000, in: *International Conference on Emerging Infectious Diseases*, Atlanta, GA.

Lin, C.T. Jordan, 1995, Demographic and socioeconomic influences on the importance of food safety in food shopping, *Agricultural and Resource Economics Review* **24**, 190–198.

Lin, C.J., Jensen, K.L. and Yen, S.T., 2005, Awareness of foodborne pathogens among US consumers, *Food Quality and Preference* **16**, 401–412.

Linnan, M.J., Mascola, L., Lou, X.D., Goulet, V., May, S., Salminen, C., Hird, D.W., Yonekura, M.L., Hayes, P., Weaver, R. et al., 1988, Epidemic listeriosis associated with Mexican-style cheese, *New England Journal of Medicine* **319**, 823–828.

Mead, P.S., Slutsker, L., Dietz, V., McCaig, L.F., Bresee, J.S., Shapiro, C., Griffin, P.M., and Tauxe, R.V., 1999, Food-related illness and death in the United States, *Emerging Infectious Diseases* **5**, 607–625.

Medeiros, L., Hillers, V.N., Kendall, P. and Mason, A., 2001, Food safety education: What should we be teaching to consumers?, *Journal of Nutrition Education* **33**, 108–113.

Oldways, n.d., *Food Issues Think Tank*, Retrieved October 4, 2006, http://www.oldwayspt.org/.

Pagotto, F., Ng, L.K., Clark, C., Farber, J. and Canadian Public Health Laboratory Network, 2006, Canadian Listeriosis reference service, *Foodborne Pathogens and Disease* **3**, 132–137.

Peeler, J.T. and Bunning, V.K., 1994, Hazard assessment of *Listeria monocytogenes* in the processing of bovine milk, *Journal of Food Protection* **57**, 689–697.

Priest, S.H., 2000, US public opinion divided over biotechnology?, *Nature Biotechnology* **18**, 939–942.

Pritchard, T., 2005, Ensuring safety and quality I: Hazard Analysis Critical Control Point and the cheesemaking process, in: P. S. Kindstedt and Vermont Cheese Council (eds.), *The Complete Guide to Making and Selling Artisan Cheeses*, Chelsea Green Publishing, White River Junction, VT, pp. 139–152.

Public Health Seattle and King County, 2002, Listeriosis in pregnant women linked to consumption of queso fresco, *Epi-Log Newsletter* **42**.

Reed, B.A. and Bruhn, C., 2003, Sampling and farm stories prompt consumers to buy specialty cheeses, *California Agriculture* **57**, 76–80.

Rimal, R.N. and Adkins, A.D., 2003, Using computers to narrowcast health messages: The role of audience segmentation, targeting, and tailoring in health promotion, in: T.L. Thompson, A.M. Dorsey, K.I. Mller and R. Parrott (eds.), *Handbook of Health Communication*, Lawrence Erlbaum Associates, Mahwah, NJ.

Sagoo, S.K. and Little, C.L., 2004, *European Commission Co-ordinated Programme for the Official Control of Foodstuffs for 2004: Microbiological Examination of Cheeses Made from Raw or Thermised Milk from Establishments of Production and Retail in the United Kingdom.*

Sanaa, M., Coroller, L. and Cerf, O., 2004, Risk assessment of listeriosis linked to the consumption of two soft cheeses made from raw milk: Camembert of Normandy and Brie of Meaux, *Risk Analysis* **24**, 389–399.

Schmidt, D., n.d., *Food Safety: A Risk Approach to Informing the Public.*

Shaw, S.A., 2000, Cheesy does it: Getting your hands on great cheese in the US means circumventing an archaic FDA regulation, *Travel & Food*, Retrieved March 2, 2006, http://www.salon.com/travel/food/feature/2000/01/28/cheese/.

Shiferaw, B., Yang, S., Cieslak, P., Vugia, D., Marcus, R., Koehler, J., Deneen, V., Angulo, F. and The Foodnet Working Group, 2000, Prevalence of high-risk food consumption and food – Handling practices among adults: A multistate survey, 1996–1997, *Journal of Food Protection* **63**, 1538–1543.

Stein, N., 2001, Would you like cheese with that? You may have to settle for Kraft. Nicholas Stein questions the FDA's investigation into whether unpasteurised cheese poses a health threat, *Fortune*, April 2, 199–200.

Stewart, L.K., 2002, Listeriosis in pregnant women linked to consumption of Queso Fresco, *Communicable Disease and Epidemiology News* **42**, 2.

Teng, D., Wilcock, A. and Aung, M., 2004, Cheese quality at farmers markets: Observation of vendor practices and survey of consumer perceptions, *Food Control* **15**, 579–587.

Teuber, M., 2000, Fermented milk products, in: B.M. Lund, T.C. Baird-Parker and G.W. Gould (eds.), *The Microbiological Safety and Quality of Food*, Aspen Publishers, Gaithersburg, MD.

Todd, E., 2007, Listeria: risk assessment, regulatory control and economic impact, in: E.T. Ryser and E.H. Marth (eds.), *Listeria, Listeriosis, and Food Safety*, Marcel Dekker, New York, Chapter 17.

US Food and Drug Administration Center for Food Safety and Applied Nutrition, US Department of Agriculture Food Safety and Inspection Service, and Centers for Disease Control and Prevention, 2003, *Quantitative Assessment of the Relative Risk to Public Health from Foodborne Listeria monocytogenes Among Selected Categories of Ready-to-Eat Foods*, United States Department of Agriculture, pp. 1–523.

US Food and Drug Administration Center for Food Safety and Applied Nutrition, and Centers for Disease Control and Prevention, 2005. *Current FDA Activities Related to the Listeria monocytogenes ActionPlan*, Retrieved May 24, 2007, http://www.foodsafety.gov/~dms/lmr2pla2.html.

Williams, P.R.D. and Hammitt, J.K., 2001, Perceived risks of conventional and organic produce: Pesticides, pathogens, and natural toxins, *Risk Analysis* **21**, 319–330.

World Health Organization and Agriculture Organization of the United Nations (WHO/FAO), 2004, *Risk Assessment of Listeria monocytogenes in Ready-to-Eat Foods*, Technical Report, Geneva, Switzerland.

Woteki, C.E. and Kineman, B.D., 2003, Challenges and approaches to reducing foodborne illness, *Annual Review of Nutrition* **23**, 315–44.

Chapter 10
Nagara River Estuary Barrage Conflict

Norio Okada[1], Hirokazu Tatano[1] and Alkiyoshi Takagi[2]
[1]*Disaster Prevention Research Institute, Kyoto University, Japan*
[2]*Faculty of Engineering, Gifu University, Japan*

Introduction

This case study is an application of IRGC's risk governance framework to an actual water resources management problem which challenged the disaster risk governance system in Japan, the construction of the Nagara River Estuary Barrage. It represents an example of a problem in which decision-makers were faced with difficult tradeoffs between protection of public safety and important water resources on the one hand, and concerns about adverse socio-economic and environmental impacts of the barrage on the other. This problem also illustrates the evolving nature of conflicts over time, where the values that dominated the decisions in the early planning stages were not those that drove public opinion toward the end.

The Nagara River Estuary Barrage was planned at the mouth of the Nagara River by the Ministry of Construction of the Japanese Government in order to develop water resources and mitigate flood disasters. The Nagara river runs through the Nagoya metropolitan region which is the third largest metropolitan area in population and a very important industrial area for Japanese economy, especially, automobile and machinery industries. However, local fishermen and eventually became opposed to the plan, and were joined in their opposition by an emerging new group of environmentalists. Crucial conflicts occurred among government officials, local people and societal groups holding diverse values. The conflicts have become compounded, evolved and lasted from 1968 to now, though some tentative resolution seems to have been reached.

The chapter consists of three parts: the first part provides the reader with the history of this project and related conflicts; the second part is a retrospective analysis of the Nagara River Estuary Barrage conflict within the context of the IRGC risk governance framework; and the third provides our conclusions about the applicability and limitations of the framework and our recommendations for the future.

O. Renn and K. Walker (eds.), Global Risk Governance: Concept and Practice Using the IRGC Framework, 221–230.

Fig. 1 Nagara River Estuary Barrage.

The Nagara River Estuary Barrage Conflict

Nagara River Estuary Barrage

The Nagara River flows out of the mountains of Gifu Prefecture situated in central Japan. For roughly a half of its 136-kilometre course, it runs southwards through narrow valleys. Then, flowing through high dykes, it makes its way across the Nobi Plain before flowing into Ise Bay (see Figure 1). Until recently, it was one of the last major rivers without a barrage in Japan.

In the 1960s, Japan was in a period the Japanese refer to as 'Kodo Keizai Seicho Ki' or the 'Era of Rapid Growth of Japanese Economy'. Japan's Water Resources Development Agency (WRDA) was established in 1962. It was anticipated that industrial water shortage could become a bottleneck for continued economic growth. As a result, major water resources development plans were made all over Japan during this period. The Kiso River System Master plan, which was officially completed in 1968, included development of the Nagara River Estuary Barrage. It was not until 1995 that the barrage project was finally completed.

Purpose of the Barrage

The Nagara River Estuary Barrage was designed for two purposes, flood control, and protection of water supply. It enables implementation of large-scale dredging projects which in turn guide safe water flow of the Nagara River in case of flooding. The barrage also prevents upstream intrusion of saltwater, thereby making it possible to make extra fresh water available as domestic water and industrial water supplies for Aichi Prefecture, Mie Prefecture and Nagoya City. Table 1 shows the progress of revising the plan.

Table 1 Water resources development plan in the Kiso River System.

	Date	Planned Volume of Water Supply	Facilities
Master Plan	1968.10.15	73 m^3/s^{*1}	Nagara River Estuary Barrage
Total Revision	1968.10.15	121 m^3/s^{*2}	Three other dams development plans are included
Total Revision	1993.3.31	34 m^3/s^{*3}	

	Target Year	Municipal Water	Industrial Water	Irrigation	Total=Target Water Supply	Capacity of Facility in Target Water Supply
*1	1985	25	42	6	73	65
*2	1995	40	60	22	121	83
*3	2000	14	6	14	34	49

Evolution of Conflict: Changes in Issues and Key Stakeholders

The project faced a long time lag between its inception in 1968 and completion in 1995. A number of changes occurred in both the original economic incentives for the project and in the societal values surrounding such projects in general. Industrial water demand dropped due to both increased use of recycled water by heavy industries and structural shifts in Japanese industry, particularly from that of materials production (e.g. iron and chemicals) to manufacturing (e.g. automobiles and electronic equipment). Though the plan remained unchanged for a long period of time, this decline in industry's water demand was paralleled by declines in both municipal and agricultural water demand; ultimately, the planned supply from the Nagara River Estuary Barrage declined accordingly (see Table 1). In the same time period, the general public began to place greater emphasis on sustainable society than on economic development. Issues like the preservation of nature increasingly gained public awareness and support, and stakeholders requested they be given a higher priority than before in public project development.

The conflict surrounding the construction of the Nagara River Estuary Barrage fell into roughly three stages, discussed below. A complete chronology of the barrage planning, construction and conflict is shown in Table 2.

First Stage: 1973 to 1981

The first stage was marked by a lawsuit against the WRDA, brought by an opposition group, demanding that the court suspend construction of the barrage. The key stakeholders bringing the suit at this stage were inhabitants in the area at the mouth of the river and fishermen. Their key concern was whether the barrage would really

Table 2 Historical development of the Nagara River Estuary Barrage Conflict.

FY1960	Preliminary survey started.
FY1963-FY1967	The Estuary Resource Survey Team for the Three Kiso Rivers ('KST') conducted survey.
FY1968	Project implementation plan survey started.
Oct. 1968	The Basic Plan for Water Resource Development in the Kiso River Water System (the 'basic plan') was decided.
FY1971	Construction began.
Dec. 1971	Policy on implementation announced.
Mar. 1973	Basic plan (partial amendment).
Jul. 1973	Implementation plan approved.
Dec. 1973	Action demanding suspension of construction filed (first suit).
Sep. 1976	Riverbank collapsed at Anpachi Town, Gifu Pref.
Mar. 1981	Suit demanding suspension of construction abandoned (first suit).
Apr. 1982	Suit demanding suspension of construction filed (second suit).
Feb. 1988	All fishermen's cooperative associations agreed on the start of the work.
Mar. 1988	Construction of the main structure of the barrage began.
Dec. 1988	Change of policy toward implementation indicated.
Feb. 1989	Change of policy toward implementation approved.
Mar. 1992	Additional survey report published.
Apr. 1992	Technical report published.
Mar. 1993	Basic plan (complete revision).
Apr. 1994	Nagara Estuary Barrage Survey conducted. (Finished Mar. 1995)
Jul. 1994	Intermediate appeal filed on the case of suspension of construction (Nagoya High Court).
Mar. 1994-Apr.	Round table conference held (8 times).
Mar. 1995	Change of implementation plan approved. Facility management policy indicated. Facility management regulation approved. Survey report submitted to Construction Minister Nosaka.
Apr. 1995	Management started.
May 1995	Construction Minister Nosaka: 'Full-scale operation from May 23'.
Jul. 6 1995	All gates began operating.
FY1995-FY1999	Nagara River Estuary Barrage Monitoring Committee (12 times).
Jul. 1997	Dredging in the mound section started.
Apr. 1998	Diversion to Nagara water supply system started; Diversion to Hokuchusei water supply system started.
Dec. 1998	Court-of-appeal decision on action demanding suspension of construction (Nagoya High Court).
Sept. 1999	5900 m^3/s of discharge observed during Typhoon No. 16. (Largest discharge after regular operation started).
Apr. 2000	Transition to follow-up survey.

contribute to flood control. After the suit was filed, industrial water demand dropped down. Shortly afterward, the appeal was dismissed.

Second Stage: 1982 to 1998

The second stage of the conflict was initiated by a second lawsuit against the WRDA, again demanding suspension of construction of the barrage. The major stakeholders against the barrage were the inhabitants in upper and middle of the Nagara river basin whose key concerns were the environmental impacts of the dam. On the other side of the debate were the WRDA and citizens in flood prone areas. The WRDA won this lawsuit. It was quite likely that the court made that judgement by overall assessment of what was then the public consensus (i.e., support of infrastructure development).

Third Stage 1999 to Present

The third has been characterised by use of arbitration and Round Table conferences. A key concern is preservation of nature put forward by nature enthusiasts living outside of the basin. The stakeholders who continue to support the project include the WRDA and citizens who live in flood prone areas protected by the barrage.

Retrospective Analysis Using the IRGC Framework

The Nagara River Estuary Barrage conflict has spanned over three decades and many economic, social and environmental changes since it was first proposed. Our analysis is therefore a retrospective assessment of how the conflict was managed and how the IRGC risk governance framework might have helped improve the process. The organisation follows the four phases of the IRGC framework: 'Pre-Assessment', 'Risk Appraisal', 'Tolerability and Acceptability Judgement', and 'Risk Management'.

Pre-Assessment

It seems quite likely that the problems faced by this water resources development and management project in Japan had their beginning in what IRGC calls the 'pre-assessment' phase. A principal purpose of the pre-assessment phase of the IRGC framework is to develop a common frame or basis for discussing the risks. This requires a 'common understanding of the risk issue(s) being addressed ... or to raise awareness amongst the parties of the differences in what is perceived as risk' (IRGC 2005). In retrospect, it appears that the Ministry of Construction, the WRDA and others responsible for this project started to deal with this large-scale construction project within a traditional frame – a top-down public sector approach typical for that period of Japan's development. Within this frame, the risks that were the focus of the project were (1) flooding and (2) insufficient water supply to meet what were then growing fresh water demands from the industrial, agricultural and municipal sectors.

The question is then: Had the authorities conducted a pre-assessment like that recommended by IRGC could they have avoided the conflicts that followed? Our answer is probably, 'No' because nobody could have anticipated the two major changes in the basic context for the project: (1) the decline in the industrial demand for water supply and (2) the crucial shift in the public's preference from economic development to sustainable growth and protection of the environment.

However, implementation of the proactive practices recommended in the pre-assessment phase of the IRGC framework might have helped avoid sowing the seeds of mutual distrust and suspicions that later gave rise to the larger conflict. Through the use of scientific workshops appealing to the public, early framing, initial screening of possible issues and discussion of the 'major assumptions, conventions and rules for assessing risk', citizens could have benefited from better access to information. The development of 'early warning' or monitoring systems whereby emerging problems could be identified and discussed as the project went forward might also have provided greater opportunity for communication. We presume that it was the lack of such communication that eventually led opponents to file the first lawsuit. Therefore, appropriate and timelier implementation of the IRGC pre-assessment phase might have helped the conflict take a different path, leading to much faster resolution.

Risk Appraisal

A complete risk appraisal process, one which included both the risk assessment and concern assessment components of the IRGC framework, was also not a part of the early planning phases of the Nagara River Estuary Project. The reality is that the mindset of the Ministry of Construction, the WRDA, and other agencies responsible for the project and their framing of the problem as a large public works project largely limited the scope of potential and emerging risks they were willing to address openly and proactively. They thus were more likely to dismiss public concerns. The authorities appeared to do a better job of listening to the concerns of more 'direct interest groups', like fishermen, and other potential stakeholders who could claim financial compensation for damages resulting from the construction of the barrage. Little care seemed to be taken to address public concerns about 'lack of participation' or the concerns of emerging 'external citizen groups' who were highly sensitised to any potential loss of habitat and of endangered species living in the Nagara River.

It seems likely that had a risk assessment and, in particular, a 'concern assessment', been conducted, they could have made the situation much better. The responsible Ministry, agencies and citizens could have reached better understanding of the issues and of each others' concerns before the plan was decided and put into practice.

Nonetheless, it is important to acknowledge that it would have remained difficult to overcome the fundamental split in concerns about the risks and values held by proponents and opponents of the barrage. One hopes however, that the public sector would have become more sensitised to the concerns of 'minority' groups who at that time had virtually no access to open public debate. The IRGC's alternative approach to governance might have changed the dismal mood of mistrust, which contributed to years of bitter conflict.

Tolerability and Acceptability Judgement

Given that the early planning for the Nagara River Estuary Barrage did not include the equivalent of the pre-assessment and risk appraisal phases of the IRGC framework, it is not surprising that explicit implementation of a 'tolerability and acceptability judgement' phase was also missing. The basic groundwork – identification of the full array of risks and concerns from a broader group of stakeholders – had not been laid. Consequently, informed discussions about the tolerability or acceptability of these issues could not be properly conducted.

Risk Management

The problems encountered in the risk management phase of the Nagara River Estuary Barrage project were perhaps rooted in the basic framing of the problem earlier on, as noted in our discussion of the 'Pre-assessment phase'. In terms of the 'risk characteristics and their implications for risk management' (see Table 6 in Chapter 1 of this volume), the Ministry of Construction, WRDA, and other public agencies primarily engaged in the project had mistakenly assumed that the risks could be classified as 'simple risk problems'. As a result, they relied on what IRGC's framework refers to as 'routine-based management strategies' for a problem that turned out to be much more complex and uncertain.

The Cyclic Nature of the IRGC Risk Governance Framework; the Risk Management Escalator and Stakeholder Involvement

The IRGC framework on risk governance envisions risk governance as a cyclic process in which a project might go through a number of stages or cycles depending on the initial characterisation of the risks involved and on the evolution of knowledge about the risks over time. It represents this concept in Figure 2, the Risk Management Escalator and Stakeholder Involvement, in which risks may fall into one of four broad categories, 'Simple', 'Complexity Induced', 'Uncertainty Induced', and 'Ambiguity Induced'.

In retrospect, the entire process of evolving conflict surrounding the Nagara River Estuary Barrage could have been divided into three stages. The timing of each stage shift could have been an appropriate time for the core governmental agents to climb up to a higher step of the 'Risk Management Escalator and Stakeholder Involvement', thus proceeding to another round of the cyclic process.

Even if the Nagara River Estuary Barrage project had been incorrectly characterised as a 'Simple' risk problem, if it had properly gone through the IRGC risk governance process (i.e. from 'Pre-assessment', 'Risk Appraisal', 'Tolerability and Acceptability Judgement', then to 'Risk Management'), the whole picture might have been totally different. Decision-makers might have become aware of the emerging shift of public values, become more sensitised to people's concerns, and therefore might eventually have reached a much better outcome.

However, it is quite likely that what appeared as a typical water resources management project at the outset, a 'Simple' risk problem actually should have been considered at least an 'Uncertainty-Induced' risk problem. It later evolved into a more 'Ambiguity-Induced' risk problem as the differences in public values emerged. If the relevant stakeholders had been able to engage in thoughtful discourse at an earlier and more appropriate time, the conflict would at a minimum have been better handled, if not finally settled down. The proposal late in the conflict to develop a system for publicly monitoring the impact of the completed barrage on the river ecosystem is evidence of a good outcome from constructive dialogue between stakeholders. This proposal was ultimately implemented.

Discussion and Conclusions

Our study of the Nagara River Estuary Barrage shows that IRGC's proposed risk governance framework can provide an effective process through which potential users can critically review an actual conflict resolution problem, identify the governance lapses that occurred, and reflect on how these lapses may have contributed to a prolonged conflict. Stakeholders on all sides of the issue can reflect on how the process could have been handled differently and achieved outcomes more acceptable to all involved.

The real challenge to the IRGC is to extend this methodology to current public projects or to prospective future projects. In order to do so, IRGC will need to address some of the distinctive difficulties in interpretation and practical application of the framework that we have identified in the course of our retrospective analysis. We raise the following questions:

1. *At which step of the 'Risk Management Escalator and Stakeholder Involvement'* (Figure 2) should a project start? The answer to this question may not be easy. In reality, stakeholders often have only limited information and knowledge to identify whether the risk is in fact a 'simple' one more 'complex', 'uncertain', or 'ambiguous'. A safer strategy might be to start with the assumption of an 'Uncertainty-Induced' risk problem which would ensure involvement of a

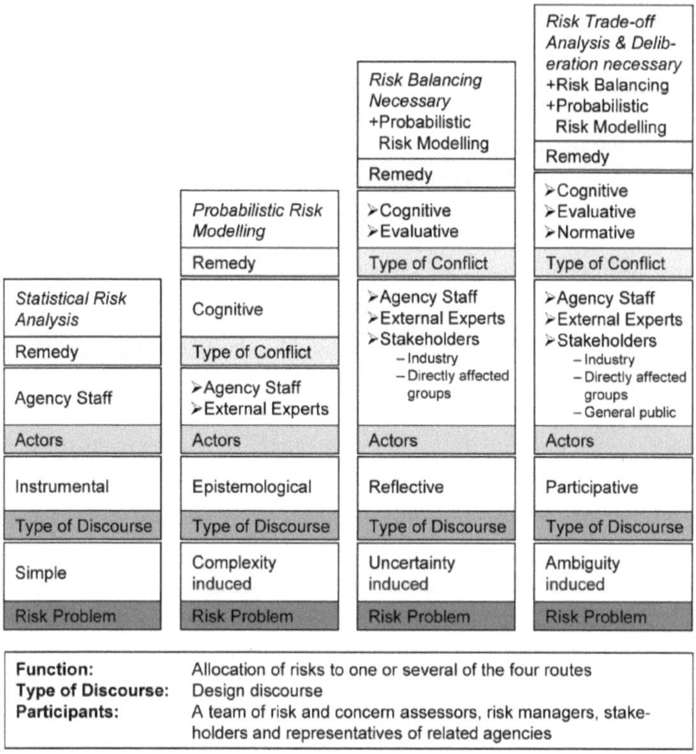

Fig. 2 The risk management escalator and stakeholder involvement (from simple via complex and uncertain to ambiguous phenomena).

broader representation of stakeholders and identification of the key risks and concerns early in the process.

2. *How can we set up sound institutional practices or guidelines that will help achieve the objectives of the IRGC's Risk Management Escalator?* Our answer is that IRGC needs to propose the tools that might help institutionalise (or otherwise realise) the theoretical risk governance goals (e.g., development of laws or ordinances for public participation, release of public information, for accountability and for transparency).

3. *What happens if conflict is considered highly adversarial or is rooted in a fundamental split in values, and thus apparently impossible to resolve?* Our tentative answer is, 'It depends'. There may be some cases where IRGC's assumption of a context in which dialogue can occur simply does not apply (e.g. religious conflicts, highly politicised international issues). If there is no room for compromise, and rational and logical reasoning is not occurring, the IRGC's methodology will probably not work until the situation becomes more sensible.

4. *Has this case study provided adequate insight into typical risk governance issues encountered in disaster risk management?* Our response is, 'Not yet overall'. The type of problem we have addressed in this case study is characterised by potential tradeoffs between disaster risk reduction on the one hand, and other public management goals such as regional economic development and environmental preservation on the other. More common types of disaster risk management challenges end to involve less conflict among stakeholders and more collaborative risk management. Such issues might be adequately treated as 'Complexity-Induced' risk problems or in some cases, as 'Uncertainty-Induced' risk problems and still require appropriate levels of discourse among potential stakeholders.

   However, the kinds of discourse envisioned by the IRGC framework are not always easily achieved. For instance, disaster experts and common citizens often find it difficult to communicate. Even experts can have difficulty understanding one another; they may use the same terms, but the terms can mean different things to different technical disciplines. Given these realities, it is often very hard to set up a common communication platform.

We would like to see the IRGC risk governance framework tested with further case studies in Japan and other countries to demonstrate more systematically how it can support prospective stakeholders, including government agencies, to better manage risks and conflicts like those faced in the Nagara River Estuary Barrage project.

Reference

IRGC, 2005, *White Paper on Risk Governance: Towards an Integrative Approach*, IRGC, Geneva.

Chapter 11
Acrylamide Risk Governance in Germany

Sabine Bonneck
Cologne, Germany

Introduction

The risk governance framework of the International Risk Governance Council (IRGC) provides an analytical structure within which to handle risks, from assessment to management. This chapter is a case study of the the events in Germany connected with the discovery of acrylamide in foodstuffs. A central question the case study hoped to answer was whether the framework could help deal with such situations as newly arising hazards from harmful substances in foodstuffs.

Advances in science and technology have led to the emergence of new sources of risk. At the same time, they have offered improved opportunities for identifying existing risks. For instance, modern measurement methods can detect substances in our air, water, and food at concentrations in the parts per billion (ppb) or even parts per trillion (ppt) range. With these improved analytical methods it is very likely that some undesirable compounds, in some cases unintended chemical by-products of production, will be detected in our foods. Since it is hardly feasible to test all foodstuffs for the approximately 100,000 known chemical substances (UBA 2001), it follows that there is some chance that some of them will be detected by chance, and these findings will then reach the public unfiltered. Prompt and careful reactions by scientific and government authorities can prevent the public from being unduly alarmed and such situations from developing into communication crises. What is important in such cases is to find solutions which meet legal health protection requirements and are acceptable to as many parties as possible.

This chapter provides an overview of the acrylamide in food crisis as it unfolded in Germany. It begins with some history on acrylamide, its uses and characteristics, followed by a summary of events which led to its discovery in foodstuffs and the reporting of these finding to the public. The next two sections present first, the institutional structures of consumer protection in Germany and second, the risk governance of acrylamide in Germany. The chapter concludes with an analysis of the events to see whether the risk could have been better handled if the IRGC frame-

O. Renn and K. Walker (eds.), Global Risk Governance: Concept and Practice Using the IRGC Framework, 231–274.

work had been applied. Since the Federal Institute for Risk Assessment (BfR)[1] has recently evaluated the communication of risk in the acrylamide case,[2] only a few observations will be made about this at this point.

Acrylamide History and Toxicity

Acrylamide was first synthesised in 1949. Since then, production of this toxic substance within the European Union has reached approximately 100,000 tonnes per year. It is used almost exclusively in the production of polyacrylamides. Polyacrylamide is used in sewage treatment, in paper and pulp processing, and in the treatment of minerals, and is an additive to cosmetics and paints (Madle et al. 2003).

Acrylamide was the object of worldwide scientific research even before its discovery in foodstuffs. The first results from animal experiments were published in 1986. The data on the incidence of tumours indicated that no 'safe dose', in the sense of a threshold value, could be assumed (BgVV 2002b).

However, it is difficult to demonstrate a direct connection between human exposure to a toxin and an increased probability of developing cancer, particularly based on animal experiments alone. Numerous chemicals have been described as carcinogens in animal experiments but to assume the threat of human cancer from this is by no means clear. Particular chemicals often induce cancer in organs which already exhibit an increased incidence of tumours. The increase due to the substance in question can only be estimated relative to the existing risk. Although epidemiological studies are preferable for evaluating the cancer risks, they do not exist for many substances.[3] The possibility that there are chemicals to which only laboratory animals, and not humans, are sensitive cannot be theoretically excluded.

These uncertainties in relating the results of animal experiments to humans are normally dealt with by convention – that when a substance is found to be carcinogenic in animals it is also treated as carcinogenic in humans, even in the absence of evidence from human studies (Henschler 1993).

If one assumes the existence of a relationship, one must deal with extrapolation from the empirically tested high doses in animals to the hypothetically estimated

[1] The chapter uses a number of abbreviations for the names of several institutions or laws. In some cases there are no official English translations available and the German names had to be translated to create a readable English text – cf. index of abbreviations and translated names at the end of the chapter.

[2] See Vierboom et al. (2007).

[3] The first understanding of the causes of cancer came from the observations of occupational medicine, since workers often experience higher exposure levels than the general public. An early milestone was the publication by the English surgeon Percival Pott regarding scrotal skin cancer among chimney sweeps in 1775. Further connections were subsequently revealed, for instance increased bladder cancer among dye workers in 1895, leukaemia due to Benzol in 1941, and lung cancer due to asbestos fibres since the beginning of the 1940s (Henschler 1993).

low doses for humans.[4] The question then arises whether the observations from high dose experiments are also relevant for low doses. The answer depends on whether or not the carcinogen is genotoxic, i.e., able to cause genetic damage. If this is not the case, the majority of scientists assume that a threshold exists, a threshold being a dose below which the risk of cancer is not increased. With genotoxic substances, which include acrylamide, the common convention is to make the conservative assumption that no threshold exists, so that even a minimal dose may lead to a rise in cancer risk (BfR 2002).

The results of scientific analyses of acrylamide have led to the following evaluations by international agencies:

- The International Agency for Research on Cancer (IARC) rated acrylamide as 'probably carcinogenic to humans' (Class 2 A) in 1994 (IARC 1994). This means that the properties of acrylamide determined through animal experiments are estimated to be relevant to and thus transferable to humans (BgVV 2002b).
- The European Scientific Committee on Food (SCF) evaluated acrylamide in 1991 as a 'genotoxic carcinogen'. The SCF re-examined this question in 2002 and left its evaluation unchanged, since it believed that the grounds for the decision remained the same (SCF 2002).
- The European Chemicals Bureau conducted an extensive risk assessment of acrylamide within the European Existing Substances Regulation. The final report contains a comprehensive statement of the state of research up to 1995. The report designates acrylamide as 'non-threshold carcinogen', and it recommends that human exposure to acrylamide be restricted as much as possible (EC/JRC 2002).

Scientific findings on acrylamide were also gathered in Germany:

- An assessment of the exposure to acrylamide in packing material was conducted at the Max von Pettenkofer Institute of the German Department of Public Health in 1993. The limit of detection for this assessment was one milligram per kilogram (Böhme/Grunow 1993).
- At the 58th Session of the Commission for Cosmetics of the Federal Institute for Consumer Health Protection and Veterinary Medicine on 29 April 1999, the Commission addressed the issue of 'genotoxic carcinogens' and demanded 'most insistently again that the residual content of monomeric acrylamide in polyacrylamide be minimised as far as possible through an appropriate choice of raw materials' (BgVV 1999: 2).

Measures to prevent unwanted human exposure to acrylamide had already been included in the recommendations of the World Health Organisation and in German legislation.

- To ensure that no residues remain in drinking water, in 1993 the World Health Organisation defined a maximum permissible value of 0.5 microgram per litre of

[4] In animal experiments, to arrive at statistically significant results, animals must be exposed to very high doses to which humans are not normally exposed.

water. This value is still in force in the current third edition of the Drinking-Water Guidelines (WHO 2004). According to Article 6 of the German Drinking Water Ordinance of 21 May 2001, the maximum permissible value of 0.0001 milligram per litre is not to be exceeded (TrinkwV 2001).

- The handling of acrylamide falls under the Ordinance on Prohibited Chemicals (Chemikalienverbotsverordnung), an ordinance to protect people and the environment from hazardous chemicals. Among other requirements, the substance is not to be passed to private end-users (ChemVerbotsV 1993).
- Appendix II of the Ordinance on Hazardous Substances of 26 August 1986, an ordinance which contains provisions for employee protection, also rates acrylamide as a carcinogen. This means that special protective measures are to be observed when handling acrylamide (GefStoffV 1986).

In short, the danger that acrylamide exposure could pose to humans was known, but the problem was believed to be limited to a few applications in which acrylamide was added, directly or indirectly, through human intervention. These situations were effectively regulated by laws and provisions. Acrylamide was certainly not of interest to the general public of Germany.

Events in Sweden up to 24 April 2002

The situation was quite different in Sweden. These events have already been analysed by Löfstedt (2003). The history of the case also plays a role in the German acrylamide case and thus, based largely on Löfstedt's report, is discussed further in this section.

Acrylamide became broadly known in Sweden in 1997 as a result of the 'Hallandsås scandal'. Hallandsås is a hilly area in Schonen, a region of southwest Sweden. It was here that, in 1992, the Swedish Railway Administration began work on an 8.6-kilometre tunnel which would reduce the journey time between Malmö and Göteborg. Construction quickly fell badly behind schedule. To prevent water leaking into the tunnel, which would further delay progress, the use of an acrylamide-containing sealant, Rhoca-Gil, was resorted to in March 1997.

The first dead fish were found in nearby fish farms in September 1997. Cattle grazing near a stream containing water that had leached from the tunnel later became lame and had to be destroyed. It was rapidly established that large quantities of acrylamide had contaminated the water. The construction site was contaminated both above and under ground. The entire vicinity was deemed to be highly dangerous. This promptly led to all products from the region being taken from the shelves of food stores. The events stoked fears in the Swedish population (Löfstedt 2003, Törnqvist 2005).

Naturally, the media took up the story. Reports appeared about the tunnel workers, who had worked with the sealant without the appropriate safety measures. The Swedish Railway Administration commissioned Margarita Törnqvist, the head of a research team at the Department of Environmental Chemistry at the University of

Stockholm,[5] to examine the blood of the cattle that had been destroyed. Törnqvist found evidence that the cows had been poisoned by acrylamide. In the tunnel workers' blood, which Törnqvist analysed next, such a high concentration of acrylamide was found that adverse effects on their health were feared. This finding also attracted a great deal of media attention. This combination of circumstances – acrylamide as the contaminant of an entire district, poisoned cattle, and unacceptable health risks for tunnel workers – led to the fact that within a few days the majority of the Swedish population knew that acrylamide was a toxic substance.

The investigations later expanded to the investigation of other sources of acrylamide when, in further experiments, Törnqvist also found unexpectedly high levels of acrylamide in the blood of a control group. The research team quickly struck on the idea of looking at food preparation, since the scientific community had long known of the formation of genotoxic chemicals during the roasting and baking of foodstuffs, known as the Maillard Reaction (Widemark 1939). To test this hypothesis, the researchers fed one group of rats normal food and a second group with baked and roasted food. Ten times as much acrylamide was found in the blood of the second group as in the first. The researchers published this result in the journal *Chemical Research in Toxicology*, arriving at the conclusion that:

> These data render it likely that cooking of food is a major source of the background dose of [acrylamide] also in humans. (Tareke et al. 2000: 1)

This article attracted hardly any media interest. The editor suggested sending out a press release, but Törnqvist refused. She wanted to await further research results and so give recommendations for a reduction of acrylamide concentrations. Otherwise, she feared, unnecessary concern to the population could be caused. Nonetheless, she published another article in the magazine of the Swedish National Food Administration (NFA), in which, among other issues, she discussed how acrylamide levels in hamburgers rose with rising frying temperatures (Tareke and Törnqvist 2001, quoted in Löfstedt 2003). This article did not create much interest either.

Törnqvist was supported in her investigations by a private laboratory, AnalyCen. In the beginning of autumn 2001, some thousands of micrograms of acrylamide could be detected in baked potatoes using a new procedure. These results were discussed with the NFA in October 2001. The NFA's agents, led by Leif Busk, suspected that the issue held a significant potential for crisis and initiated their own experiments at AnalyCen. Törnqvist and the laboratory continued to work on improving their analytical method and were soon able to confirm Törnqvist's original results. Her scientific paper reporting these was rejected by the journal Nature in February 2002. At the same time, the NFA agents had confirmed Törnqvist's results with their own experiments and wanted to bring these to the public's attention. Bertil Norbelie, the NFA's director, later defended the decision:

> The Food Administration has to work in the interests of the consumers and this includes going public with the information that we had as we take the view that any secrecy is not defensible. (Norbelie 2002, quoted in Löfstedt 2003: 410)

[5] Törnqvist's team had developed a method at the beginning of the 1990s with which the blood of Chinese factory workers had been tested for acrylamide.

Törnqvist had asked Leif Busk of the NFA at this stage to wait until her paper had undergone the regular peer review process, and Busk had agreed. In Löfstedt's view, 'however, at this stage things began spiralling out of control' (2003: 410), because information had already been leaked from two sources:

1. NFA employees had begun to inform their European colleagues of the results. Löfstedt gives two reasons for this. Firstly, they wished to spare their colleagues being confronted by the publication of the research results without preparation, and, secondly, there was the possibility of receiving research funds from the European Union. The second argument was decisive; the NFA scientists suspected that, following possibly serious consumer reaction, they should have pointed out they had initiated Europe-wide research activities.
2. The people in charge at AnalyCen laboratory, without consulting either Törnqvist or the NFA, published a two-page article in their customer magazine about their expertise in the analysis of acrylamide. The laboratory had already received enquiries about this. The despatch of the magazine could not be stopped in time to prevent the leaking of information.

In the middle of April 2002, Törnqvist's article was accepted for publication in the *Journal of Agricultural and Food Chemistry*. In correspondence with the author of this case study in 2006, Törnqvist described the situation thus:

> Yes, the paper was indeed accepted. That meant that several referees had looked at it and have had viewpoints that we had met in a revised manuscript. According to the editor, with whom we checked by email, *the paper was accepted and should be sent for printing.*

At this point, the NFA wanted to take their information to the public. A press conference was called for 11.00 on 24 April. The invitations for this were written jointly by the NFA and the PR department of the University of Sweden and contained the following passage:

> Researchers at Stockholm University have found an element that can cause cancer and which is formed during cooking a wide range of foodstuffs. The National Food Administration has in a pilot study found the substance in many food staples. The levels (of the substance) are high and new research findings will have international importance with regard to risk valuation, food production and consumption. You are therefore invited to receive this information at a press conference arranged by Stockholm University and the National Food Administration. (Translated from the Swedish, quoted in Löfstedt 2003: 411)

This text was issued on the afternoon of 23 April, some 20 hours before the conference. This period was to allow representatives of the national and specialist press to attend. In addition, the NFA held an internal meeting with Swedish food manufacturers beforehand to discuss the research findings. After this meeting the NFA and the University's representatives decided to give out no more information before the press conference.

However, the press's reaction was immediate. Journalists began to harass the individuals involved: Leif Busk of the NFA had 40 calls to his mobile phone within three hours, and his wife and children were also harried with numerous calls. Törnqvist's team was traced in the same way as the publisher of the Journal of

Agricultural and Food Chemistry. Nonetheless, the journalists did not succeed in discovering which results were to be presented on the following day.

On the morning of 24 April the editor of the *Journal of Agricultural and Food Chemistry* informed Törnqvist that none of the results from her paper could be made public, since this would qualify as a publication and so would prevent publication in the journal.

Hundred-fifty journalists appeared at the press conference; Sweden had not seen an event of comparable magnitude since the murder of Olaf Palme. In addition, it was also broadcast live on television.

The NFA explained which products and brands possessed the highest acrylamide contents. It is Busk who is quoted here:

> I have been in this field for 30 years and I have never seen anything like this before. The discovery that acrylamide is formed during the preparation of food, and at high levels, is new knowledge. It may now be possible to explain some of the cases of cancer caused by food. (See, for example, reports from BBC News 2002)

Törnqvist, however, did not regard herself as able to present any details from her article, which had a negative effect on her credibility. A Swedish science reporter who made no secret of her irritation with the fact that Törnqvist's article was not available asked whether there were any publication of the results being presented which had undergone a peer review. In retrospect, Törnqvist says:

> I think she wanted to show that she knew that scientific papers should be peer-reviewed. But the paper was peer-reviewed but not published. This was too complicated to explain in this situation. (Personal communication with author, 2006)

The NFA reported that it did not yet have any plans to take measures such as removing particular products from the market, since further research was necessary first. The statement led to the recommendation to eat low-fat products. In the final question-and-answer session, researchers and NFA representatives were asked for their personal recommendations, for instance, what they would advise their own children about eating potato crisps (Löfstedt 2003).

International Response to the Press Conference

The press conference attracted a great deal of attention from the whole western world although immediate reaction was very ambivalent.

The BBC's response was relatively restrained. It was even observed that 'the research was deemed so important that scientists took the unusual step of going public with their findings before the details had been officially published in an academic journal'. Subsequently, however, some experts' opinions were given, such as that of David Phillips of Cancer Research UK, who described the study's findings as 'highly significant' (BBC News 2002).

The New York Times addressed the issue in its editorial of 29 April 2002, arriving at the conclusion that the findings would certainly raise concerns, but that

> the Swedes were not so sure of themselves that they took any immediate action to change their own food supply or processes. They simply urged the European Union, other food safety agencies and the food industry to explore the issue. (New York Times 2002)

A sharper tone was taken on the following day under the headline 'Scientists Cautious on Report on Cancer from Starchy Foods': 'The scientists have not published a paper on their small study. Instead, they made their announcement at a news conference last week' (Kolata 2002). While the WHO had already announced that it would be holding a meeting of experts as soon as possible, the opinion of 'many experts' was 'that it made no sense to be alarmed over unpub lished data on a chemical that was very unlikely to have a measurable impact on cancer rates'. The toxicologist Stephen Safe from the A&M University in Texas was quoted as saying: 'it's just dumb, dumb, dumb. There are carcinogens in everything you eat'. In addition, it was pointed out that humans were exposed to numerous natural substances which could cause cancer in rodents. The risk of cancer due to acrylamide had yet to be proven. The article ended with the polemic comment that many scientists would be happy to read the Swedish findings if they were published in a reputable scientific journal (Kolata 2002).

Sceptical reactions also appeared in the German and Swedish press. The *Frankfurter Allgemeine Zeitung* referred to Swedish scientists who saw 'the danger of cancer for consumers dramatised in an irresponsible manner' (FAZ 2002). This article also pointed out here that the NFA had reported on acrylamide in certain foodstuffs 'before the customary publication of re search findings in the specialised press'. The NFA's declaration that acrylamide in foodstuffs in Sweden was responsible for 'several hundred deaths' per year was challenged by a comment in the Swedish newspaper *Dagens Nyheter*: 'the information released is quite insufficient. Without the facts, trust evaporates' (FAZ 2002). The *Frankfurter Rundschau* opened its coverage of 26 April 2002 with the words: 'crisps and chips probably do *not*, in the view of researchers, pose the substantial risk of cancer that the Stockholm Food Authority has warned of' (FR 2002).

In Sweden, a discussion of the actions of the NFA and the researchers began immediately after the press conference. Foremost among the criticisms was the accusation that the press conference had simply stoked fears, since no advice was given about how the danger could be reduced. Communications experts even speculated about the motivations of the NFA, specifically that the NFA had sought a way to draw attention to itself or to obtain research funding. Other voices, however, asserted that the greatest error lay in the fact that the media had not responded to Törnqvist's findings earlier.

Just one day later, on 25 April 2002, criticism in the Swedish media was more restrained. Löfstedt (2003) found the reasons for this to be that Swedish politicians had not criticised the NFA and that the WHO had rated the results as so important that they could not have been withheld from the public. On 26 April the WHO issued its own press release which announced that it would be holding a meeting of experts to discuss the significance of the results (WHO 2002).

Four weeks after the press conference, a survey of 250 Swedish households showed that the population was well aware of the fact that crisps and chips con-

Table 1 International expert events concerning acrylamide.

Date	Event	Level
25.06.2002	FAO/WHO Consultation	WHO
03.07.2002	Scientific Committee on Food	EC
15.10.2002	EU Commission: Stakeholder Meeting	EC
28.10.2002	JIFSAN[a]/NCFST[b] Workshop: 'Acrylamide in Food: Scientific Issues, Uncertainties and Research Strategies'	USA
16.03.2003	FAO/WHO 'Seminar on Acrylamide in Food: Current State of Affairs – Exchange of Views – Update on Ongoing Research – Identification of Gaps'	WHO
28.03.2003	EFSA Workshop on Acrylamide Research	EC
28.04.2003	Joint Research Centre (JRC): Analytical Methods Workshop	EC
21.10.2003	EU Commission Stakeholder Meeting	EC
17.11.2003	EFSA Workshop on Acrylamide Formation in Food	EC
22.03.2004	JECFA Seminar	EC
13.04.2004	JIFSAN Workshop on Acrylamide in Food	USA
08.02.2005	64th JECFA Meeting in Rome	EC

[a]Joint Institute for Food Safety and Applied Nutrition, established between the United States Food and Drug Administration (FDA) and the University of Maryland (UM).
[b]National Centre for Food Safety and Technology (NCFST), Chicago, Illinois. A research consortium among the FDA Center for Food Safety and Applied Nutrition (CFSAN), Illinois Institute of Technology (IIT) and the food industry (cf. http://www.ncfst.iit.edu/main/home.html, accessed 31 May 2006).

tained acrylamide (82% and 63% respectively). This knowledge, however, was not reflected in changes to consumption patterns: just 8% of those questioned said that they had reduced their intake of crisps, 13% planned to consume less in the future.

Nonetheless, the events did have economic consequences; the shares of Chips, the largest Swedish manufacturer of crisps, fell 15% on 24 April 2002. Sales of crisps fell 40% in the week following the press conference but recovered steadily thereafter, until in the third week after the press conference they had settled at the level of the previous year (Löfstedt 2003).

The German market for crispbreads fell some 30% in 2002. Whereas the traditional Swedish firm Wasa, which has been part of the Italian company Barilla since 1999, reported a downturn in sales in Germany of 15%, several low-cost producers were faced with much more drastic falls. In Sweden, average sales were down some 5% on the previous year (Kruse 2003, confirmed by the Barilla-Wasa press office).

The surprise the news created internationally also had the consequence that, within a short time, numerous committees of experts had been assembled and research projects launched to learn more about the formation of acrylamide and to discuss the possibilities of reducing the content in foodstuffs. Table 1 lists the conferences and events on acrylamide which have taken place since April 2002 with the European Commission, the WHO and in the USA. This overview is by no means complete, but does clearly demonstrate the many activities undertaken to address the new problem.

By 2005, the European Union was already conducting studies in 10 areas in connection with acrylamide, for instance on toxicological and epidemiological issues, on methods of analysis, and on the extent of human exposure (EC 2005). The manufacturers of affected foodstuffs were also active; in Sweden, for instance, the manufacturers of crispbreads formed a research association (Kruse 2003). Germany also saw the launch of a research association by producers. Coordinated by their umbrella organisation, the German Federation of Food Law and Food Science (BLL) and provided with a budget of 1.7 million Euro, of which 1.6 million came from the Federal Ministry for Economic Affairs and Employment, between 2003 and 2005 issues related to analysis as well as toxicological and technological topics were dealt with (BLL/FEI 2005).[6]

Evaluation of the Events in Sweden

The course of events in Sweden was very largely determined by weaknesses in the communications process before and during the press conference. Löfstedt (2003) in his analysis arrives at the following conclusions:

1. Since the information had already started to leak, there was no alternative to making it officially public. The decision to use the instrument of a press conference was only made following long discussions between the NFA and the University. Their objective was to avoid misunderstanding. The journalists should therefore have the opportunity to pose questions directly, and it was assumed that the researchers would find it easier to explain their complex results in such a context.

   No further consideration was given to the fact that the NFA had almost no practice in arranging press conferences, and that Margareta Törnqvist and her team had no training how to deal with the media.

   In addition, certain salient questions were not addressed, such as what kind of information the public required at this juncture. Löfstedt put forward the hypothesis that some very simple information would have sufficed, for instance that the NFA was working to verify some preliminary indications of the presence of acrylamide in certain foodstuffs. Holding a press conference was possibly overhasty. Communications experts would have advised a press statement to specialised journals instead.

   Whether the public's trust in the NFA fell after these events, as many observers assume, has not been investigated until now.[7]

---

[6] See also http://www.ilu-ev.de/acrylamid/acrylamid.htm, accessed 31 May 2006.

[7] Löfstedt mentions in his study that Leif Busk of the NFA planned an evaluation of the communications measures. This has not, however, yet been implemented.

2. There were also long discussions about the wording of the invitation. On this point it is particularly significant that with the NFA and the University of Stockholm, two actors with entirely different interests were involved. While the University was concerned about addressing scientists by emphasising Törnqvist's findings, the NFA favoured striking formulations which would awaken media interest and reach as many consumers as possible. The text was, therefore, the result of a compromise which:

   > ... gave the worst of both possible worlds with a somewhat sensationalist press invitation with a significant delay leading to information vacuums causing rumour and speculation. (Löfstedt 2003: 420)

   This information vacuum was worsened by the refusal of the organisers to move the press conference forward to the evening of 23 April or to answer the enquiries of the media directly. This was why media representatives began to look for other sources of information. Löfstedt's research shows that both organisers considered it too complicated to depart from the procedure once it had been decided on. Both would have moved the press conference forward had they been able to work independently.

   The central question concerns the nature of the news to be presented. Was it the release of research findings, or an important piece of information for consumers? The University and the NFA should have resolved this issue before all others.

3. In the text of the invitation to the press conference, and during the press conference itself, words which function as social amplifiers were used, such as 'cancer', 'base foodstuffs', and 'international concern;. Naturally the media picked these words up and made headlines of them, e.g. 'Cancer poison found in food' (*Dagens Nyheter*, quoted in Löfstedt 2003: 420), or 'Alarm about cancer poison in common foodstuffs' (*Svenska Dagbladet*, quoted in Löfstedt 2003: 420). Due to the events in Sweden described above, 'acrylamide' was still in the thoughts of journalists and the population as a whole, and functioned as an additional social amplifier.

   However, the commotion in the media waned quickly. Instead, the NFA was blamed for fuelling fears and exaggerating risks. Löfstedt names the following reasons for this development:

   a) Acrylamide is not mixed into the food, but forms there in natural process. Natural dangers are accepted as facts, in contrast to technological dangers, for which someone can be blamed.
   b) The risk appears familiar. It seems safe to assume that, since humanity has used fire, acrylamide has been able to form during roasting. It is not, therefore, a risk whose novelty need concern people.
   c) The risk is reckoned to be controllable. If one wishes to reduce the risk, one needs simply to reduce consumption of products with the highest acrylamide content, chiefly crisps and chips. If people see a risk as being controllable,

they often regard it as less dramatic. This is also the case for smoking or the consumption of alcohol.

4. At the press conference particular brand names were disclosed. The NFA was obliged to do this because of the legal situation, or the media could have called on the 'principle of transparency' (*Offentlighetsprincipen*), under which all government documents are to be made available to public scrutiny. The NFA decided in this case to take a proactive approach which was intended to win the public's trust and to prevent the media making accusations about the withholding of information. Naturally, the NFA was fiercely criticised by the manufacturers for this decision. What was particularly problematic was that only one test had been conducted for each brand. It has since become known that the acrylamide content can vary widely between different samples of the same brand.

5. There have been a series of food scares in Europe since the 1990s, of which the most prominent have been BSE, foot and mouth disease, and dioxin in chicken food and hens' eggs. Research has shown that people are not normally able to distinguish these risks with which they are confronted in their daily lives or through media coverage. Single risks are normally present in people's consciousness until they are threatened by a new risk. People who are worried about acrylamide in food at a particular point in time may have forgotten that a few months ago they were concerned about aflatoxin in dried figs. And the issue of acrylamide will probably remain in most people's minds only until such time as the revelation of a novel risk in foodstuffs causes new worries.

   It is, in Löfstedt's (2003) opinion, especially problematic if the scientific community is not in agreement in its evaluation of such a circumstance and if the controversy is played out in public. Regarding the questions whether acrylamide caused cancer in humans at all, and to what extent the results of experiments on mice and rats could be applied to humans, heated debates took place between epidemiologists and toxicologists in the Swedish press. Such open disagreements generate a mistrust of science (Löfstedt and Renn 1997).

   Sharp (2003) contests this. He sees a contradiction between, on the one hand, promoting transparency and proactive procedures and, on the other hand, permitting selected controversies to take place behind closed doors. He also asks if there is any way in which public disputes between scientists could be directed by a third party.

6. If the information sources in the Swedish acrylamide case had not been seen as credible, the topic would never have been taken up by the press to the extent that it was. The high level of trust that the NFA and scientists in Sweden enjoyed contributed to an increase in the drama surrounding the events (Löfstedt 2003).

In addition, the question of the peer review also played a significant role. Even in favourable coverage after the press conference it was pointed out that the research findings had not been published in a journal of scientific standing. As late as autumn 2005, this criticism was made in the context of an expert conference in Germany

(Wolf 2006). The credibility of the findings therefore suffered badly. However, the topic had been already identified as explosive by employees of the NFA in autumn 2001. An explicit obligation to confidentiality on the parties would probably have generated less pressure to make the findings public at a particular time and allowed time to wait for the publication of Törnqvist's article.

Summary of the Characteristics of the Acrylamide Case: Relevance for Risk Governance

Renn (2003) mentions the following peculiarities of the acrylamide case, which help to define it as a 'systemic' risk for society:

- Acrylamide was already known as a constituent of sealant and rated as a carcinogen. There is still no clear and conclusive study linking its presence in food and an increased risk of cancer.
- In Sweden, acrylamide was already known to large portions of the population as a hazardous substance.
- At the time of the discovery neither the extent of exposure was known, nor could a reliable dose-response relationship be specified.
- Practically every person in the world is exposed to acrylamide, since it is generated in the cooking of potatoes and cereals.
- Acrylamide forms in a natural process without any human intervention.

The OECD (2003) describes a risk as systemic when reaction to it entails significant economic, social and even political consequences. Systemic risks require a holistic approach which takes account of this complexity (Renn 2003). The IRGC framework provides an approach to dealing with cases of systemic risks, because it goes far beyond the simple cause-effect risk analyses and stresses the importance of understanding the interdependencies between the different dimensions of risks, networks of actors, and the public in the governance of risks (IRGC 2005).

The occurrence of acrylamide in foodstuffs possesses other particularities which are fundamental to further challenges to risk governance approaches. The first measurements done in Sweden made it evident that different levels of acrylamide are present not just in different foodstuffs but in identical foodstuffs. Not even the individual crisps in a bag are contaminated to the same extent. That means that the establishment of a threshold value would not solve the problem, but that a solution must be very complex to take account of these various facts. Furthermore, every attempt to lower the quantity of acrylamide by changing the manufacturing method or ingredients also has potential effects on the quality and flavour of a product, and thus also its acceptance with consumers.

The foodstuffs affected also involve foods which are cooked at home, such as roast potatoes and chips. However, the private sphere cannot be influenced by such measures as the establishment of threshold values.

The Institutional Structures of Consumer Health Protection in Germany

Before discussing the case of acrylamide in Germany, the structures of consumer health protection will be described, including a first short glance at possible weak points in risk governance (see review in Dressel et al. 2006). As described earlier, the use of acrylamide in Germany in the 1990s was legally regulated in several areas. However, far-reaching changes occurred in the structures of the institutions concerned with the ultimate regulation of acrylamide in Germany between that time and when acrylamide was first discovered in foodstuffs in 2002.[8] Figure 1 depicts the chronological sequence of changes in which the institutions relevant to this discussion appear in grey.

With the restructuring of 1994, the Federal Public Health Department's fields of responsibilities were transferred to three successor organisations, one of which was the Federal Institute for Consumer Health Protection and Veterinary Medicine (BgVV). The BgVV had been moved from the domain of the Federal Ministry of Health to the then Federal Ministry of Food, Agriculture and Forestry now the Federal Ministry of Food, Agriculture and Consumer Protection (BMELV).

In 2002, a new arrangement was introduced which, amongst other changes, institutionalised the organisational separation of risk assessment and risk communication from risk management.[9] The requirement for a separation of risk assessment and risk management had already received international support in the report of the Codex Alimentarius Commission (Codex Alimentarius 2001). In the aftermath of the BSE crisis, Germany also took up this requirement (von Wedel 2001) and in August 2002 implemented in the Act on the Reorganisation of Customers' Health Protection and Food Safety. On this statutory basis, the duties of consumer health protection were divided between two new authorities, the Federal Institute for Risk Assessment (BfR) and the Federal Office of Consumer Protection and Food Safety (BVL). The separation of the fields of risk assessment and risk management promised more transparency and independence of consumer health protection and was directed towards regaining the trust of the population. In addition, the separation fa-

---

[8] See Act on Successor Institutions of the Federal Public Health Department and Act on the Reorganisation of Customers' Health Protection and Food Safety.

[9] At European level in relation to food safety according to chapter 1 article 3 of EC Regulation 178/2002, the following expressions are used as defined:
Risk Assessment: 'a scientifically based process consisting of four steps: hazard identification, hazard characterisation, exposure assessment and risk characterisation'.
Risk Communication: 'the interactive exchange of information and opinions throughout the risk analysis process as regards hazards and risks, risk-related factors and risk perceptions, among risk assessors, risk managers, consumers, feed and food businesses, the academic community and other interested parties, including the explanation of risk assessment findings and the basis of risk management decisions'.
Risk Management: 'the process, distinct from risk assessment, of weighing policy alternatives in consultation with interested parties, considering risk assessment and other legitimate factors, and, if need be, selecting appropriate prevention and control options'.

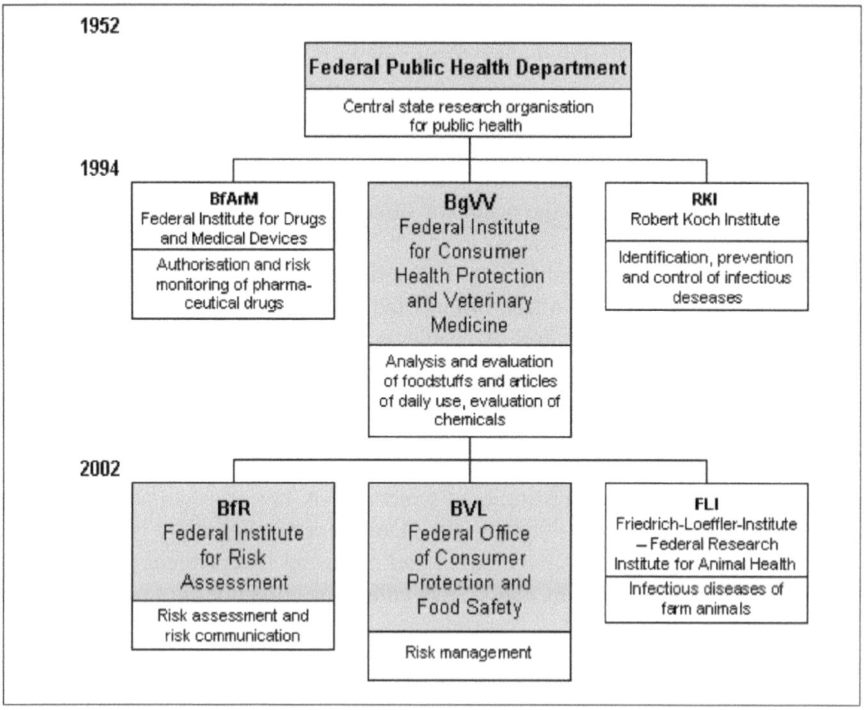

Fig. 1 Changes in the institutional structures and main roles in consumer health protection in Germany since 1994.

cilitated the cooperation between the European Union, the Federal government and the regional governments in this area (Dressel et al. 2006).

The core duties of the BfR are scientific risk assessment in consumer protection and risk communication. In order that these assessments could take place free of interested parties, the BfR was constituted as a legally capable public corporation (Deutscher Bundestag 2002a). The BfR is active in the context of regulatory procedures, on its own initiative, at the request of the BVL and the BMELV, as well as in cooperation with EU institutions (Wissenschaftsrat 2006).

The main role of the BVL is to take preventive and protective measures in the fields of food safety and consumer protection in regard to, amongst others, foodstuffs, cosmetics, and other articles of daily use. In addition, it is involved in the design and review of the corresponding monitoring programmes for the regional governments and is the interface for the European early warning system (Deutscher Bundestag 2002a).

A similar organisational separation has also been implemented in France (Ministère de l'Agriculture 2005), which, like Germany, adopted the structures created at the European level in the wake of the BSE scandal (Dreyer et al. 2006). This approach has not been followed in all the member countries. In England, these tasks are administered by various departments of the Food Standards Agency.[10] In Sweden, risk assessment and management – and even legislation – reside with the NFA.[11]

The chosen organisational form in Germany is not uncontroversial. This is a topic which, because of its complexity, can only be touched on here. However, as early as 2002, criticism of the arrangement was being expressed by Böschen et al. (2002). They suggested that the problems of contemporary societies, by reason of their increasing complexity, can no longer be solved only by scientists providing the knowledge and then handing it to the politicians. Decisive situations are ever more frequently affected by a lack of knowledge. Scientific debate occurs increasingly in public, and descriptions of problems arise as the result of many-layered public-political discourse. Moreover, due to the underlying belief in the objectivity of science, the strict division of duties prevents the search for possible approaches to handling possible differences of opinion productively (Böschen et al. 2002).

Furthermore, Böschen et al. (2002) voiced specific concerns about the fate of risk communication within the institutional structures. A fair representation and consultation of societal interests cannot take place, because the public is seen simply as the addressee of risk communication and not as an independent pole between political administration and business. Risk communication does not take place as a dialogue but as a monologue.

This last criticism is particularly troubling because monologues are not appropriate for bridging the frequent gap between the assessment of a hazard by science and its appraisal by the public. Current research strongly suggests that risk assessments should involve not only quantifiable scientific findings but also societal perceptions and ideals. Furthermore, the OECD sees the education of the population as a basic prerequisite for the management of systemic risks. It is only in this way that broad-based risk prevention can take place in a decentralised and market-based society (OECD 2003).

As part of the planning for the establishment of these new authorities, it had been agreed that the BfR should undergo regular inspection by an external body to ensure the quality of its work (BMVEL 2001). The Science Council submitted its first report in May 2006. It found a lack of plans in regard to risk communication. The Science Council criticised the fact that the department was overloaded with other work.

In the view of the Science Council, BfR's establishment of an in-house research department fulfilled a major prerequisite for its 'high-quality administration' of statutory duties (Wissenschaftsrat 2006: 7). The quality of the research effort was rated overall as good, in some fields even as very good. However, the lack of strategies for identifying topics for research in anticipation of possible future risks was criticised.

The Science Council has confirmed that the BfR enjoys a good reputation amongst the relevant EU institutions. This reputation is reflected in the chairmanships and other leading roles held by BfR staff on EU committees, panels and work-

---

[10] See http://www.food.gov.uk/aboutus/how_we_work/, accessed 31 May 2006.

[11] See http://www.slv.se/templates/SLV_Page.aspx?id=2051, accessed 31 May 2006.

ing groups to the BfR's management, as well as the adoption of methods developed by the BfR in European standards and guidelines.[12]

Criticism was directed at the way in which the research work of the BfR has been hindered by protracted testing and approval procedures and the long-delayed allocation of research funds. In addition, the Science Council expressed concern that division of responsibilities between the BfR and BVL have not been sufficiently well demarcated. Counterproductive friction has resulted.

The BfR is the only national institution that handles the assessment of risks arising from human and animal foodstuffs, chemicals, and items of daily use, a responsibility that involves testing and monitoring roles. The Science Council does not actually regard these tasks as appropriate for a departmental research institution; however, two reasons favour this current arrangement. Firstly, these permanent roles cannot be executed to the necessary scope and continuity at a university. Secondly, risk assessments performed by a departmental research institution like the BfR are considered to be particularly authoritative. Maintaining control over the administration and quality of these functions raises the trust placed by consumers and business in the outcomes of BfR's assessments (Wissenschaftsrat 2006).

Furthermore, state-sponsored private institutions in Germany do exist which provide information to consumers. This task is not something that must necessarily be administered by the state, but it should be ensured that consumers 'can take part in the development of the market as an equal partner' (BMVEL n.d.: 16) and so can increase the personal responsibility that they take. In the acrylamide case, these institutions were chiefly the *aid-Infodienst – Verbraucherschutz, Ernährung, Landwirtschaft – e. V.* (aid), as well as the Federation of German Consumer Organisations (vzbv), the umbrella organisation of the 16 regional states' consumer centres and 23 other consumer-oriented associations.[13]

These remarks have already given a first sketch of the strengths and weaknesses of these structures in Germany, which are also significant in the acrylamide case.

Risk Governance in the Acrylamide Case in Germany

From IRGC's perspective, risk governance includes the totality of actors, rules, conventions, processes, and mechanisms concerned with how relevant risk information

---

[12] The first experts' discussion on the acrylamide issue, which took place at the WHO in Geneva from 25 to 27 June 2002, was chaired by Dieter Arnold, the then president of the BgVV (BgVV 2002e).

[13] *aid-Infodienst* was founded in 1950. It was intended to inform the needy post-war population about correct nutrition (http://www.aid.de/allg/geschichte.php, accessed 31 May 2006). The vzbv has developed from working group of the German Association of Consumer Organisations (*Verbraucherverbände e.V. AgV*). This was founded in 1953 and conducted the first comparative product test in 1961, now seen as one of the central tasks of consumer protection (http://www.vzbv.de/start/index.php?page=wir&pagelink=geschichte, accessed 31 May 2006).

is collected, analysed and communicated and management decisions are taken.[14] It describes an integrative approach through which the examination and management of risks can be systematised, regardless of the nature of the risk in question. The framework is divided into four phases: Pre-Assessment, Risk Appraisal, Tolerability and Acceptability Judgement, and Risk Management (IRGC 2005).

In the following sections, the course of events in the German acrylamide case is compared to the individual phases of the Framework. The events are reconstructed with the use of media reports, the results of an Internet search, and interviews with experts. Acrylamide is particularly suitable as a case study for examining the risk governance model, since systematic risks require an integrated approach. The goal of this comparison is to determine whether the Framework can contribute to improved risk governance outcomes.

Pre-Assessment

Even before a society is confronted by a risk, two fundamental questions need to be addressed:

1. What is understood by the term risk? and,
2. What indicators are there for the existence of a risk?

The scarcity of resources obliges societies to select particular topics to undergo the process of risk governance. This choice is not always easy, because opinions may vary widely. What counts as a risk to someone may be an act of God to someone else or even an opportunity for a third party (IRGC 2005).

The IRGC governance framework identifies four elements of pre-assessment: problem framing, early warning, screening and scientific conventions. Three are specifically addressed in the context of the German acrylamide case.

Problem Framing

Tversky and Kahneman (1981) demonstrated that how a decision is framed is ethically significant. The psychological model that leads to decisions can be influenced, depending on how the question is formulated. The authors showed that a course of action is more likely to be chosen when it is positively formulated and linked with benefits. The same option is rejected when the costs are named, even though the distribution of costs and benefits are absolutely identical in both cases. For instance, it makes a great difference whether the results of a vaccination is given as '90 out of 100 vaccinated were saved' or '10 out of 100 vaccinated died' (Tversky and Kahneman 1981).

---

[14] Within the following section a number of phrases and sentences have been taken directly fom the IRGC's White Paper No. 1. In every case the source has been referenced as IRGC (2005) within the paragraph (for easier readability these quotations are not put in quotation marks, however).

Whether a consensus evolves about what requires consideration as a relevant risk depends on the legitimacy of the selection rule. The acceptance of selection rules rests on two conditions. First, all actors need to agree with the underlying goal. These are often legally prescribed, such as prevention of adverse health impacts, purity laws for drinking water, etc. Second, they need to agree with the implications which the identified hazard can have on the desired goal (IRGC 2005).

The goal in the present case, the protection of human health from adverse impacts from food, is by no means controversial, but legally established. According to paragraph 1 of the German Food and Feed Code, the purpose of this law is:

> ... to ensure the protection of consumers by preventing or defending against hazards to human health in human and animal foodstuffs, cosmetics and articles of daily use. (§1 LFGB)

The question of the potential consequences from acrylamide in foodstuffs could not be answered in the early stages of the acrylamide case, because everyone had been equally surprised by the Swedish research findings. However, acrylamide had been rated by various institutions around the world as carcinogenic and hazardous to humans. Its presence in drinking water is thus considered undesirable. By extension, it seems likely that a broad consensus would have existed for the restriction of acrylamide in other foodstuffs[15] to negligible or extremely low concentrations.

Early Warning

Löfstedt (2003) had already quoted critics from Sweden who said that the greatest error in the development of the acrylamide case was that no attention had been paid to the scientific publications of Tareke and Törnqvist in 2001. Wiedemann et al. (2002) argues that this error is an indication of the need for the early identification of risks so that problems and gaps in existing knowledge can be addressed in good time.

The international specialist community did not recognise early warning signals, but learned of the new problem at the same time as the general public. Public perception of the following events led to great pressure on authorities and stakeholders to act which certainly frustrated the search for jointly agreed solutions.

Scientific Conventions

As a further element of Pre-Assessment, the IRGC framework suggests agreement on scientific conventions to be used in assessing and evaluating risks. As already partially explained, and as the following discussion will indicate, it would have been helpful in the acrylamide case if these issues, at least, had been decided on:

- the social definition of what is to be regarded as adverse, for example by defining the 'No Adverse Effect Level' in food (NOAEL);

---

[15] Drinking water is a kind of foodstuff (see EC 2002b).

- the selection rule determining which potentially negative outcomes should be considered in the risk governance process knowing that an infinite number of potential negative outcomes can be theoretically connected with almost any substance, activity or event;
- the selection of the testing and detection methods;
- the selection of valid and reliable methods for measuring perceptions and concerns;
- the determination of models to extrapolate high dose effects to low dose situations, for example linear, quadra-linear, exponential or other functions or assumptions about thresholds or lack of thresholds in dose-response relationships;
- the extrapolation of the results of animal studies to humans;
- the assumptions made about exposures and definition of target groups.

Decisions on these issues, developed by consensus between experts and the involvement of relevant regulatory institutions, are indispensable for the later step of Risk Appraisal (IRGC 2005). Without agreement on these conventions, discord between scientists ran like a thread through the acrylamide case.

One example of a key disagreement was about the findings of two epidemiological studies, one by Mucci et al., published in the *British Journal of Cancer* in January 2003, and the other by Pelucchi et al., published in the *International Journal of Cancer* in July 2003. The objective of both these studies was to investigate the relationship between the intake of acrylamide and the probability of developing cancer. Neither study detected a statistically significant relationship. The BfR issued full statements in response to both studies, which cannot be discussed in detail here, but which noted numerous failings in both studies. The inadequate size of the population samples alone, for example, made it difficult for either study to demonstrate such a relationship (BfR 2003a, 2003b).

The smaller the risk in question, the larger the size of the sample studied must be to provide evidence of the risk. Mucci et al. had 591 participants in their study. According to the BfR, this sample size would have only permitted the detection, with sufficient statistical certainty, a 50% rise in the risk of bladder or kidney cancer (BfR 2003a). In the risk governance context, the convention that must be agreed upon what weight or credibility to give to epidemiological studies like these that yield 'negative' results.

According to the promotion on their websites, the *British Journal of Cancer* 'publishes high quality original papers and reviews that make a significant contribution to increasing understanding of the causes of cancer and to improving the treatment and survival of patients'.[16] Like the *International Journal of Cancer*, it undertakes a peer review of every article submitted.[17]

However, by deciding to publish the papers, both journals appeared to disregard the sample size and other problems. The journals could be sure that the studies could be frequently cited and in fact, both studies were reported widely in the mass media

[16] See http://www.nature.com/bjc/index.html, accessed 31 May 2006.

[17] See http://www3.interscience.wiley.com/cgi-bin/jabout/29331/ForAuthors.html, accessed 31 May 2006.

(see, for example, Stern Online 2003a). This had a considerable impact: even today both papers are cited in support of arguments against the risk of cancer associated with human exposure to acrylamide.

This experience, in addition to the controversy surrounding Törnqvist's unpublished research findings, raises the question whether a paper's appearance in an academic journal in fact constitutes a guarantee of sound scientific work. It seems likely that specialist journals also are under a certain pressure to be able to publish contributions that are relevant to current, and in particular, controversial discussions.

In the early stages of the acrylamide case, an agreement about dealing with disputes between experts could not be reached. However, suggestions for dealing with scientific controversies in general had been proposed by the Helmholtz Society or the Max Planck Institute (cf. Wiedemann et al. 2002).

A pre-assesssment, as proposed by the IRGC framework, was only fulfilled in a limited way for the acrylamide in food crisis, which contributed to difficulties encountered in the risk governance process. An early warning system for the detection of new risks like acrylamide might have given authorities and manufacturers more time to develop their response to the crisis, but it did not exist. The absence of scientific conventions for assessing the seriousness of the risk further hampered risk governance decisions. In effect, however, the presence of acrylamide in foodstuffs was framed as a risk that was undesirable in society, one that ultimately needed to be addressed in some way.

Risk Appraisal

In the IRGC framework, the phase of risk appraisal consists of two components, risk assessment and concern assessment. The aim of this phase is to gather the information that is necessary to assess the size and likelihood of a risk and societal concerns about them in a serious way, crucial steps that must be taken before society can decide whether it wishes ultimately to agree to a risk. The information thus includes the scientific assessment of both the risks (to human health in the case of acrylamide) and the concerns stakeholders may have regarding social and economic implications (IRGC 2005).

Risk Assessment

When a society is confronted by a systemic risk such as acrylamide, complexity, uncertainty and ambiguity have first of all to be clarified. The impacts of these three types of problems for risk assessment and ultimately, risk management, must be made clear (IRGC 2005).

1. *Complexity*
   The acrylamide problem exhibited substantial complexity. The difficulties of firmly establishing whether acrylamide is carcinogenic have been described

earlier. The chain of cause and effect, from exposure to acrylamide to increases in the risk of cancer is not easy to define. There may be a multitude of potential causal agents, interactions between them, as well as a number of specific observed cancer effects many of which may take years to develop.

Complexity can sometimes be resolved if all available knowledge is brought together in what IRGC calls epistemological discourse, for instance, with the help of Delphi interviews (IRGC 2005). No such systematic approach was used to attempt to deal with the complexity of the acrylamide problem, though numerous scientific institutions around the world had begun to research in this area following the Swedish findings.

2. *Uncertainty*
In the context of assessing risks, particularly new risks, human knowledge is almost always incomplete and thus contingent on uncertain knowledge and assumptions. The acrylamide case was also dominated by a great deal of uncertainty, in particular the uncertainty arising from the extrapolation of data from animal experiments to humans. At the time that acrylamide problem arose in foodstuffs, the policy convention in place was that substances that were definitely carcinogenic in animal experiments were also treated as carcinogenic in humans. However, as described above, this convention was questioned by some epidemiologists and toxicologists, resulting in considerable media response.

Uncertainty affects risk evaluation and the development of management options. The classic question of 'how safe is safe enough' is replaced by the question of 'how much uncertainty and ignorance are the main actors willing to accept in exchange for some given benefit'. In such cases, risk managers are well advised to include the main stakeholders in the risk management process and ask them to find a consensus on approaches to issues like the extra margin of safety in which they might be willing to accept in exchange for avoiding potentially catastrophic consequences. This type of deliberation, called 'reflective discourse', relies on a collective reflection about balancing the possibilities for over- and under-protection. If too much protection is sought, innovations may be prevented or stalled; if we go for too little protection, society may experience unpleasant surprises. It is recommended that policy makers, representatives of major stakeholder groups, and scientists take part in this type of discourse. The reflective discourse can take different forms: round tables, open forums, negotiated rule-making exercises, mediation or mixed advisory committees including scientists and stakeholders (IRGC 2005).

3. *Ambiguity*
The result of a risk assessment is normally evaluated differently by different actors. High complexity and uncertainty favour the emergence of ambiguity. Ambiguity arises when values, priorities, or limitations cannot be agreed upon. Two forms of ambiguity may be distinguished:

a) *interpretative ambiguity* ⟶ The result of a risk assessment provokes differing interpretations because there is a lack of clarity whether an effect is adverse or not, e.g., low concentrations of genotoxic substances.
b) *normative ambiguity* ⟶ This arises as a consequence of different concepts of what can be regarded as tolerable in terms of e.g. ethics, quality of life parameters, distribution of risks and benefits, and so on. Examples of this would include passive smoking, nuclear power, pre-natal genetic screening.

Both forms of ambiguity are identifiable in the events surrounding acrylamide in Germany. Firstly, as discussed above, no agreement could be found on the issue whether the quantity of acrylamide that a person would normally absorb in their diet should be considered a cause for concern. Secondly, the economic interests of the manufacturers of the affected foodstuffs played a role. The fronts between the different interest groups seem here to have hardened particularly. Since the general public took less part in the debate over the interpretation of toxicological data, it would surely have been sensible to include the issue of the ambiguity in the public discourse.

In problems of high ambiguity, as in the acrylamide case, IRGC does not see a simple demonstration of open-mindedness about publicly expressed concerns to be sufficient. Such circumstances require an opening of the process of risk evaluation to the participation of the public and new forms of consultation in the form of a participative discourse. These offer affected parties the possibility to exchange arguments, and openly discuss assumptions and ethical values. This form of discourse allows numerous opportunities to resolve conflicts since, for instance, in the course of the discourse common values can be identified, or ways found in which different societal groups can realise their own concepts without interfering with others (IRGC 2005).

The Beginning of the German Acrylamide Case

The problem of complexity, bringing together available knowledge on acrylamide, was easily addressed in Germany. The German authorities were informed of the Swedish findings through the European Union's early warning system on 23 April 2002 (Deutscher Bundestag 2002b). The BgVV expressed itself promptly with a press release (BgVV 2002a). Subsequently, remaining unanswered questions were addressed in a series of official activities, as detailed in Table 2.[18]

A first important step in the acrylamide risk assessment was surely the expert discussion at BgVV on 14 May 2002. Scientists, consumer and business associations, experts from the regional governments, and representatives of the BVL dis-

[18] The multitude of activities undertaken by the BgVV/BfR cannot be discussed in detail at this point. For further information on the BgVV's/BfR's activities in this regard and on acrylamide, please see the Instititute's homepage. All the documents of the BgVV and the BfR discussed in this text are to be found at: http://www.bfr.bund.de.

Table 2 Official activities in Germany relating to the risk assessment of acrylamide in foodstuffs.

14.05.2002	BgVV	1st public expert discussion at BgVV
19.06.2002	BgVV	1st meeting of the acrylamide analysis working group
02.07.2002	BgVV	Statement on FAO/WHO consultation from 25.06 to 27.06.2002
01.08.2002	BgVV	BgVV: Proposal for the introduction of an action value of 1000 Microgram per kilogram foodstuff
August 2002	BgVV	Beginning of proficiency tests
17.09.2002	BgVV	Determination of signal values
01.10.2002	BgVV	Publication of the process for determining acrylamide in solid and pasty foodstuffs
28.11.2002	BfR	Publication of a document on the updated risk assessment of acrylamide in foodstuffs
09.12.2002	BfR	Press statement from BfR: baby foods can also contain acrylamide
16.12.2002	BfR	Results of proficiency tests
27.01.2003	BMVEL	Hearing of the consumer committee of the German parliament on acrylamide in cosmetics
30.01.2003	BfR	Statement on Mucci et al.'s study
10.02.2003	BfR	Publication of results of proficiency tests
15.07.2003	BfR	Results of spot survey of young people on absorption of acrylamide
29.10.2004	BfR	Expert colloquium on research activities (part I)
07.02.2005	BfR	Expert colloquium on research activities (part II)
15.03.2005	BfR	Information on the outcome of the 64th JECFA meeting

cussed the significance of the Swedish findings for Germany. The development of validated methods of analysis was seen as a precondition for the determination of the acrylamide contents of products on the German foodstuffs market. Great uncertainty remained regarding the exposure of humans because no current data was available on German eating habits. The experts unanimously supported the risk assessment undertaken by the EU which named acrylamide as a carcinogen. Among other items confirmed as open questions and next steps were that a comparison of laboratories (proficiency test) had to be organised to guarantee uniform measurement, that affected foodstuffs should be investigated in a coordinated fashion, and that human consumption be estimated (BgVV 2002b).

Simultaneously with the report on the expert discussion, a statement was issued with advice for consumers on how absorption of acrylamide could be lowered by changing eating habits (BgVV 2002c).

At the first meeting of the acrylamide analysis working group, findings from current research were presented. For instance, data showed that there are foodstuffs with their own potential for the formation of acrylamide, that the highest acrylamide levels are in potato products and crispbreads, and that the lower the amount of water present, the greater the formation of acrylamide (BgVV 2002d).

An international expert debate on the significance for human health of acrylamide in foodstuffs took place from 25 to 27 June 2002. The organisers of the event were the Food and Agriculture Organisation of the United Nations (FAO) and the World Health Organisation (WHO). In the meantime, the Swedish findings on the topic had been confirmed by experiments in various countries, including Norway, Switzerland, Great Britain, the USA and Germany.

Provisional exposure estimates suggested that consumers take in less than 1 microgram per kilogram bodyweight per day of the substance over the long term. Experts viewed the question of the genotoxicity and mutagenicity of by acrylamide as crucial. There was as yet no precise assessment of human carcinogenicity available, because data from studies of occupational exposures were, due to the low numbers of cases involved, not appropriate for capturing small changes in cancer risk. However, acrylamide was to be found in higher quantities in foodstuffs than any other carcinogen. The conclusion of the experts was, therefore, to consider acrylamide in foodstuffs as a 'major concern' (WHO/FAO 2002: 1 and 20). The process of the formation of acrylamide in foodstuffs was still not understood although the methods for its detection were considered valid. Given the incomplete state of the data, no specific recommendations regarding consumption of specific foodstuffs could be given, only tentative advice (WHO/FAO 2002).

The event reinforced the BgVV's stance and approach up to that point that 'the very thorough discussion between internationally recognised experts did not reach any conclusions which were substantially divergent from the BgVV's previously held opinion' (BgVV 2002e: 5).

On 1 August 2002, the BgVV issued a statement on the current situation. The Agency stated that a scientifically established conclusion regarding acrylamide in foodstuffs could not be expected in the foreseeable future. Internationally, the only recommendation being made was to keep exposure 'as low as reasonably achievable' (ALARA). The BgVV perceived a necessity to improve the situation as fast as possible, and suggested as a first measure the introduction of an 'action value' of 1,000 micrograms, i.e., 1 milligram per kilogram foodstuff. With more meaningful data, maximum values for specific foodstuffs could perhaps be established later. In accordance with the current state of knowledge, the action value was applied above all to crisps and other similar snacks, biscuits, and crispbreads (BgVV 2002f).

In December, the Swiss health authorities issued an estimation regarding the absorption of acrylamide through foodstuffs. The diet of 27 participants had been analysed over two days. A fairly low level of acrylamide intake was ascertained – just 0.28 micrograms per kilogram body weight per day. A significant result arose from the observation of relative acrylamide absorption from different foodstuffs: 36% of the daily dose came from coffee. It was assumed that this value was so high because potato products were underrepresented in the experiment. Nonetheless the conclusion remained that foodstuffs which contained relatively low levels of acrylamide could make a significant contribution to the total absorbed if consumed in large quantities (BAG 2002).

In December 2002, discussions began between the manufacturers of affected foodstuffs and the authorities at BMVEL (BLL 2003).

A particular hazard from acrylamide for young people was identified, since it was supposed that they would have a higher consumption of crisps and chips (BfR 2002). In July 2003 the results of a study evaluating the exposure of Berlin schoolchildren in the 10th grade of general schools were released. Until then assessments of the exposure to acrylamide had rested largely on data from the national consumption study of 1989 and the National Health Survey of 1998. From these old data, it was concluded that young male adults between 19 and 24 years of age would, on average, take in 50 micrograms of acrylamide per day. An outcome reported from the more recent study was that the average intake was 69 micrograms of acrylamide, or 1.1 micrograms per kilogram bodyweight per day. Among 5% of the schoolchildren, this value was as high as 3.2 micrograms per kilogram bodyweight per day (Mosbach et al. 2003).

On 19 March 2004 the BfR issued a statement on 'two years of acrylamide' and gave a list of activities analysing the risk up to that point in time:

- The formation mechanism could be explained to a large extent. Validated analytical methods were available for many foodstuffs.
- A daily intake of 0.5 to 1.0 micrograms per kilogram bodyweight for adults was estimated.
- Through blood tests it could be shown that human exposure actually occurred.
- No indication existed that even low doses of acrylamide could be considered without risk of cancer.
- Experts estimated that an average daily exposure to acrylamide of one microgram per kilogram bodyweight over a lifetime would result in 6 to 100 additional cases of cancer per 10,000 individuals.

Using the state of knowledge current at the time, the content of acrylamide in commercially produced and prepared foodstuffs could be reduced by varying the factors of temperature, time and water content. At the same time, it became clear that the preparation of a product could not be altered beyond certain limits without the product losing its characteristic qualities. Further findings were still to be collected (BfR 2004).

The manufacturers cooperated in a project researching acrylamide. Various analytical, toxicological and technological issues were addressed with the aim of reducing the quantity of acrylamide in foodstuffs. Participants included the German Research Centre for Food Chemistry, the Institute for Food and Environment Research, the German Institute of Food Technology, the Federal Research Centre for Nutrition and Food, and the University of Kaiserlautern. The project's final report was submitted in December 2005 (BLL/FEI 2005).

The unanimous opinion among experts is still that a conclusive quantification of the risk arising from acrylamide in foodstuffs will not become available in the foreseeable future. The risk assessment was quite comprehensively carried out, despite the complexity of the problem. The further development of risk governance approaches took account of the fact that uncertainty is very high in the acrylamide case. Nevertheless, it is striking that the manufacturers were far more involved in the attempt to minimise uncertainty and ambiguity than the other stakeholders, in particular the consumer protection organisations.

Concern Assessment

The IRGC framework proposes that the concern assessment phase should include: research on concerns about and perceptions of risk by interested societal groups; an understanding of potential economic effects; and the courses of action for addressing them. The rationale for a concern assessment phase is rooted in well-described differences in the way individuals and scientists assess and perceive risks which are described briefly in this section.

A central finding of risk perception research is that the estimation of a hazard by the majority of the population frequently does not correspond with the scientific assessment of the risk.[19] Schütz and Peters (2002) see an explanation of this in the fact that different rules underlie the formation of risk realities for the population, for scientists, and also for the media as the link connecting them.

'Lay people's' perceptions of risk are not based on methodically sophisticated and systematic procedures, but incorporate several qualitative aspects of the risks. For example, risks are judged to be less hazardous if they are seen as being controllable or if they are undertaken voluntarily. Risks are viewed as many times more hazardous when an institution viewed as responsible for imposing the risk can be identified and targeted for recrimination, particularly when they might be viewed as making profit from the situation (Renn and Kastenholz 2000). Furthermore, hazards are held to be less worrying if they have a limited potential to create catastrophes, because the simultaneous death of many people in one place is perceived as significantly more threatening than the deaths of the same number of people over a period of time or in several locations (Jungermann and Slovic 1993). Additional influences include the state of scientific knowledge, the personal effect of the risk, degree of familiarity with the risk, the personal utility of the source of the risk, the balance of risk and utility within society including its possible effects on future generations. But lay people also weigh up probabilities subjectively; for instance, the probability of an event is reckoned to be higher if it can more easily be imagined or if it has already been experienced (Schütz and Peters 2002).

These assumptions also hold true for cancer risks, since the perceptions of the population do not correspond with the scientific assessment of the causes of cancer. This can be illustrated by discussing three examples from the USA in the 1990s. The pesticide Alar, used in the apple industry, was a cause for great concern. The US Environmental Protection Agency estimated that the increase in the risk of cancer to an individual from a lifetime's consumption of apples treated with Alar was 0.00045%. Nonetheless, people developed a great aversion to Alar, and sales of apples fell enormously. In the second case, an article about the relationship between drinking coffee and an increase in the incidence of cancer of the pancreas was publicised. When this finding became known, the consumption of coffee fell for a number of days, but then soon returned to its former level. (The relationship reported could not be verified.) In the third instance, the US Food and Drug Administration planned to withdraw the sweetener saccharin from the market following its classification as

[19] An overview of the current state of risk perception research can be found in Schütz and Wiedemann (2005).

a carcinogen. However, protest from the population against this measure was so strong that it was not implemented (Brody 1999).

These examples illustrate that people are indeed prepared consciously to accept some risk of cancer with particularly popular or useful foodstuffs. The opposite reaction in the case of the pesticide Alar may well be explained by the involuntary nature of the risk, that is, that people viewed that the apples were treated with Alar by the producers and that they could only avoid the risk by stopping eating apples. In contrast to coffee, the slighter preference for apples in individuals' diets probably actually resulted in a permanent boycott of apples.

People's views of risks are also influenced by media coverage (Renn and Kastenholz 2000). The topic of acrylamide was a particular object of media interest in 2002. According to a search of the Lexis-Nexis online archive, over 150 media entries on the theme appeared between April and November 2002 (Wiedemann et al. 2002). These reports appeared to lead to least a temporary alteration in consumer behaviour, because sales of crispbreads and potato crisps, both types of product which were at the centre of media coverage along with chips, fell markedly in 2002 and 2003.[20] At the beginning of February 2003, the opinion research institute Polis ascertained that 15% of Federal citizens had changed their eating patterns as a result of media coverage of acrylamide (Stern Online 2003b). In 2003, total sales of savoury snacks had actually risen 2.7% the previous year, although sales of potato crisps specifically were down 4.8%. Instead, more peanuts, pretzel sticks, *Erdnussflips* (extruded specialities), and savoury and cheese biscuits were bought (Handelsmagazin 2004).

Manufacturers and consumer protection organisations are the two groups whose concerns should have been incorporated in a risk appraisal phase for acrylamide. Survey findings show that consumer protection organisations enjoy the highest levels of trust within a population when information about food risks is in question (vzbv 2001; EC 2006). The German acrylamide case was characterised by substantial disagreements between the two parties. Consumer protection organisations claimed that the manufacturers would not do enough against possible health risks for consumers, and felt that they had not been adequately involved in the risk assessment process (vzbv 2002).

The manufacturers initially felt that they had been made solely responsible for the acrylamide problems (e.g., GDCh 2003, confirmed in expert interviews). They signalled their willingness to cooperate on solutions, but emphasised the need for

[20] An interesting point in this context is that sales of crispbreads were apparently already falling in 2002. However, the decline in sales of potato crisps began, according to the statements of market leaders Intersnack, only in November 2002, following a new wave of media coverage on estimations of incidences of death due to acrylamide in foodstuffs. This could, firstly, be because crispbreads tend to be eaten by people who are more interested in health issues and so react more sensitively to such media reports. Alternatively, it could be that there are many people who are more ready to give up crispbreads than potato crisps, just as the Americans gave up apples more quickly than coffee. A further analysis of the figures wold be of interest to clarify whether the Germans actually reacted more cautiously than their European neighbours. With potato crisps in particular, it should be recalled that a soccer world cup took place in 2002, so a certain decline in sales might have been expected in 2003, even without acrylamide.

research before the delivery of assessments and criticised the first assessments of the BgVV about human exposure as 'unscientific'. The introduction of an action value was viewed 'very critically', since it did 'not contribute to a drastic reduction of acrylamide absorption through foodstuffs' (BLL 2002b).

The dispute, conducted in public, reached a crisis when, during an information event at the BgVV in August 2002, a manufacturers' association attempted to prohibit the choice of moderator for the closing debate. The doctor of chemistry, who had been appointed moderator, was accused of being a professed communist. 'There is a strong suspicion that, through the choice of moderator, a clear opposition of the BgVV to the food industry is being engineered' (Baking Goods and Materials Producers' Association, quoted in Schrum and von Aster 2003).

In 2003, after a balance had been drawn up in some media on the occasion of the anniversary of the announcement of the Swedish findings, the scope of media coverage fell markedly. In April 2003 the last press statement on the topic was issued by the BfR.

The discord remained. In 2005, the consumer organisation Foodwatch expressed the opinion that the acrylamide case was an example of 'how frivolously commerce and politicians deal with risk substances' (Foodwatch 2005: 2). The manufacturers, on the other hand, saw the topic's handling by the media as 'barely responsible', complained that their largest loss of sales internationally were in Germany, and emphasised the danger that such crises represent for jobs (Wolf 2006).[21]

These episodes/incidents show that no systematic concern assessment was conducted in the German acrylamide case. Instead the controversy between the chief stakeholders, the manufacturers and consumer protection bodies, was conducted in public. The media coverage partly contributed fuel what was already heated debate.

Tolerability and Acceptability Judgement

Judging whether a risk is seen as acceptable or tolerable involves two steps, risk characterisation and risk evaluation. This phase is often the most controversial part of the risk governance process, particularly when, as in the case of acrylamide, the

[21] An article on p. 1 of the *Kölner Express* of 25 November 2002 was seen as particularly lurid:

> Frying-Death: More Deaths Than Traffic. It's hiding in chips, crisps, and popcorn – yes, even in Mum's roast potatoes: the secret cancer agent acrylamide. Boffins have now worked out that the chip poison could cause more than 8,000 deaths a year in Germany. Is there anything left we're allowed to eat? After the horror of BSE (mad cow disease), and the trauma of foot and mouth disease the latest findings have stoked more fears. What is this stuff that leaves mealtimes sticking in our throats? (Renz 2002)

Newspapers have to be filled, even on days when there is little news. Then journalists turn to less momentous themes and try to make news of them. The results can then make such an article. Stephan Russ-Mohl, Director of the European Journalism Observatory, is critical: 'Even under trying editorial circumstances one should always consider whether one is causing unnecessary fears' (Stute 2006).

risk in question is accompanied by complexity, uncertainty, and ambiguity. Deciding that a risk is 'acceptable' means that no further measures need be taken to reduce it. A 'tolerable' risk, in contrast, is one that is viewed as reasonable, but further measures to reduce the risk are deemed necessary. The IRGC framework describes in detail the processes and factors to be considered in evaluating the tolerability or acceptability of a risk (IRGC 2005).

In the German acrylamide case, it is certain that no such systematic and comprehensive judgement of the acceptability and tolerability of the presence of acrylamide in foodstuffs took place. Stakeholders were consulted, and it was attempted by the ministry to conduct a tolerability judgement in the first instance with the representatives of the manufacturers. However, no agreement between the participants was reached, so that each group developed its own judgements based on the available information and its own concerns. In its statement of 19 March 2004, the BfR reached the conclusion that with acrylamide it was dealing with a risk that was 'to be regarded as significant in comparison with other material risks', and demanded 'that the exposure be drastically reduced as rapidly as possible' (BfR 2004: 1). This decision corresponds to a judgement of the risk as *tolerable* in IRGC's terminology. In its 'Acrylamide Status Report', the BLL described the 'problem' as being 'taken very seriously by all participants, since acrylamide appeared to be carcinogenic in animal experiments', yet simultaneously emphasised that acrylamide was also formed during the domestic preparation of food, and had been in the foods in question since the discovery of fire (BLL 2003: 1). This description of the situation tends far more to a judgement of the risk as *acceptable*. It is noticeable that the term acrylamide appears in quotation marks in the title of the document and partially also in the document. The quotation from Kulling in the journal of the Society of German Chemists sounds similar: 'the daily consumption of fried foods brings an entirely different health problem to the fore: one-sided, imbalanced nutrition, together with a high energy intake, above all through a large proportion of fats' (Kulling 2002: 1104).

In contrast, consumer advocates drew on the appraisals of scientists that regarded 'acrylamide as far more dangerous than all other chemicals yet found in foodstuffs' (bio verlag 2002). Such statements pushed the acrylamide risk towards one which was no longer tolerable – although no evidence has been found during this research that consumer advocates or other actors wanted to ban affected foodstuffs. However, as shown below, consumer advocates called for the most far-reaching measures for the reduction of the acrylamide levels in foodstuffs. Apart from this, the German Advisory Council on the Environment came to the conclusion that 'the intake of acrylamide in food lies beyond the realm of the tolerable' in its environmental survey of 2004 (Deutscher Bundestag 2004). However, in the question session of the German Bundestag on 13 November 2002, the then parliamentary secretary to the Federal Minister for Consumer Protection, Nutrition and Agriculture announced that 'Let me refer to a similar process: Benzopyrene is produced by barbecueing. This is accepted, because the barbecued food tastes better. Nonetheless, the public has the right to be informed of the risk to health' (Deutscher Bundestag 2002b: 482). This multitude of judgements then accompanied the introduction of measures in risk management with which the problem of acrylamide in foodstuffs should be met in Germany.

Risk Management

When all the phases of the Risk Governance Framework discussed above have been applied and the necessary information processed and collected, risk management is faced with three possible situations:

- *Intolerable*: the risk source (such as a technology or a chemical) needs to be abandoned or replaced or, in cases where that is not possible, for example natural hazards, vulnerabilities and exposure need to be reduced.
- *Tolerable*: the risks need to be reduced or handled in some other way within the limits of reasonable resource investments (ALARA). This can be done by private actors such as corporate risk managers, or public actors such as regulatory agencies, or by public-private partnerships (IRGC 2005). The precautionary principle might also be applied in selecting actions to take.[22]
- *Acceptable*: the risks are so small, perhaps even regarded as negligible, that any risk reduction effort is unnecessary. However, risk-sharing via insurance and further risk reduction on a voluntary basis both present options for action which can be worth pursuing.

The IRGC framework discusses the risk management approaches that might be considered, as well as appropriate measures for involving relevant stakeholders depending on the type of risks, the results of the assessments, and the extent to which the risks remain complex, uncertain, or ambiguous.

The acrylamide case represents a situation in which the German authorities determined that the risk was at some level tolerable, but was one that needed to be managed carefully. The BgVV, as discussed earlier, at this time still combined the role of risk assessment and risk management in the same organisation. After the WHO and the EU had classified acrylamide as a cause for concern and recommended the fastest possible reduction of exposure, the BgVV regarded setting a limit on acrylamide levels in food as essential. They made the risk management decision to apply the precautionary principle and established an 'action value' of 1,000 micrograms per kilogram foodstuff. The level chosen for acrylamide corresponded to the

[22] The European Commission has also endorsed, under certain conditions, the application of the precautionary principle, for instance where a phenomenon, product, or process potentially leads to hazardous consequences, but the risk cannot be defined with sufficient certainty. The recommendation also acknowledges the difficulty of 'weighing the rights and freedoms of individuals, companies and associations on one hand with the need to reduce the danger of negative consequences for the environment and the health of people, animals, or plants on the other' (EC 2002a: 1). However, the precautionary principle has attracted criticism because this approach does not permit risk managers a differentiation by priorities. In addition, what 'reasonably' can be taken to mean within the term ALARA is open to question (BfR 2005).

limit for Benzopyrene in the regulations concerning flavourings and cheese (BgVV 2002f).

The manufacturers' reaction was to reject this. In their view, the level had no toxicological foundation. The BLL reacted with a press release stating that the level had absolutely no effect on the marketability of products (BLL 2002a).

Following the information event of 29 August 2002 the minimisation concept was presented. According to the BMVEL's press release, the idea had been proposed by the newly established BVL and 'was reviewed with the highest foodstuffs monitoring authorities of the regional governments' (BMVEL 2002). The minimisation concept comprised the idea that foodstuffs be classified into defined commodity groups which would then sampled for acrylamide. In each commodity group, a 'signal value' was established which was defined as the lowest acrylamide level found in the top 10% of foods within a commodity group (in other words, at the 90th percentile of all values). In no case could the signal value be set at a level greater than 1,000 micrograms per kilogram of foodstuff. Signal values, once set, could not be raised (BVL 2005c). So-called 'minimisation dialogues' were then held with the manufacturers whose products exceeded the signal values in order to introduce measures for the reduction of acrylamide concentrations.

The first calculation of signal values was made in September of 2002. Shortly before the introduction of the minimisation concept, only a very few analytical findings were available: 63 for bread, 42 for crispbreads and 87 results for potato crisps (Galle-Hoffmann 2002). By the publication of the last calculation in October 2005, 10,000 tests had been evaluated. The results of the five published calculations are presented in Table 3.

The aim of the minimisation concept is that the difference between the levels measured in a commodity group be made as narrow as possible, or that the signal value be as near as possible to the average. But the levels do not vary to the desired degree.

In 2003–2005 observed levels of acrylamide were at or above the signal value from the preceding year or above the maximum 1,000 micrograms limit in seven to ten commodity groups. Acrylamide levels appeared to decline in 5 out of the 13 commodity groups, including chips, shortcrust pastries, baked goods for diabetics, almond biscuits, and children's rusks. However, even in these commodity groups, the trend was not consistent.

The BfR welcomed the minimisation concept but argued that data collection needed to be significantly improved (BfR 2004). It was confirmed in experts' discussions that the process of data collection was unstructured and that there was no control of any changes from the previous year's values; for example, different products were tested from year to year which could account for some of the inconsistencies in results observed. In any case, the monitoring of foodstuffs within the regions was reaching its limits, because the organisations involved also had to deal with other risks associated with foodstuffs.

The BfR reached the conclusion 'that a representative quantification of the reduction of the exposure is not possible at present' (BfR 2004) because acrylamide exposure from the domestic preparation of the affected products could not be as-

Table 3 Overview of the signal values for the period 2002–2005.

Commodity Group	**Date of Analysis**				
	September 2002	January 2003	November 2003	November 2004	October 2005
Acrylamide in micrograms per kilogram of foodstuff					
Shortcrust pastries	800	660	575	575 (760)*	300
Breakfast cereals (Cornflakes and Muesli)	260	260	200	200 (240)*	180
Roast coffee	370	370	370 (520)*	370 (420)*	370 (537)*
Potato crisps	1000 (1500)*	1000 (1200)*	1000 (1470)*	1000 (1029)*	1000 (1333)*
Crispbreads	610	610	610 (1260)*	610 (640)	590
Chips, prepared	770	570	570	540	530
Potato fritter, prepared	1000	1000 (1300)*	1000 (1080)*	1000 (1215)*	1000 (2520)*
Gingerbread and pastries containing gingerbread	1000	1000 (1370)*	1000 (1460)*	1000 (1020)*	1000 (1270)*
Almond biscuit	1000	710	710 (760)*	560	560 (706)*
Rusks or biscuits for babies and toddlers	n.c.	n.c.	360	360	245
Long-life baked goods for diabetics	n.c.	n.c.	1000 (1740)*	1000 (1010)*	545
Soluble coffee	n.c.	n.c.	1000 (1110)*	1000 (2380)*	1000 (1030)*
Coffee substitute	n.c.	n.c.	1000 (2080)*	1000 (2910)*	1000 (2341)*

*Observed value
**n.c. = not calculated
Source: BVL (2005a)

sessed (BfR 2004). The BVL also held a critical view of the results of the minimisation concept, because 'acrylamide levels in foodstuffs have only been fractionally reduced', and called for 'a closer relationship' with manufacturing practice (BVL 2005b).

No official critical evaluations of the minimisation concept from the manufacturers' side could be identified. However, unofficial discussions suggest that it was predominantly viewed as a measure for mollifying consumers. The BfR's appraisal – that the tests were conducted unsystematically and that thus the results did not reflect reality – was affirmed. In addition, the fact that foreign manufacturers had not been involved in this process was seen as a problem.

In contrast, the consumer protection organisations complained that the minimisation concept did not go far enough. Foodwatch criticised the fact that the worst levels were taken as benchmarks. To arrive at best practice, Foodwatch argued that it would be more sensible to take lower levels as reference values. Above all, there was a lack of pressure on manufacturers. The test results were to be published to force the manufacturers to take steps to reduce acrylamide levels (Foodwatch 2006a). Foodwatch gave, as an example, the Dutch firm FZ Organic Food, a manufacturer of organic potato crisps amongst whose products the Tra'fo brand was found to contain extremely high levels of acrylamide. After Foodwatch had made this public, FZ Organic Food had to accept a 30% drop in sales. The company was subsequently able to alter its production and significantly reduce the acrylamide levels in its crisps (Foodwatch 2005).[23] A further central demand of Foodwatch was the labelling of products with a scale which would indicate the 'exposure class' of a product (Foodwatch 2006b). Dissatisfaction with the regulations was clearly expressed (cf. 'Künast bows to the snacks industry', '*Künast kuscht vor Knabberindustrie*', bio verlag 2003).

The German Food and Feed Code prohibits authorities in Germany to name manufacturers publicly.[24] However, in the near future a law relating to consumer information[25] would be enacted to take account of the food scares of the recent past and strengthen the rights of consumers to information. The current draft, however, still includes grounds for exempting information. This includes cases in which 'through the information requested, company or commercial secrets or other information relevant to competition which is comparable to a company or commercial secret in its significance for the company would be made public' (VIG Entwurf). This restriction also attracted strong criticism since this issue is certainly handled in a less restrictive way in other countries, for instance in Sweden with the public principle mentioned above.

This prohibition on divulging the names of manufacturers is not entirely advantageous to them all. A massive image problem was created for the crispbread manufacturer Wasa, whose products were from the beginning shown to contain very low levels of acrylamide. Their sales also declined sharply due to the acrylamide levels found in their competitors' products (Kruse 2003).

---

[23] Foodwatch's test results are available on the company's homepage, accessed 31 May 2006, http://www.fzorganicfood.com.

[24] However, there is an exception to the prohibition on naming manufacturers which arises from the consumer information law of the federal state of North Rhine-Westphalia. In Paragraph 4, 'right to information', it is stated: 'Every natural person has, due to the provisions of this law against the points named in §2, the right to access to official information available at that point' (IFG NRW). On this legal basis the individual acrylamide test results are indeed available on the internet page of the Ministry of the Environment and Conservation, Agriculture and Consumer protection of North Rhine-Westphalia (cf. http://www.munlv.nrw.de/sites/arbeitsbereiche/verbraucherschutz/produkte.htm, accessed 31 May 2006). But the findings are presented in a very opaque fashion and provide no practical help to consumers with their purchasing decisions. According to the employee responsible for this at the ministry, however, this is due purely to a lack of finances and personnel. Unfortunately, the number of visits to the site had not been counted. In conversations the assumption was expressed several times that such publications would affect the consumer as confidence-building measures which would hardly alter the long-term purchasing decisions of average consumers. In this context it would have been very interesting to gather evidence for this assumption.

[25] Verbraucherinformationsgesetz (VIG).

The German minimisation concept was praised internationally many times, but not adopted. One reason may be that nobody wanted to give new life to the topic of acrylamide in neighbouring European countries. Most people with whom the topic was discussed reckon that it will also soon be forgotten in Germany.

In Switzerland, the German signal values were even taken as a reason to ban a children's biscuit made by Milupa (Nachrichten.ch 2005). In Germany, no attempt has been made to enforce the minimisation concept, e.g. withdraw a product from the market due to a high acrylamide level.

This overview of the most important events in risk management in Germany has shown that a solution was found which corresponds to the recommendations of IRGC. For risks which are evaluated as tolerable, IRGC suggests a management strategy be considered which is based on the precautionary principle. Since, as the events summarised in this chapter have indicated, this risk could be dealt with as a tolerable one, the minimisation concept was considered to be an appropriate measure for reducing that risk. It provided an opportunity to examine individual exposures to acrylamide, which can differ markedly depending on the foodstuff.

However, for the majority of people with whom the issue was discussed, the minimisation concept was judged to be a failure from the outset, because it was not combined with incentives or sanctions. It was developed by regulatory authorities and the affected industries without input from other stakeholders or evaluation of how it met other important criteria advocated by IRGC. For example, the minimisation concept did not really provide a valid monitoring programme because of problems with inconsistencies in the sampling program from year to year.

Summary and Conclusion

The purpose of this document has been to examine the case of acrylamide found in foodstuff in Germany within the context of the IRGC risk governance framework. The IRGC framework should improve the societal process of risk governance, especially for systemic risks such as acrylamide. The acrylamide case presented a particular challenge for risk governance as it was a situation in which the emergence and recognition of the potential risk occurred rapidly and caught many unprepared. The news that acrylamide had been discovered in foodstuffs reached the public under unusual circumstances. The way in which scientific information was released to the public contributed to ongoing debates about the quality of the research findings. Although much time has been spent on the investigation of acrylamide, a conclusive assessment of the hazard to humans is not likely to be available for the foreseeable future.

The rapidity with which the crisis developed made it difficult for a full pre-assessment, as advocated by IRGC, to be completed. Two factors might have helped. No early warning system existed with whose help the risk governance process could have begun under less pressure. The situation was further described by a lack of scientific conventions which could define the appropriate basis for publicly conducted disputes. For sensitive topics like this, it would also have been appropriate and helpful to have some conventions with which the specialist and popular press might have had to comply without, of course, interfering with the fundamental freedom of the press, but to avoid creating more of a problem. At least the third and crucial factor was met; the factor that acrylamide in foodstuffs presented a type of risk to which society would consider it necessary to apply the the process of risk governance.

The second phase of the framework involves risk appraisal, in which all the available information which society needs to decide whether to take a risk is collected. For the acrylamide case, the risk assessment was conducted largely as the framework suggests. However, the opinions of stakeholders were not considered to the extent recommended. Only a few activities were undertaken to include stakeholder concerns regarding the effects of the risk and the possible risk management options.

For the third phase of the framework, no systematic judgement took place whether the acrylamide risk was to be viewed as acceptable or tolerable on the basis of the knowledge gained. Instead, a number of voices were raised, by direct stakeholders and other public actors, who had developed their own judgements. It is clear from official statements, that government authorities regarded this as a tolerable risk, a risk which needed to be reduced. Manufacturers, however, were inclined to view acrylamide as an acceptable risk, one for which little or no action should be required. The diversity of opinions about the tolerability/acceptability accompanied the introduction of risk management measures.

In the last section of the framework, the risk management phase, the government authorities implemented the minimisation concept, a tool which, could have been perfectly suitable for leading to a reduction of acrylamide levels in foodstuffs. However, since the government relied on voluntary cooperation of the industry, the minimisation concept was not combined with any incentives or sanctions. This half-hearted implementation led to a risk management strategy, that, in reality, would have been only suitable for handling an acceptable risk.

At the same time, the manufacturers in Germany and in Europe made considerable efforts to develop improvements in production methods which corresponded more closely to the management of a tolerable risk.

Whether the manufacturers would have been more proactive in building trust if an early warning system had existed and given them more time to prepare themselves to deal with the problem remains an open question. It has not been possible within this study to confirm whether the manufacturers were made fully responsible for the problem. However, it appears that they could have made far more use of the opportunity to publicise their activities to monitor and reduce acrylamide levels.

Had the IRGC framework been followed, it would have been necessary to establish a consensus between manufacturers and consumer protection groups. However, it is doubtful that inclusion of the public in the debate over the management of acrylamide would have been suitable for the acrylamide problem in Germany. For a while, the public avoided some popular products, perhaps waiting for a sign that a solution had been found or that something had happened. Since nothing did happen, and the manufacturers were not obligated to take any measures, it was certainly no longer clear to the public why they should restrict themselves from food products they desired and so they returned to their former eating patterns.

Although the importance of separating risk assessment from risk management is often stressed in various risk governance frameworks, including IRGC's, the organisational division of risk assessment and risk management in Germany between the BfR and the BVL, respectively does not appear to have played a particularly positive role in the course of the events discussed. In private conversations, it was lamented that the old BgVV had had greater powers of enforcement and implementation. The BfR could only issue risk assessments but without any enforcement powers, while the BVL had relatively little independence to implement effective risk management measures.

The result of this case study is that if the IRGC framework had been applied it would probably have improved the societal process of risk governance for acrylamide in food in Germany. The framework helped to give a structure to this evaluation of the acrylamide case and to identify clearly the weak points where the process could have been improved, thereby leading to results which would have had greater benefits for society.

Index of Abbreviations and Translated Names

	Gesetz über Nachfolgeeinrichtungen des Bundesgesundheitsamtes	Act on Successor Institutions of the Federal Public Health Department
	Gesetz zur Neuorganisation des gesundheitlichen Verbraucherschutzes und der Lebensmittelsicherheit	Act on the Reorganisation of Customers' Health Protection and Food Safety
	Verband der Backmittel- und Backgrundstoffhersteller e. V.	Baking Goods and Materials Producers' Association
	Wissenschaftsrat	Science Council
BfArm	Bundesinstitut für Arzneimittel und Medizinprodukte	Federal Institute for Drugs and Medical Devices
BfEL	Bundesanstalt für Ernährung und Lebensmittel	Federal Research Centre for Nutrition and Food
BfR	Bundesinstitut für Risikobewertung	Federal Institute for Risk Assessment
BGA	Bundesgesundheitsamt	Federal Public Health Department
BgVV	Bundesinstitut für gesundheitlichen Verbraucherschutz und Veterinärmedizin	Federal Institute for Consumer Health Protection and Veterinary Medicine
BLL	Bund für Lebensmittelrecht und Lebensmittelkunde e. V.	German Federation of Food Law and Food Science

BMELF	Bundesministerium für Ernährung, Landwirtschaft und Forsten	Federal Ministry of Food, Agriculture and Forestry (from 20 September 1949 to 12 January 2001)
BMELV	Bundesministerium für Ernährung, Landwirtschaft und Verbraucherschutz	Federal Ministry of Food, Agriculture and Consumer Protection (since 22 November 2005)
BMVEL	Bundesministerium für Verbraucherschutz, Ernährung und Landwirtschaft	Federal Ministry of Consumer Protection, Food and Agriculture (from 12 January 2001 to 18 October 2005)
BMG	Bundesministerium für Gesundheit	Federal Ministry of Health
BVL	Bundesamt für Verbraucherschutz und Lebensmittelsicherheit	Federal Office of Consumer Protection and Food Safety
ChemVerbotsV	Chemikalienverbotsverordnung	Ordinance on Prohibited Chemicals
DFAL	Deutsche Forschungsanstalt für Lebensmittelchemie	German Research Centre for Food Chemistry
DIL	Deutsches Institut für Lebensmitteltechnik	German Institute of Food Technology
EFSA	Europäische Behörde für Lebensmittelsicherheit	European Food Safety Authority
FAO	Welternährungsorganisation	Food and Agriculture Organisation
FLI	Friedrich-Loeffler-Institut – Bundesforschungsinstitut für Tiergesundheit	Friedrich-Loeffler-Institute – Federal Research Institute for Animal Health
GDCh	Gesellschaft Deutscher Chemiker	German Chemical Society
GefStoffV	Gefahrstoffverordnung	Ordinance on Hazardous Substances
ILU	Institut für Lebensmittel- und Umweltforschung	Institute for Food and Environment Research
JECFA		Joint FAO/WHO Expert Committee on Food Additives
LFGB	Lebensmittel-, Bedarfsgegenstände- und Futtermittelgesetzbuch	German Food and Feed Code
MUNLV	Ministerium für Umwelt und Naturschutz, Landwirtschaft und Verbraucherschutz NRW	Ministry of the Environment and Conservation, Agriculture and Consumer Protection of North Rhine-Westphalia
RKI	Robert-Koch-Institut	Robert Koch Institute

SRU	Sachverständigenrat für Umweltfragen	German Advisory Council on the Environment
TrinkwV	Trinkwasserverordnung	Drinking Water Ordinance
vzbv	Verbraucherzentrale Bundesverband e. V.	Federation of German Consumer Organisations

References

BAG, Bundesamt für Gesundheit, 2002, Preliminary communication. Assessment of acrylamide intake by duplicate diet study, http://www.bag.admin.ch/dokumentation/medieninformationen/01217/index.html?lang=de&msg-id=4095 (accessed 31 May 2006).

BBC News, 2002, Bread and crisps in cancer risk scare, *BBC News*, 25 April 2002, http://news.bbc.co.uk/1/hi/health/1949413.stm (accessed 31 May 2006).

BfR, Bundesinstitut für Risikobewertung, 2002, Aktualisierung der Risikobewertung von Acrylamid in Lebensmitteln, Stellungnahme des BfR vom 28. November 2002.

BfR, Bundesinstitut für Risikobewertung, 2003a, Bedeutung der Studie von Mucci et al. für die Risikobewertung von Acrylamid in Lebensmitteln, Stellungnahme des BfR vom 25. Februar 2003.

BfR, Bundesinstitut für Risikobewertung, 2003b, Bedeutung der Arbeit von Pelucchi et al. für die Risikobewertung von Acrylamid in Lebensmitteln, Stellungnahme des BfR vom 8. Juli 2003.

BfR, Bundesinstitut für Risikobewertung, 2004, Zwei Jahre Acrylamid – Eine Bilanz aus Sicht der Risikobewertung, Stellungnahme des BfR vom 19. März 2004.

BfR, Bundesinstitut für Risikobewertung, 2005, Risikobewertung genotoxischer und kanzerogener Stoffe soll in der EU harmonisiert werden, Stellungnahme Nr. 029 des BfR vom 18. Mai 2005.

BgVV, Bundesinstitut für gesundheitlichen Verbraucherschutz und Veterinärmedizin, 1999, Bericht über die 58. Sitzung der Kommission für kosmetische Mittel des Bundesinstituts für gesundheitlichen Verbraucherschutz und Veterinärmedizin (BgVV) am 29. April 1999 in Berlin.

BgVV, Bundesinstitut für gesundheitlichen Verbraucherschutz und Veterinärmedizin, 2002a, Schweden weisen Acrylamid in Lebensmitteln nach, Pressemeldung Nr. 10/2002, 25. April 2002.

BgVV, Bundesinstitut für gesundheitlichen Verbraucherschutz und Veterinärmedizin, 2002b, Zum Vorkommen von Acrylamid in Lebensmitteln, Bericht des Bundesinstituts für gesundheitlichen Verbraucherschutz und Veterinärmedizin über das Expertengespräch vom 14. Mai 2002.

BgVV, Bundesinstitut für gesundheitlichen Verbraucherschutz und Veterinärmedizin, 2002c, Einfluss der Ernährung auf die Aufnahme von Acrylamid. Durch Änderungen im Ernährungsverhalten kann der Verbraucher die Acrylamid-Aufnahme deutlich reduzieren, Stellungnahme des BgVV vom 4. Juni 2002.

BgVV, Bundesinstitut für gesundheitlichen Verbraucherschutz und Veterinärmedizin, 2002d, Analytik von Acrylamid in Lebensmitteln, Protokoll einer Sitzung im BgVV vom 2. Juli 2002.

BgVV, Bundesinstitut für gesundheitlichen Verbraucherschutz und Veterinärmedizin, 2002e, Zur gesundheitlichen Bedeutung des Vorkommens von Acrylamid in bestimmten zubereiteten Lebensmitteln, Kommentar zu den Ergebnissen einer internationalen Expertenberatung vom 2. Juli 2002.

BgVV, Bundesinstitut für gesundheitlichen Verbraucherschutz und Veterinärmedizin, 2002f, Höchstmengen für Acrylamid in Lebensmitteln, Stellungnahme des BgVV vom 1. August 2002.

bio verlag gmbh, 2002, Erste Fortschritte bei Acrylamid in Lebensmitteln, Bericht im ARD-Wirtschaftsmagazin Plusminus, Aktuelle Meldung vom 30.10.2002, http://www.naturkost.de/meldungen/021030ev1.htm (accessed 31 May 2006).

bio verlag gmbh, 2003, 'Künast kuscht vor Knabberindustrie'. Ein Jahr Acrylamid, Aktuelle Meldung vom 23.04.2003, http://www.naturkost.de/meldungen/2003/030423e2.html (accessed 31 May 2006).

BLL, Bund für Lebensmittelrecht und Lebensmittelkunde e. V., 2002a, Acrylamid-Forschung auf gutem Wege, Pressemeldung, 29. August 2002.

BLL, Bund für Lebensmittelrecht und Lebensmittelkunde e. V., 2002b, BLL Stellungnahme zu Acrylamid, Pressemeldung, 2. September 2002.

BLL, Bund für Lebensmittelrecht und Lebensmittelkunde e. V., 2003, 'Acrylamid'-Sachstandsbericht.

BLL/FEI, Bund für Lebensmittelrecht und Lebensmittelkunde e. V./Forschungskreis der Ernährungsindustrie e. V., 2005, *Development of New Technologies to Minimize Acrylamide in Food*, AiF-FV 108 ZBG.

BMVEL, Bundesministerium für Verbraucherschutz, Ernährung und Landwirtschaft, n.d., *Gesundheitlicher Verbraucherschutz – Wer macht was?*

BMVEL, Bundesministerium für Verbraucherschutz, Ernährung und Landwirtschaft, 2001, Bericht der Arbeitsgruppe Reorganisation des gesundheitlichen Verbraucherschutzes.

BMVEL, Bundesministerium für Verbraucherschutz, Ernährung und Landwirtschaft, 2002, Künast: Minimierungskonzept zu Acrylamid vorgelegt, Pressemeldung Nr. 279, 4. Juni 2002.

Böhme, Chr. and Grunow, W., 1993, Monomere in Kunststoffen mit Lebensmittelkontakt. Datenübersicht zur gesundheitlichen Beurteilung wichtiger Monomere, *MvP Hefte* **2**.

Böschen, S., Dressel, K., Schneider, M. and Viehöver, W., 2002, Pro und Kontra der Trennung von Risikobewertung und Risikomanagement – Diskussionsstand in Deutschland und Europa – Gutachten im Rahmen des TAB-Projektes 'Strukturen der Organisation und Kommunikation im Bereich der Erforschung übertragbarer spongiformer Enzephalopathien (TSE)', TAB-Diskussionspapier Nr. 10, Büro für Technikfolgenabschätzung.

Brody, J.E., 1999, Communicating cancer risk in print journalism, *Journal of the National Cancer Institute Monographs* **25**, 170–172.

BVL, Bundesamt für Verbraucherschutz und Lebensmittelsicherheit, 2005a, Übersicht über die Signalwerte der ersten bis fünften Berechnung, letztes Änderungsdatum: 21.10.2005, http://www.bvl.bund.de/cln_007/nn_493416/DE/01_Lebensmittel/03_UnerwStoffeUndOrganismen/04_Acrylamid/SW_Berrechnungen/5te_SW_Berrechnung/Uebersicht_5SWBerrechnung.html (accessed 31 May 2006).

BVL, Bundesamt für Verbraucherschutz und Lebensmittelsicherheit, 2005b, Acrylamidgehalte in Lebensmitteln sinken nur geringfügig, Pressemeldung, 21. Oktober 2005, http://www.bvl.bund.de/cln_007/nn_493378/DE/08_PresseInfothek/01_InfosFuerPresse/01_PI_und_HGI/Rueckstaende/Acrylamid_2005.html (accessed 31 May 2006).

BVL, Bundesamt für Verbraucherschutz und Lebensmittelsicherheit, 2005c, Minimierungskonzept zur Senkung der Acrylamidgehalte in Lebensmitteln, letztes Änderungsdatum: 28.11.2005, http://www.bvl.bund.de/cln_007/nn_493378/DE/01_Lebensmittel/03_UnerwStoffeUndOrganimen/04_Acrylamid/00_Minimierungskonzept/minimierungskonzept_node.html_nnn=true (accessed 31 May 2006).

ChemVerbotsV, 1993, Verordnung über Verbote und Beschränkungen des Inverkehrbringens gefährlicher Stoffe, Zubereitungen und Erzeugnisse nach dem Chemikaliengesetz, Fundstelle: BGBl I 1993, 1720.

Codex Alimentarius Commision, 2001, *Procedural Manual: Twelfth Edition*, Joint FAO/WHO Food Standards Programme, Rome.

Deutscher Bundestag, 2002a, 14. Wahlperiode, Gesetzentwurf der Bundesregierung: Entwurf eines Gesetzes zur Neuorganisation des gesundheitlichen Verbraucherschutzes und der Lebensmittelsicherheit, Drucksache 14/8747.

Deutscher Bundestag, 2002b, Plenarprotokoll 15/9, Stenografischer Bericht, 9. Sitzung, Berlin, Mittwoch, den 13. November 2002, pp. 479–495.

Deutscher Bundestag, 2004, 15. Wahlperiode: Umweltgutachten 2004 des Rates von Sachverständigen für Umweltfragen: Umweltpolitische Handlungsfähigkeit sichern, Drucksache 15/3600.

Dressel, K., Böschen, S., Schneider, M., Viehöver, W., Wastian, M. and Wendler, F., 2006, Food Safety regulation in Germany, in: E. Vos and F. Wendler (Eds.), *Food Safety Regulation in Europe: A Comparative Institutional Analysis*, Series Ius Commune, Intersentia Publishing, Antwerp, pp. 287–330.

Dreyer, M., Renn, O., Borkhart, K. and Ortleb, J., 2006, Institutional re-arrangements in European Food Safety Governance: A comparative analysis, in: E. Vos and F. Wendler (Eds.), *Food Safety Regulation in Europe: A Comparative Institutional Analysis*, Series Ius Commune, Intersentia Publishing, Antwerp, pp. 9–64.

EC, European Commission, 2002a, Mitteilung der Kommission – Die Anwendbarkeit des Vorsorgeprinzips. KOM (2000) 1, endgültig, Brüssel, den 2.2.2000, http://europa.eu.int/eur-lex/lex/LexUriServ/site/de/com/2000/com2000_0001de01.pdf (accessed 31 May 2006).

EC, European Commission, 2002b, Regulation (EC) No. 178/2002 of the European Parliament and of the Council of 28 January 2002 laying down the general principles and requirements of food law, establishing the European Food Safety Authority and laying down procedures in matters of food safety, *Official Journal* **L 031**, 1–24.

EC, European Commission, 2005, Acrylamide – EU Summary of Activities, http://europa.eu.int/comm/food/food/chemicalsafety/contaminants/acryl_database_en.htm (accessed 31 May 2006).

EC, European Commission, 2006, Risk Issues, Special Eurobarometer 238/Wave 64.1, 2006, http://europa.eu.int/comm/public_opinion/archives/eb_special_en.htm (accessed 31 May 2006).

EC/JRC, European Commission and Joint Research Centre, 2002, European Union Risk Assessment Report Acrylamide, CAS No. 79-06-1, EINECS No. 201-173-7, http://ecb.jrc.it/documents/Existing-Chemicals/risk_assessment/report/acrylamidereport011.pdf (accessed 31 May 2006).

FAZ, 2002, Verdacht auf Krebserreger in Pommes frites und Chips, *Frankfurter Allgemeine Zeitung*, 25. April 2002.

Foodwatch, 2005, Bio-Kartoffelchips: Der Acrylamid-Test von Foodwatch im Juli 2005, 05.12.2005, http://foodwatch.de/themen_aktivitaeten/acrylamid/kartoffelchips/index_ger.html (accessed 31 May 2006).

Foodwatch, 2006a, Acrylamid – Ein unnötiges Krebsrisiko, http://foodwatch.de/themen_aktivitaeten/acrylamid/index_ger.html (accessed 31 May 2006).

Foodwatch, 2006b, Orientierung für Verbraucher durch Kennzeichnung, http://foodwatch.de/themen_aktivitaeten/acrylamid/kennzeichnung/index_ger.html (accessed 31 May 2006).

FR, 2002, Krebsalarm bei Chips empört Forscher, *Frankfurter Rundschau*, 26. April 2002, No. 97, p. 38.

Galle-Hoffmann, U., 2002, Acrylamid in Lebensmitteln – Untersuchungsergebnisse aus Deutschland und anderen europäischen Ländern, Folienpräsentationen zur Informationsveranstaltung des BgVV vom 29. August 2002: Acrylamid in Lebensmitteln – ernstes Problem oder überschätzte Gefahr, http://www.bfr.bund.de/cd/1732 (accessed 31 May 2006).

GDCh, Gesellschaft Deutscher Chemiker e. V., 2003, Die chemische Seite von Mutter Natur – Veränderungen in Lebensmitteln, Pressemeldung Nr. 41/2003.

GefStoffV, 1986, Gefahrstoffverordnung vom 26. August 1986, BGBl. I p. 1470.

Gesetz über Nachfolgeeinrichtungen des Bundesgesundheitsamtes vom 24. Juni 1994, Fundstelle: BGBl I 1994, 1416.

Gesetz zur Neuorganisation des gesundheitlichen Verbraucherschutzes und der Lebensmittelsicherheit vom 6. August 2002, BGBl I Nr. 57, ausgegeben zu Bonn am 14. August 2002.

Handelsmagazin, 2004, Für gesalzene Umsätze, *Handelsmagazin*, **4**, 44–45, http://www.markant-gmbh.de/download/hm_2004/handelsmagazin_04_04.pdf (accessed 31 May 2006).

Henschler, D., 1993, Krebsrisiken im Vergleich. Forderungen für Forschung und politisches Handeln, *GSF mensch + umwelt* **8**, March, 65–72.

International Agency for Research on Cancer, IARC, 1994, *Acrylamide*, IARC Monographs on the Evaluation of the Carcinogenic Risk of Chemicals to Humans, Vol. 60, IARC, Lyon, France.

IFG NRW, 2001, Gesetz über die Freiheit des Zugangs zu Informationen für das Land Nordrhein-Westfalen (Informationsfreiheitsgesetz Nordrhein-Westfalen – IFG NRW), vom 27. November 2001.

IRGC, 2005, *Risk Governance: Towards an Integrative Approach*, International Risk Governance Council, White Paper No. 1, IRGC, Geneva, 2005.

Jungermann, H. and Slovic, P., 1993, Charakteristika individueller Risikowahrnehmung, in: Bayerische Rückversicherung (Ed.), *Risiko ist ein Konstrukt. Wahrnehmungen zur Risikowahrnehmung*, Gesellschaft und Unsicherheit, Vol. 2, Knesebeck, München, pp. 89–107.

Kolata, G., 2002, Scientists cautious on report of cancer from starchy food, *The New York Times*, 30 April.

Kruse, H., 2003, Ein hartes Brot, *Die Welt*, 14. Oktober.

Kulling, S.E., 2002, Krank durch Lebensmittel: Oder was wir selbst tun können, *Nachrichten aus der Chemie* **50**, Oktober, 1103–1105.

LFGB, 2006, Lebensmittel-, Bedarfsgegenstände- und Futtermittelgesetzbuch (Lebensmittel- und Futtermittelgesetzbuch – LFGB), Bundesgesetzblatt Jahrgang 2006, Teil I Nr. 20, ausgegeben zu Bonn am 27. April 2006.

Löfstedt, R.E., 2003, Science communication and the Swedish acrylamide 'alarm', *Journal of Health Communication* **8**(3), 407–432.

Löfstedt, R.E. and Renn, O., 1997, The Brent Spar Controversy: An example of risk communication gone wrong, *Risk Analysis* **17**(2), 131–136.

Madle, S., Broschinski, L., Mosbach-Schulz, O., Schöning, G. and Schulte, A., 2003, Zur aktuellen Risikobewertung von Acrylamid in Lebensmitteln, *Bundesgesundheitsblatt – Gesundheitsforschung – Gesundheitsschutz* **46**, 405–415.

Ministère de l'Agriculture de l'Alimentation de la Pêche et des Affaires Rurales, 2005, Das französische System für Lebensmittelsicherheit, http://www.securitesanitairedesaliments.agriculture.gouv.fr/sections/guide-sa-curita/ta-chargez-guide-dans/sections/guide-sa-curita/ta-chargez-guide-dans/ (accessed 31 May 2006).

Mosbach-Schulz, O., Seiffert, I. and Sommerfeld, C., 2003, Abschätzung der Acrylamid-Aufnahme durch hochbelastete Nahrungsmittel in Deutschland, http://www.bfr.bund.de/cd/3862?index=65&index_id=4185 (accessed 31 May 2006).

Mucci, L.A., Dickmann, P.W., Steineck, G., Adami, H.-O. and Augustsson, K., 2003, Dietary acrylamide and cancer of the large bowel, kidney, and bladder: Absence of an association in a population-based study in Sweden, *British Journal of Cancer* **88**, 84–89.

Nachrichten.ch, 2005, Krebserregender Stoff in Kinder-Biskuit?, 6. September, http://www.nachrichten.ch/detail/221086.htm (accessed 31 May 2006).

New York Times, 2002, Fried carcinogen, *The New York Times*, 29 April.

Norbelie, B., 2002, Att tiga är inte alltid guld [To be silent is not always golden], *Vår Föda* **3**, 3 (quoted in Löfstedt 2003).

OECD, Organisation for Economic Co-Operation and Development, 2003, *Emerging Risks in the 21st Century. An Agenda for Action.*

Pelucchi, C.. Franceschi, S., Levi, F., Trichopoulos, D., Bosetti, C., Negri, E. and La Vecchia, C., 2003, Fried potatoes and human cancer, *International Journal of Cancer* **105**, 558–560.

Renn, O., 2003, Acrylamide: Lessons for risk management and communication, *Journal of Health Communication* **8**(3), 435–441.

Renn, O. and Kastenholz, H., 2000, *Risk Communication for Chemical Risk Management, An OECD Background Paper*, Bundesinstitut für gesundheitlichen Verbraucherschutz und Veterinärmedizin (BgVV).

Renz, C., 2002, Fritten-Gift: Mehr Tote als im Verkehr?, *Express*, 25. November.

Risikokommission, 2003, *Abschlussbericht der Risikokommission*, Aktionsprogramm Umwelt und Gesundheit.

Scientific Committee on Food, SCF, European Commission, Health & Consumer Protection Directorate-General, Directorate C, 2002, Opinion of the Scientific Committee on Food on New Findings Regarding the Presence of Acrylamide in Food, http://europa.eu.int/comm/food/fs/sc/scf/out131_en.pdf (accessed 31 May 2006).

Schütz, H. and Peters, H.P., 2002, Risiken aus der Perspektive von Wissenschaft, Medien und Öffentlichkeit, *Aus Politik und Zeitgeschichte*, **10-11**, 40–45.

Schütz, H. and Wiedemann, P., 2005, Risikowahrnehmung: Forschungsansätze und -ergebnisse, in: Strahlenschutzkommission (Ed.), *Abschätzung, Bewertung und Management von Risiken*, Veröffentlichungen der Strahlenschutzkommission, Vol. 56, Bonn, pp. 228–256.

Schrum, A. and von Aster, E.-L., 2003, Ess-Kalation im Schlaraffenland – Wie sicher sind unsere Lebensmittel?, *SWR2 Wissen*, Manuskriptdienst, Broadcast Wednesday, 15 January, 8.30 a.m., http://db.swr.de/upload/manuskriptdienst/wissen/wi0120031794.rtf (accessed 31 May 2006).

Sharp, D., 2003, Going public on acrylamide, *Journal of Health Communication* **8**(5), 433–434.

Stern Online, 2003a, Schwedische Studie gibt Entwarnung, 28. Januar, http://www.stern.de/wissenschaft/gesund_leben/503208.html?nv=cb (accessed 31 May 2006).

Stern Online, 2003b, Deutsche scheren sich nicht um Acrylamid, 9. Februar, http://www.stern.de/wissenschaft/gesund_leben/503685.html?nv=cb (accessed 31 May 2006).

Stute, D., 2006, Aufbauschen und zuspitzen – die Vogelgrippe in den Medien, *Deutsche Welle*, dw-world.de, 23. Februar, http://www.dw-world.de/dw/article/0,2144,1913433,00.html (accessed 31 May 2006).

Tareke, E., Rydberg, P., Karlsson, P., Eriksson, S. and Törnqvist, M., 2000, Acrylamide: A cooking carcinogen?, *Chemical Research in Toxicology* **13**, 517–522.

Tareke, E. and Törnqvist, M., 2001, Akrylamid – inte bara i Hallandsåsen utan även i stekta hamburgare [Acrylamide – not only in the Hallandsås but also in fried hamburgers], *Vår Föda* **2**, 28–29 (quoted in Löfstedt 2003).

Törnqvist, M., 2005, Acrylamide in food: The discovery and its implications: A historical perspective, in: M. Friedman and D. Mottram (Eds.), *Chemistry and Safety of Acrylamide in Food. Advances in Experimental Medicine and Biology* **561**, 1–19.

TrinkwV, Trinkwasserverordnung, 2001, Verordnung über die Qualität von Wasser für den menschlichen Gebrauch, BGBl I Nr. 24, ausgegeben zu Bonn am 28. Mai 2001.

Tversky, A. and Kahneman, D., 1981, The framing of decisions and the psychology of choice, *Science* **211**(4481), 453–458.

UBA, Umweltbundesamt, 2001, *Nachhaltigkeit und Vorsorge bei der Risikobewertung und beim Risikomanagement von Chemikalien*, Texte 30/01, Berlin.

Vierboom, C., Härlen, I. and Simons, J., 2007, *Acrylamid in Lebensmitteln – Ändert Risikokommunikation das Verbraucherverhalten?*, A. Epp, R. Hertel, G.-F. Böl (Eds.), BfR Wissenschaft, Berlin.

VIG Entwurf, Gesetz zur Verbesserung der gesundheitsbezogenen Verbraucherinformation, 2006, Verbraucherinformationsgesetz – VIG – Entwurf, http://www.bmelv.de/cln_044/nn_751678/DE/02-Verbraucherschutz/Verbraucherinformationsgesetz.html_nnn=true (accessed 31 May 2006).

von Wedel, H., 2001, *Organisation des gesundheitlichen Verbraucherschutzes (Schwerpunkt Lebensmittel)*, Schriftenreihe der Bundesbeauftragten für Wirtschaftlichkeit in der Verwaltung, Band 8., Stuttgart.

vzbv, Verbraucherzentrale Bundesverband e. V., 2001, Prüfsteine. Zur Glaubwürdigkeit der Wende in der Verbraucherpolitik, Berlin/Bonn.

vzbv, Verbraucherzentrale Bundesverband e. V., 2002, Hersteller sollen Acrylamid-Belastung offenlegen, Pressemeldung, 12. September 2002.

WHO, World Health Organisation, 2002, WHO to hold urgent expert consultation on acrylamide in food after findings of Swedish National Food Administration, Press Report No. 32, 26 April.

WHO, Word Health Organisation, 2004, *Guidelines for Drinking-Water Quality, Vol. 1*, Third Edition, WHO, Geneva.

WHO/FAO, World Health Organisation in collaboration with the Food and Agriculture Organisation of the United Nations, 2002, Health implications of acrylamide in food, Joint FAO/WHO Consultation, Geneva, Switzerland, 25–27 June.

Widemark, E.M.P., 1939, Presence of cancer producing substances in roasted food, *Nature* **143**, 984–985.

Wiedemann, P.M., Mertens, J., Clauberg, M., Hennings, W. and Schütz, H., 2002, *Umweltstandards. Neuordnung der Verfahren zur Risikobewertung und Standardsetzung im Umwelt- und Gesundheitsschutz*, Abschlussbericht zum BfS-Projekt StSch 4337.

Wissenschaftsrat, 2006, *Stellungnahme zum Bundesinstitut für Risikobewertung (BfR)*, Berlin, Drs. 7258-06.

Wolf, W., 2006, Acrylamid in Deutschland: Wie Hysterie zur Umsatzbremse wird, in: Bundesinstitut für Risikobewertung (Ed.): *Was kostet eine Krise? – Fakten, Erfahrungen, Handlungsmöglichkeiten*, Tagungsband zur BfR-Status-Konferenz am 05. September 2005 im Bundespresseamt, pp. 97–101.

Chapter 12
Energy Security for the Baltic Region[1]

D. Warner North
NorthWorks, Inc., Belmont, and Department of Management Science and Engineering, Stanford University, California, USA

Introduction

Since the events of January 2006, in which natural gas supplies to Ukraine and Georgia were interrupted, energy security for Europe has become a 'hot issue', with many national leaders calling for changes in policies and actions (see Appendix 1).

Achieving energy security for the Baltic region is particularly difficult because of the way this region has evolved with the end of the Cold War. The energy infrastructure was created when this region was part of the Soviet Union. Now Estonia, Latvia, and Lithuania are independent nations that have joined the European Union. As part of the process that led to its membership in the European Union, Lithuania has agreed to shut down the two nuclear power plants at Ignalina (3000 MW of installed capacity) that have been providing most of the electric generation for Lithuania and a large amount for export to neighbouring countries. Unit #1 was shut down at the end of 2004, and the agreement specifies that Unit #2 will cease to operate by the end of 2009. While proposals have been made for expanded use of renewable energy supplies, it seems clear that most of the replacement for this nuclear generation will, in the near term, have to come from fossil fuel generation, such as use of natural gas from Russia, plus perhaps some limited use of heavy oil ('orimulsion') from Venezuela.

[1] The author's interest and involvement in energy security for the Baltic region began with a discussion with Dr. Christian Kirchsteiger of the European Commission staff in the fall of 2004 and a Stanford University class project during the winter of 2005. During the past year and a half the author has had extensive discussions with experts in Lithuania, in Moscow, and in Brussels (North 2005). The effort became a case study for the IRGC following the Annual Meeting of the Society for Risk Analysis in December, 2005. The author recently gave a presentation on this case study at a meeting of the Organisation for Security and Cooperation in Europe (OSCE). The technical program for this OSCE meeting, on April 18–19, 2006, was organised in large part through the work of Jean-Pierre Contzen and others from IRGC. At the meeting's end, a resolution was adopted by OSCE, that it would carry out activities aimed at achieving improved risk governance, with IRGC in support.

O. Renn and K. Walker (eds.), Global Risk Governance: Concept and Practice Using the IRGC Framework, 275–299.

In January, representatives from 'relevant ministries, energy industry, energy associations, regulators' from the three Baltic nations plus Finland, Poland, and Sweden met at a workshop in Vilnius, Lithuania and reached agreement supportive of building a new nuclear power plant in Lithuania. They agreed to work together to 'prepare common energy strategy of the Baltic during 2006' ('Declaration' 2006). A few days afterward, leaders of the Lithuanian Energy Institute (LEI), the author, and Dr. Kirchsteiger of the European Commission all attended a risk management meeting sponsored by Gazprom, Russia's large, state-controlled natural gas company and major supplier of natural gas to the Baltic and Europe, and its operating subsidiary, Vniigaz, in Moscow. The author presented an invited paper at a plenary session at this conference, on energy security for the Baltic region, which is attached as Appendix 2 and is a good background document on energy security issues and the technical tools for addressing them.[2] The following discussion presents a case study on Baltic energy security as seen from the perspective of the IRGC's risk governance framework, particularly as an illustration of the framework's 'pre-assessment phase'. It represents the author's views on issues and opportunities for further work in this area.

Baltic Energy Security Viewed from the IRGC Framework: Uncertainty, Complexity, and Ambiguity

Energy policy in the Baltic and elsewhere is characterised by considerable uncertainty. The prices for oil and oil products, natural gas, and electric power production from these fossil fuels can vary considerably over time, as has been demonstrated in the past year and also in previous periods of 'energy crisis' (e.g., as seen with the sharp rise of international oil price in the 1970s). Energy policy is also quite complex; a complex supply chain links the economics of extracting energy materials from the ground and transporting and transforming these materials into the forms and locations needed to meet demand for energy end uses, such as heat in a homes, propulsion of a vehicle such as an automobile, airplane, or electric train, or operation of a television set, hot water heater, or cooking appliance. Furthermore, modern industrial societies rely on multiplicity of energy sources to meet this multiplicity of energy needs. As energy prices change, both consumers and energy companies adapt by changing their behaviour in the purchase of energy equipment and materials as well as in end uses for energy demand. In the short term, measured in days, weeks, or months, demand for energy materials is highly inelastic – modern societies need energy to function, and people cannot maintain their life styles without it,

[2] The reader unfamiliar with the energy security issues and the technical tools proposed for addressing them may wish to read Appendix 2 before going on. The author's paper from the European Commission meeting in Brussels (North 2005), plus papers at this conference by former US Ambassador to Lithuania, Keith C. Smith (2005), Dr. Juozas Augutis (2005), and Dr. Arvydas Galinis and Dalius Tarvydas (2005) provide additional details on the background of planning energy security for the Baltic regions.

so they pay the higher prices that usually result from a change that reduces supply. In the longer term, measured in periods of several years to several decades, purchases of energy equipment adapt to the altered patterns of energy price, and the response can be large increases in supply for some energy materials and large reductions in demand – especially for energy materials at a price that turns out to be significantly more expensive than the competing energy materials and technologies.

Understanding these uncertainties and the complexity of energy markets is a domain that has generated much specialised analysis, especially by companies that deal in energy materials, by government agencies with responsibilities in the energy sector, and by centres of scholarship that have focused on energy policy issues. Because of the importance of the energy sector, energy issues can become important issues of national policy. These issues may differ among nations and among regions. This case study of energy security in the Baltic region addresses the upcoming change from the closure of Ignalina Unit #2 in Lithuania, the implications of this action in making the region more dependent on imported materials, particularly natural gas from Russia, the probability of interruption in the supply of such energy materials, and the economic and other impacts of supply interruption.

There are many complex value-laden issues associated with choices among energy technologies and policies. These issues represent the 'ambiguous' dimension of this particular risk problem, as discussed in IRGC's framework on risk governance. The acceptability of nuclear power generation, including management for safe operation without accidents leading to substantial release of radioactivity, and the management of high-level and long-lived radioactive waste materials resulting from nuclear power generation, have been the subject of discussion and debate for decades in Europe, the United States, and other countries. A second major issue is possible tradeoffs between the risks of nuclear power generation and the accumulation of greenhouse gases such as carbon dioxide from fossil fuel combustion, which may cause alteration in earth's climate.

The decision on replacement energy supply after the closure of Ignalina will affect Lithuania, its neighbours in the Baltic region, and the European Union as a whole. It is a *relatively simple* problem in international planning and governance for a region of Europe. The need is immediate, because Lithuania's agreement with the European Union specifies that Ignalina Unit #2 is to be closed by the end of 2009 – which gives very little time for the construction of new energy facilities. The need for analysis of risks to inform decisions on assuring energy security for the Baltic region should not depend on resolution of broader disagreements regarding the future use of nuclear power generation and measures to reduce emissions of carbon dioxide from fossil fuel combustion. These important issues may be dealt with on a longer time scale. The immediate risk facing Lithuania and its neighbours is an interruption in natural gas supply since these nations import natural gas from one supplier – the Russian Federation, via its state-controlled gas company, Gazprom.

The Baltic region obtains most of its natural gas from Russia. There are small domestic supplies in some countries, but far less than is needed to meet domestic demand. The Baltic countries are currently connected to the electricity grid for Russia, and not to the grid that serves Europe from Portugal to Poland, or to the one that serves the Scandinavian countries such as Sweden and Finland. While there are plans for such interconnection in the future, building high- capacity connections to the electrical grid to the north or to the west will be expensive. On the other hand, extensive excess capacity for electric generation now exists in the Baltic region, and new, more efficient gas-fired generation can be installed relatively quickly. The dilemma is that the available fuel supply is limited and the best candidate fuel is natural gas from Russia. Increased dependence on Russia for vital energy supply is of great concern to the three Baltic countries and to their southern neighbour, Poland. Sudden interruption of energy supplies is at a minimum an inconvenience. It can be highly disruptive to economic activity, and in cold weather it can be life-threatening.

There can be multiple contributors to disruptions of energy delivery, some foreseeable, other not. Energy companies plan for equipment failure, and for periods of extreme weather that can cause elevated levels of demand and also equipment failure. Acts of terrorism or civil unrest can disrupt energy supply. Deliberate interruption by a national authority can also cause sudden shortages.

Events in recent years provide a strong basis for these concerns about overreliance on Russian gas supplies. In April of 2004, Gazprom cut gas supplies to Belarus and to Lithuania when Belarus failed to pay its gas bill on time. In early January 2006, Gazprom briefly curtailed gas supply to the Ukraine. In late January of 2006, two natural gas pipelines from Russia into Georgia and several electric power transmission towers were destroyed by acts of terrorism with the result that many areas of Georgia were without electricity or natural gas. People withstood the cold weather as best they could.

One alternative energy solution for Baltic countries might be to take advantage of existing pipeline projects. Construction of a natural gas pipeline under the Baltic Sea is already underway, after a decision to proceed in September 2005 by Germany and Russia. While the current plan is to build this pipeline from Russia (near Saint Petersburg) to Germany, the plan could be amended to create a spur line to the Baltic region, either to Latvia or to the Kaliningrad region of Russia, which is geographically cut off from the remainder of the Russian Federation by Poland and Lithuania. At the present time, natural gas to Kaliningrad comes from the rest of the Russian Federation through Belarus and then through Lithuania to Kaliningrad. A spur line from the Baltic Sea pipeline to Kaliningrad would enable gas to flow eastward to Lithuania, Latvia, and Belarus through existing pipelines, should there be an interruption in supply from the Yamal pipeline now bringing gas from Russia to Belarus, Lithuania, and Kaliningrad.

Establishment of extensive emergency supplies of natural gas is another approach. In the 1970s, when the United States faced possible interruptions in the oil supply from the Middle East, analysis showed that it was important to distinguish between alternatives that could provide a quickly accessible reserve of up to several months and those that relied on technologies that might provide liquid fuels, but which required large investments and many years for development and construction. The result was the conception, and then the implementation, of the Strategic Petroleum Reserve, the placement of large amounts of crude oil in underground salt dome structures. This crude oil could be quickly and easily retrieved to serve as an emergency supply in the event of a crude oil shortage.

A similar alternative for the Baltic Region would be the development of underground natural gas storage. Latvia has favourable geology for the construction of such facilities, which might hold amounts of gas that could meet the needs for up to several months if supplies through the existing pipelines from Russia were to be interrupted. Interruption for a much longer period would require an alternative source of supply – for example include a terminal facility for the importation of liquefied natural gas (LNG) from countries such as Qatar in the Middle East, which have very large supplies of natural gas and no nearby markets for it. Some international commerce in LNG is already occurring and worldwide commerce in LNG using large LNG tankers is expected to develop rapidly within the next decade. Liquefied natural gas delivered to the Baltic region would be expensive compared to historical prices for natural gas, but it might be economically competitive with other sources of natural gas and with fuels derived from petroleum.

For reasons set forth in the 'Declaration' (2006) from the workshop held in Vilnius, Latvia in January 2006, it may also be advantageous to proceed toward the construction of a new nuclear power plant in Lithuania, probably near the existing units at Ignalina, which has existing infrastructure, including cooling water and professional personnel trained in nuclear technology. But construction of a new nuclear unit that could operate by the end of 2009 seems highly unlikely. Construction of a new nuclear plant on the scale of one of the existing units (Unit #2 has been derated to 1350 MW) will require capital investment of many billions of dollars. It is unlikely that capital of this quantity can be obtained without financing from outside the Baltic region.

Even if it were possible to raise such capital, it would be necessary to renegotiate Lithuania's agreement with the EU so that Ignalina Unit #2 could remain in operation for a longer time. But such extension of the existing agreement between Lithuania and the rest of the European Union may be quite controversial. Discussions with one official of the European Commission suggest that Lithuania will be held to the existing agreement that Unit #2 will cease operation by the end of 2009.

In the short term, perhaps the best overall technical and economic solution might be increased cooperation between the Baltic nations and the Russian Federation and its state-owned gas company, Gazprom. Baltic natural gas infrastructure can be made more reliable through the addition of underground gas storage and redundancy in pipelines, and the threat of sudden price increases or curtailment of supply might be eliminated through new agreements between Gazprom and its customers. While Gazprom has potential access to very large known natural gas fields, its infrastructure is aging, many of its existing gas producing areas are depleting, and some of its own experts believe it may have trouble meeting its commitments for export and domestic supply during the next 15 years. Investment capital and technology from outside the Russian Federation may be very valuable to Gazprom (as well as to the Russian oil industry; see Kramer 2006) for expanding production from existing gas fields, for bringing new fields into production, and for upgrading its existing pipeline infrastructure, much of which was built in the 1970s.

Within the context of the IRGC's risk governance framework, the Baltic energy security problem requires initially that both uncertainty and complexity in the issue be addressed. The tools for dealing with uncertainty – probabilistic risk analyses – are available and routinely applied in the assessment and management of risks from equipment failure and/or from occurrence of extreme climatic events. It could also be applied to the risk of terrorist attacks. The use of decision trees and influence diagrams to characterise uncertainty in sequences of political events leading to war, civil unrest, or disruption of energy supply was pioneered by the author and colleagues during the energy crisis period of the 1970s (Howard et al. 2006). An application of decision analysis carried out for the US government in the 1970s demonstrated how uncertainty in future energy prices might be assessed for analysis of a major energy policy decision (see Appendix 2 and its references).

The tools for dealing with the complexity of energy systems are also well-developed and available from leading centres of scholarship and energy planning. The recent work of the Lithuanian Energy Institute demonstrates a high level of sophistication in analysing alternative energy supply and equipment configurations. The Stanford University Energy Modeling Forum was established in the 1970s as a centre of scholarship for the evaluation and comparison of energy models and modeling methodologies. There appears to be relatively little planning capability to assist decision making by the European Commission and the leadership of the European Union in dealing with near-term energy supply planning.

Consideration of the future role of nuclear power in Lithuania introduces ambiguity into the evaluation of energy solutions for the Baltic region. Furthermore, additional nuclear generation is a longer term solution rather than one that will meet the short-term need for energy in the event of disruption of natural gas supplies. The best approach would be to carry out a risk analysis in two stages: a first stage aimed primarily at providing assured gas supply, and a second stage that addresses the addition of new nuclear capacity. The first stage might best be characterised as 'complexity induced' and thus require the kind of analysis and stakeholder involvement suggested in the IRGC's risk management escalator (see Figure 4 in Chapter 1). The second stage might correspond more closely to an 'ambiguity' induced problem and thus require the more extensive approach envisioned by IRGC's framework.

For the first stage, the shut-down of Ignalina by the end of 2009 is taken as given. The replacement of electric generation will come primarily from existing and new thermal power plants fueled by natural gas (with perhaps a substantial contribution from orimulsion, other heavy oil, and/or biomass). The analysis should examine ways to make the supply system for natural gas into the Baltic region appropriately secure. Participants in the discussion for this analysis should include Gazprom, representatives of the St. Petersburg region of the Russian Federation, as well as Kaliningrad and neighbouring nations such as Belarus, Lithuania, Latvia, and Poland. The evaluation would focus on the probability of failure to have adequate energy supply because of interruption caused by equipment failure, weather, terrorist attack, or deliberate acts to cut off or restrict supply. The time period for this analysis would be 2010 to 2020 or slightly beyond.

A second stage of analysis could address a longer time scale – perhaps to 2050 or even further – and address a broader set of issues associated with new nuclear power generation and with steps beyond those in the Kyoto Treaty for management of greenhouse gases emitted in the combustion process. Expanding the use of renewable energy sources such as biomass, phasing out relatively carbon-rich fossil fuels such as coal and orimulsion, and the creation and management of additional nuclear waste materials all should be part of the agenda for this expanded analysis. There are strong value judgements – issues of ambiguity in the terminology of IRGC's framework – held by many individuals and nations on the acceptability of nuclear power, which have been a great source of controversy within Europe. Germany and some other countries have policies in place specifying that nuclear generation must be phased out, whereas France is highly dependent on nuclear and plans to continue to develop and deploy advanced nuclear technology. In the US, some who have been skeptical and opposed to further expansion of nuclear power generation are shifting to favour it (Kennedy 2005, *New York Times* 2006).

Every reasonable effort should be made to have the first stage of analysis precede the consideration of issues for the second. The primary issue for the first stage involves cooperation with Russia. It is not at all unusual in controversial international risk governance situations to have national leaders trading strong conflicting statements and questioning each other's credibility and intentions (Appendix 1). Within Russia and within Gazprom, there are many technical experts who would like to work out a set of arrangements that will assist Russia in developing its extensive gas resources and selling natural gas for a fair and reasonable price to export markets in European countries. Underground storage would provide resiliency, and multiple gas pipelines or an LNG terminal might alleviate the current limitation, that Lithuania and other Baltic countries are dependent on gas from a single supplier through a single natural gas pipeline.

Baltic Energy Security; IRGC's Four Phases of Risk Analysis and Management

The IRGC framework describes four phases for its process of risk analysis and management: pre-assessment, risk appraisal, tolerability/acceptability assessment, and risk management. This case study is essentially at the pre-assessment phase. We have described above the 'framing' of the problem, and the 'early warning' on the importance of increasing dependence of the Baltic Region on gas supplies from Russia. An understanding of the importance of the problem is clearly set forth in EC Green Papers (see Appendix 1) and in the writings of political leaders and widely read columnists (see Appendix 1). But as yet there has been little integrated analysis, especially in a form that includes knowledge and perspectives from the Baltic countries, their western European neighbours, and from the Russian Federation.

This case study has attempted to carry out a pre-screening and a preliminary selection of assumptions, conventions, and methodologies and procedures for progressing forward into risk appraisal and risk management. It represents essentially a reconnaissance by one US-based academic who has had the opportunity to travel in Lithuania, in other Baltic countries, and in the Russian Federation, to attend conferences in Brussels and Moscow, and to discuss the problem with a small number of leading experts. What is needed to go forward is both adequate funding and a charter from the national governments and international organisations to support a risk appraisal effort on a much larger scale than what has been done to date. A reasonable analogy for this effort exists in the planning and analysis done for the US government on synthetic fuels in the mid-1970s, involving about a dozen agencies of the US government and an analysis staff of tens of people working for a period of six months or more, supported by energy analysis models and risk methodology that took years for development (see Appendix 2 for details).

The need for greatly improved communications to support 'Risk Management' and 'Tolerability and Acceptability Judgements' also seems clear. The IRGC's framework sets forth many ideas for how to accomplish effective dialogue among the parties. IRGC recently released a White Paper on critical infrastructures (IRGC 2006) that presents the results of a preliminary investigation of coupled infrastructures, of which natural gas transport is one of five kinds of infrastructure considered. An overarching concern expressed in that analysis is that privatisation and internationalisation of such infrastructures may have increased the risk of catastrophic service interruption, that is, failure to meet demand. The paper concludes that

> Security of service supply and the impacts of extensive service interruption should be made a high-level priority for further legislation, planning, and evaluation.

The text goes on to note:

> A framework needs to be created aiming to achieve a better balance between conflicting social objectives, such as, for example, in the trade-off between economic objectives and the provision of sufficient redundancy in systems or of redundant back-up systems and reserve supplies. The IRGC Framework on Risk Governance provides one starting place for such analysis. (IRGC 2005: 12)

This case study of Baltic energy security, even in combination with IRGC's White Paper on critical infrastructures, has not yet been developed to the point where it can be taken as validation of the IRGC's risk governance framework (IRGC 2005). However, the IRGC framework still seems highly applicable to the need to address uncertainty, complexity, and ambiguity with modern analytical and risk analysis tools in support of improving risk governance. The importance of the problem of assuring reliable energy supply to the Baltic Region, and more generally, to Europe and other areas of the world, is very high. At this time, much more analysis and international dialogue are needed. The approach described in IRGC's framework seems promising for meeting this great need. Implementation of the approach, however, requires both ample funding for analysis and political leadership in support of risk management decision making.

Summary

The author views that a pre-assessment reconnaissance for Baltic Energy Security is far along, perhaps essentially completed, with respect to a first stage of analysis. The technical tools needed, energy modeling capability and probabilistic risk analysis methods, are readily available – but there appears to be little experience in using such analysis tools to support risk governance on this problem in the European Commission, or in Gazprom and the Russian Federation. As such, this first stage of risk analysis would be a challenging and pioneering effort in international dialogue and cooperation as well as in technical analysis.

The author concluded his presentation to the Organisation for Security and Cooperation in Europe (OSCE) in April 2006 with the following statement:

> We need proactive national and international leadership in pursuit of economic growth, environmental protection, and social well-being, as well as energy security. Scientists and analysts like me can help. Our community has excellent tools for planning in the face of large uncertainty and great complexity. But enlightened, proactive national and international political leadership is crucial for enabling progress. I hope OSCE can help in providing such leadership.

The author believes Baltic energy security provides an excellent case study for IRGC and for international organisations, such as the European Union and the OSCE. A fully developed case study on this subject could demonstrate what can be done with available technical tools to develop a risk analysis that can lead to improved risk governance, and thereby, to improved energy security to those in the Baltic region.

References

Augutis, J., 2005, Lithuanian power network reliability analysis, Paper presented at the SEIF-IV Conference, Brussels, November.

'Declaration' of the Workshop 'Development of the Electricity Markets and Security of Supply in the Baltic Sea Region', Vilnius, Lithuania, 26–27 January 2006, 1 page, http://www.president.lt/en/news.full/7510.

Galinis, A. and Tarvydas, D., 2005, Power systems planning for Lithuania and the Baltic Region, Paper presented at the SEIF-IV Conference, Brussels, November.

Howard, R.A., Matheson, J.E., Merkhofer, M.W., Miller, A.C. and North, D.W., 2006, Comment on influence diagram retrospective, *Decision Analysis* **3**(2), 117–119.

IRGC, 2005, *Risk Governance: Toward an Integrative Approach*, International Risk Governance Council, Geneva, Switzerland, September.

IRGC, 2006, *Managing and Reducing Social Vulnerabilities from Coupled Critical Infrastructures*, International Risk Governance Council, Geneva, Switzerland, October.

Kennedy, D., 2005, Risks and risks, Editorial, *Science* **309**, 30 September, 2137, http://www.sciencemag.org/cgi/reprint/309/5744/2137.pdf (access required).

Kramer, A.E., 2006, Russia decides it's time to try modern oil drilling, *New York Times*, May 13, p. B9, http://www.nytimes.com/2006/05/13/business/worldbusiness/13russogas.html.

New York Times, 2006, Editorial, The greening of nuclear power, May 13, p. A28, http://nytimes.com/2006/05/13/opinion/13sat1.html.
North, D.W., 2005, Analysis of risk of supply interruption to Lithuania from disruption of natural gas supply from Russia, Paper presented at the SEIF-IV Conference Brussels, Belgium, November 14–16, http://www.northworks.net/w_pub_seif-iv.htm.
Smith, K.C., 2005, Political issues arising from Central European dependency on Russia, Paper presented at the SEIF-IV Conference, Brussels, November.

Appendix 1

Quotations from Leaders and Leading News Media Writers on Energy Security with Respect to the Use of Russian Natural Gas in Europe, 2006

1. Commissioner Andris Piebalgs' Speech to EU Parliament, January 17, 2006:

> ... this dispute has underlined not only the importance but also the necessity of a clearer, more cohesive and pro-active EU-wide energy security policy. While it is true that the Commission and Parliament have argued for this for many years, the time is now right to make real progress on this.

http://europa.eu.int/comm/commission_barroso/piebalgs/doc/media/2006_01_17_gas_crisis_speakings_plenary.pdf.

2. The new European Commission Green Paper, 'A European Strategy for Sustainable, Competitive, and Secure Energy', released March 8, 2006:

- ... the Baltic States, which remain an 'energy island', largely cut off from the rest of the Community (p. 6).
- The EU needs to complete the internal gas and electricity markets. (Rec. #1, p. 18) ... Review 'could propose clearly identified priorities for the upgrading and construction of new infrastructure' (p. 15).
- A common external energy policy (Rec. 6, p. 19). As part of it, 'a new energy partnership with Russia' (p. 20). 'A new initiative is particularly opportune with regard to Russia' (p. 15).

http://ec.europa.eu/energy/green-paper-energy/doc/2006_03_08_gp_document_en.pdf.

3. President Putin's Opinion/editorial article, which appeared in the *Wall Street Journal* and in the *Moscow Times*, March 1, 2006, p. 10:

> The new policy of the leading countries should be based on the understanding that the globalisation of the energy sector makes energy security indivisible. Our common future in the area of energy means common responsibilities, risks, and benefits. ... Generally speaking, all of us should recognise and admit that 'energy egoism' in a modern and highly interdependent world is a road to nowhere. ... We [Russia] will strive to create an energy security system sensitive to the needs of the whole international community ... international cooperation opens all avenues for that.

4. Reaction, via a letter about March 9 to the *Wall Street Journal*, from former Estonian Prime Minister Mart Lahr:

> Somehow Mr. Putin's article reminded me of speeches of former Soviet leaders, when peace was praised but in reality preparations for war were made. Russia's recent actions against Ukriane, Moldova, and Georgia have made absolutely clear that Moscow has decided to use energy deliveries as a political weapon. ... His intentions became obvious when Russia unilaterally withheld Ukraine's gas supply following their price dispute, and in Moscow's actions against Moldova and Georgia. As long as Mr. Putin is inclined to use oil and gas as a foreign policy weapon, a rules-based energy system will be impossible – regardless of how many articles Mr. Putin chooses to pen in the international media.

Estonian Review, http://www.vm.ee/eng/kat_137/7390.html.

5. Valdus Adamkus, President of Lithuania, interviewed by reporter Stefan Wagstyl, *Financial Times*, London, UK, May 4:

> President Valdus Adamkus has called for a common European Union front in response to Russia's willingness to use its energy supplies to secure political influence over its neighbours. Speaking to the *Financial Times* on the eve of an international pro-democracy meeting in Vilnius, Mr. Adamkus condemned Germany for backing Russia's controversial planned Baltic Sea gas pipeline, which will circumvent transit countries including the Baltic states, Ukraine, and Poland. He said, 'I can understand the Russian position but I can't understand Germany's position. As a member of the EU, they acted without even extending the courtesy of advising the Baltic states [about their plans].' ...
>
> 'I don't want to use the word blackmail', said Mr. Adamkus in referring to Moscow's efforts to extend its influence through energy policy, but he made it clear he was very concerned about Russia's economic and political pressure.

6. Vice President Richard Chaney's speech in Vilnius, May 4, 2006 included the following characterisation of Russia:

> America and all of Europe also want to see Russia in the category of healthy, vibrant democracies. Yet in Russia today, opponents of reform are seeking to reverse the gains of the last decade. In many areas of civil society – from religion and the news media, to advocacy groups and political parties – the government has unfairly and improperly restricted the rights of her people. Other actions by the Russian government have been counterproductive, and could begin to affect relations with other countries. No legitimate interest is served when oil and gas become tools of intimidation or blackmail, either by supply manipulation or attempts to monopolise transportation. And no one can justify actions that undermine the territorial integrity of a neighbour, or interfere with democratic movements.
>
> Russia has a choice to make. And there is no question that a return to democratic reform in Russia will generate further success for its people and greater respect among fellow nations. Democratisation in Russia helped to end the Cold War, and the Russian people have made heroic progress in overcoming the miseries of the 20th century. They deserve now to live out their peaceful aspirations under a government that upholds freedom at home, and builds good relations abroad.
>
> None of us believes that Russia is fated to become an enemy. A Russia that increasingly shares the values of this community can be a strategic partner and a trusted friend as we work toward common goals.

http://www.whitehouse.gov/news/releases/2006/05/20060504-1.html.

7. Keith Smith, former US Ambassador to Lithuania, in a speech April 24, 2006 at the European Policy Exchange, London:

> European dependency on Russian gas and oil has become a hot topic of discussion in Brussels and Washington following January's Ukrainian-Russian 'gas war'. Nevertheless, there are few signs that leaders in either capital are prepared to develop a coordinated strategy to deal with the mix of opportunities and threats stemming from our greater energy dependency on an ever more aggressive and authoritarian Russia.
>
> Europe has a new Green Paper, filled with recommendations, but no enforcement power on the vital issues of energy diversity. EC President Barroso has traveled to and from Moscow with little to show for his appeal to President Putin for more business transparency, energy market reciprocity and pipeline competition.

8. Author and columnist Thomas L. Friedman, in the *New York Times*, May 10, 2006, in a response to Vice President Cheney's May 4 Vilnius speech, wrote:

> In the post-cold war world, European integration and economic reform seemed irreversible and certain. But in the post-post-cold war, Europe can't unite on anything – even on an energy policy – and so it is being pushed around by Russia.

9. Simon Zekaria, in AFX Europe, Brussels, Belgium, Wednesday May 10, 2006 wrote:

> Former prime minister of Russia Mikhail Kasyanov said that Moscow's standoff in January over gas prices that cut off supplies to Ukraine by state-owned Gazprom, which also cut off gas supplies to western Europe, was a 'big mistake'.
>
> Speaking in Brussels, Kasyanov – prime minister between 2000–2004 and tipped to be a candidate in Russia's 2008 presidential election – said: 'That was a big political mistake. It must never happen again'. He said the dispute raised questions over the 'reliability of supplies' and added that it was wrong that gas was used as a 'political weapon' over the issue.
>
> Late last month, the European Commission called on Gazprom to stick to its contractual commitments and warned it against threatening crucial European energy supplies. The commission reacted after Gazprom warned the EU not to 'politicise' terms for Russian gas supplies, implicitly threatening to sell its product elsewhere. ...
>
> Kasyanov said that it is unlikely a full energy deal on the issues could be struck in the 'near future'. 'I don't think that during this year or next year something revolutionary in energy relations will happen'.

10. Author and columnist Thomas Friedman, in the *New York Times*, October 25, 2006, in a column entitled 'The Really Cold War', wrote:

> When Europeans tell you that they fear a new 'cold war', this time they really are talking about the temperature – and the fear that Russia, if it wanted to turn off the gas, could make Europeans very cold. About 40 percent of Europe's natural gas imports come from Russia, and that is expected to grow to 70 percent by 2030.

11. *Der Spiegel* International, online edition in English, October 30, 2006, wrote about the meeting in Berlin of Polish Prime Minister Kaczynski with German Chancellor Angela Merkel:

> Germany's neighbour [Poland] is especially unhappy about the Baltic Sea pipeline which bypasses Poland on its way from Russia to Germany – a pipeline which the Polish Defense Minister Radek Sikorski compared to the 1939 Molotov-Ribbentrop pact that divided Poland up between Nazi Germany and the Stalinist Soviet Union.
>
> Kaczynski says his country is concerned about being overly reliant on Russia for its energy supplies and wants to be able to pipe in energy from Western Europe too if necessary. However, Merkel did not budge from her support for the pipeline, saying it is important to establish a Europe-wide energy market. She promised she would make Poland's access to European gas markets a priority during the German EU presidency during the first half of 2007.

Appendix 2

Assessing Risks in Long-Term Planning: Probabilistic Scenario Analysis with Generalised Equilibrium Energy Models[3]

D. Warner North
NorthWorks, Inc., Belmont CA, and Department of Management Science and Engineering, Stanford University, CA, USA northworks@mindspring.com

Paper Presented at the RIMS-2006 VNIIGAZ/GAZPROM Conference, Moscow, Russia, February 1–2, 2006
(Note: minor editorial revisions have been made subsequent to this conference.)

Abstract

This paper provides a brief introductory summary of methodologies that have been used in the United States for long-term planning of large energy investments that require extensive capital investment and development of technology. It is expected that such methods may be useful to GAZPROM and to its export customers in connection with planning and financing development of additional gas fields in the high arctic area of Russia and pipeline construction to bring such gas to export markets, such as the Baltic countries and other portions of Western Europe. There are two methodologies involved: (1) the use of probabilistic scenario analysis to examine changes in market conditions, including political events that may limit or disrupt energy supply to customers; (2) the use of large-scale energy models that project how market conditions including energy prices and the mix of energy materials may evolve under specific scenarios. The example presented will be the analysis prepared for the US Presidential Task Force in 1975, when the US Government was considering a massive investment in new facilities to make liquid and gaseous fuels from coal and oil shale. This program was subsequently implemented, proved to be

[3] Available on the web: http://www.northworks.net/w_pub_vniigaz.htm.

a market failure, and was therefore discontinued. The reasons for the market failure were clearly evident in the 1975 analysis. The author will present a retrospective review of this 1975 analysis, in which he was a participant, and an update on the energy modeling and probabilistic scenario methodologies as these have evolved in thirty years of subsequent use by US energy and risk specialists.

I. History and Background

At the midpoint of the first decade of the 21st century, oil and gas resources are depleting. Western Europe will need increasing amounts of imported oil and gas to meet its energy needs. There are also special situations. For example, the government of Lithuania has agreed as a condition for its membership in the European Union to shut down the Ignalina Nuclear Power Plant (NPP) [1]. This plant has two 1500 MW nuclear generating units of the graphite channel (RBMK-2) type, the same design as for the Chernobyl NPP in the Ukraine. Ignalina NPP was built during the time of the Soviet Union to provide electricity for the Baltic region, and it became a part of Lithuania when Lithuania became an independent country. Its 3000 MW generation capacity has allowed Lithuania to meet most of its own needs for electricity from this NPP and also to export large amounts of electricity to Latvia, Estonia, Belarus, and the Russian Federation. Unit one was shut down at the end of 2004. The second unit is scheduled to be shut down in 2009. Shutting down these two nuclear units will require that this electric generation be replaced by other energy sources, such as natural gas from Russia or a heavy oil/water mixture called orimulsion from Venezuela.

Increased reliance by European countries on natural gas from Russia has positive features for these countries and also for Russia. Russia has extremely large natural gas resources that can be developed and transported to European countries for costs that should be competitive with other energy sources. Natural gas does not contain sulfur, nitrogen, metals, or complex hydrocarbons, so control of air pollutants such as sulfur oxides, nitrogen oxides, and particulate matter is inherently much more easily accomplished; pollution control equipment is needed only for management of oxides of nitrogen formed from atmospheric nitrogen in the combustion process. The low carbon content of natural gas compared to oil and coal implies lower levels of carbon dioxide emissions into the atmosphere, compared to burning coal or oil. More use of natural gas, instead of coal or oil, therefore reduces global climate alteration. For these reasons natural gas will increasingly be viewed as a premium fuel, for which customers are willing to pay a higher price. Production areas of natural gas in Western Europe such as in the North Sea are depleting, and large new gas resources are unlikely to be discovered. Russia is known to have very large gas resources that are only now beginning to be developed, such as the Shtokman field. These resources could provide ample supplies for European consumers for much of the 21st century, and may permit export of liquefied natural gas (LNG) to North America as well [2].

Development of Russian natural gas for export to European countries will require considerable capital expenditure, including the construction of new pipelines and development of gas fields in the arctic region. While agreement was reached between GAZPROM and German companies last year to construct a new gas pipeline to provide Russian gas to Germany, the Baltic countries and Poland expressed concern that this pipeline would bypass them and leave them dependent on single pipelines from Russia. The countries would have preferred an alternative overland route. The 2004 interruption of gas supply to Belarus and the New Year's Day interruption of gas supply to the Ukraine, and the recent terrorist attacks on the pipelines to Georgia have increased concerns among Europeans that the supply of gas from Russia may not be reliable. Events such as equipment breakdowns and extreme weather can lead to supply interruptions. The planning of the multibillion dollar investments in gas field development and pipeline construction will depend on perceptions that (1) the price of the natural gas to customers will be competitive with other energy sources, and (2) that supplies will be reliable. There must be assurances that neither politically motivated shutdowns, equipment failures, extreme weather, or acts of terrorism or war will disrupt the transport of gas essential for heating and continued function of the economy in countries depending on natural gas imports. Alternative supplies energy are available to Europe through North Africa, the Middle East, and possibly from Central Asia, especially as new pipelines or LNG facilities are constructed.

What is needed for planning is methodology to deal with two issues: (1) a way of assessing the competition in price between gas from Russia and other energy materials for meeting the needs of European Countries, looking forward for decades (2) a way to assess the uncertainties arising from weather, equipment failures, and political events to plan adequate redundancy in the energy supply system to that the probability of significant supply interruption can be made acceptably low. Multiple natural gas pipelines connecting gas fields to customers, underground gas storage located in customer countries, and provision to obtain and use other supplies under upset conditions may be needed to assure adequate supply reliability. The cost of these facilities needed for adequate supply reliability should be included in calculating what it will cost to provide gas from Russia to serve export markets in Europe. Therefore, the two issues must bc linked.

II. Methodology Needed

How can these planning needs be met? In this paper the author shares experience from applications within the United States of risk and decision analysis, in the belief that such experience may be useful for those in Russia and in Europe responsible for planning GAZPROM's future and a reliable supply of energy to meet European energy needs. In the 1970s the author was involved in a number of energy planning studies in the United States, including the analysis of a major Presidential initiative to create a one million barrels-per-day synthetic fuel capability within ten years [3,

4], and a comparison of the economic and social costs of coal and nuclear generation [5]. The author was also involved in a pioneering application of decision analysis to political events in the Persian Gulf area that might impact on the reliability of energy supply from that region [6]. The influence diagram methods developed in that project are described in a recently published pair of journal articles [7, 8]. The author also chaired the review of the first set of reports to the US Congress on global climate change, for the Science Advisory Board of the US Environmental Protection Agency [9]. In October of 2004 the author substituted on short notice for his Stanford University colleague, Professor Stephen Schneider, in giving a presentation on global climate alteration in Stuttgart, at a meeting organised by the Minister of Transport and Energy of the State of Baden-Württemberg [10]. Especially in the roundtable discussion that followed the presentations, which included leaders of German industry and representatives from the German Bundestag, the author became persuaded that formal analysis of the kind done in the United States in the 1970s would be very helpful in Europe.

Analytical Tools Area #1 – Probabilistic Risk Analysis: A shortfall or failure in a complex system often can be described as a scenario, a set of events leading to this failure. Probabilistic models of such event sequences have a long history in statistics and reliability theory. Such approaches were developed in the aerospace and nuclear power industries and then widely applied in these industries and elsewhere. Decision analysis evolved out of statistical decision theory in the aftermath of World War II. In decision analysis, sequences are examined of decisions and uncertain events. A decision tree is often used as a visual display and as a computing device for computing the probability distributions corresponding to different decision alternatives, and then evaluating these to find the best decision [11].

Influence diagrams [6, 7, 8] evolved from decision trees to deal with situations in the sequence of decisions and events has a complex structure of conditionality. For example, a series of ten to twenty uncertain events precedes the system failure of interest, and the probabilities assigned to one of these events may depend on some (say, 2 to 6), but not all of the preceding events and decisions in the sequence. With binary (two possible outcomes) events, a sequence of ten events leads to a thousand end points or scenarios, and 20 leads to a million. An illustration of such a tree showing all the possible outcomes becomes impractical except in a generic form. The influence diagram is a schematic form that shows by means of arrows connecting decisions and events (collectively referred to as 'nodes') the conditionality structure. Influence diagrams have achieved widespread use in decision analysis, operations research [12] reliability [13] and also in computer science applications, sometimes under the term, Bayesian nets. A recent application to protection against terrorism by the author's department chairman at Stanford and one of her students is found in [14]. Software packages for influence diagram computations are available from several sources. For the illustrative example below, the author has used Decision Advisor, a propriety software package for R&D management marketed by SmartOrg of Menlo Park, CA.

Failure of the natural gas supply system to meet gas demand might result from four types of events:

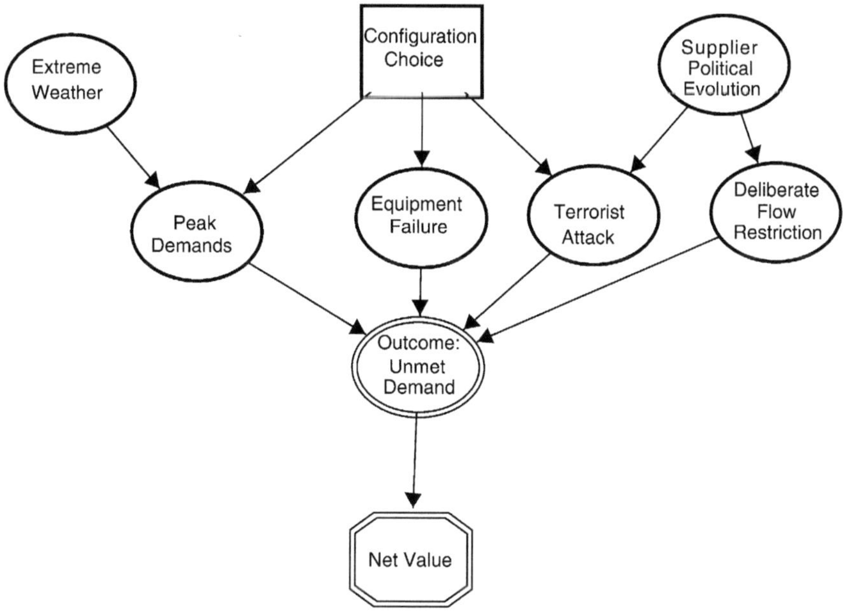

Fig. 1 Influence diagram for gas supply – Simple illustration. Diagram produced by Decision Advisor software.

1. high peak demand and/or reduced supply resulting from extreme weather
2. equipment failure for reasons other than extreme weather
3. terrorist attacks
4. deliberate interruption by the supplying country or by the operator of a pipeline, for political reasons.

An influence diagram is shown in Figure 1.

The arrows show conditionality for dependence of the uncertain events on preceding events. For example, we assume that a failure of the system to meet peak demands as the result of extreme weather depends on whether the extreme weather occurs (and perhaps to what extent) and on the configuration of the pipeline and storage system. An analysis might consider a base case and several alternatives in which additional investment has been made to have storage facilities and one or more additional pipelines from the supplier country. For each system configuration and each of the weather outcomes, a probability is assigned to the outcome of a supply shortfall. We show below in Figure 2a decision tree representation for a portion of the influence diagram: two outcomes for extreme weather (yes and no), three systems configurations (base and two alternatives) and two possibilities for shortfall (shortage and none).

Moving from left to right and then down the influence diagram in Figure 1, probabilities are assigned to 'terrorist attack', conditional on which of these outcomes occurs and the system configuration alterative. 'Deliberate flow restriction' is shown

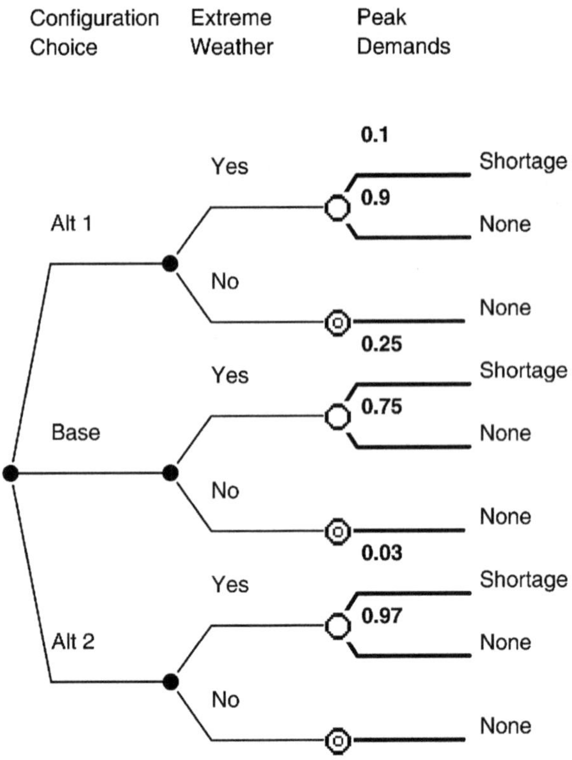

Fig. 2 Tree representation showing probability assignments for 'peak demand' node in the Influence Diagram. Tree as shown is a display produced by Decision Advisor.

conditioned only on 'supplier political evolution'. Once probabilities are assigned to all the possible outcomes for each event and decision, a composite probability can be computed for each possible path or scenario. Figure 3 shows the probability distribution computed for illustrative numbers in this highly simplified influence diagram.

The approach is capable of using a much more complex structure of decisions and uncertain events conditionally dependent on preceding decisions and events (nodes) in the influence diagram. For the other three terms, we have conditioned equipment failure only on the configuration decision, and we condition 'terrorist attack' on 'supplier political evolution'. 'Deliberate flow restriction' is also conditioned on what happens in 'supplier political evolution'. This node is intended to describe what might happen politically in eastern European countries during the next 5–15 years. A very simple characterisation of three mutually exclusive, collectively exhaustive outcomes for one or more countries might be: continued steady progress in democracy, protracted civil unrest, and reversion to authoritarian imperialism. Probabilities may be assigned to each of these outcomes, based on the judgement of experts in the politics of the area. Disagreement among such experts can be expected. Which judgements are most important? Sensitivity analysis can help determine which judgements are most important in influencing the overall probability of supply interruption and its economic and social consequences.

Once the set of conditional probability assignments are complete and outcomes for 'unmet demand' resulting from the four sets of causes are described, then the software can compute probability distributions for each configuration, so that costs and risks of supply interruptions over different time periods and impact magnitudes can be compared. An illustrative cumulative distribution is shown below in Figure 3. It shows a moderate probability (about 7%) of a limited shortfall judged to be worth about €1 billion, and a much smaller probability (0.7%) of a much larger shortfall, valued at €60 billion. Such a large disaster would be comparable to the impacts the US experienced in 2005 from the Gulf Coast hurricanes. It might plausibly result from a major, extended interruption, perhaps from a series of successful terrorist attacks, civil unrest, or an extreme action or series of actions to reduce supply from the gas sources and deliberate shutdown of pipelines.

The structure shown in Figure 1 here is intended only as a highly simplistic description of the methodology, for those not familiar with influence diagrams and probabilistic risk analysis. Applications of probabilistic risk analysis to weather and

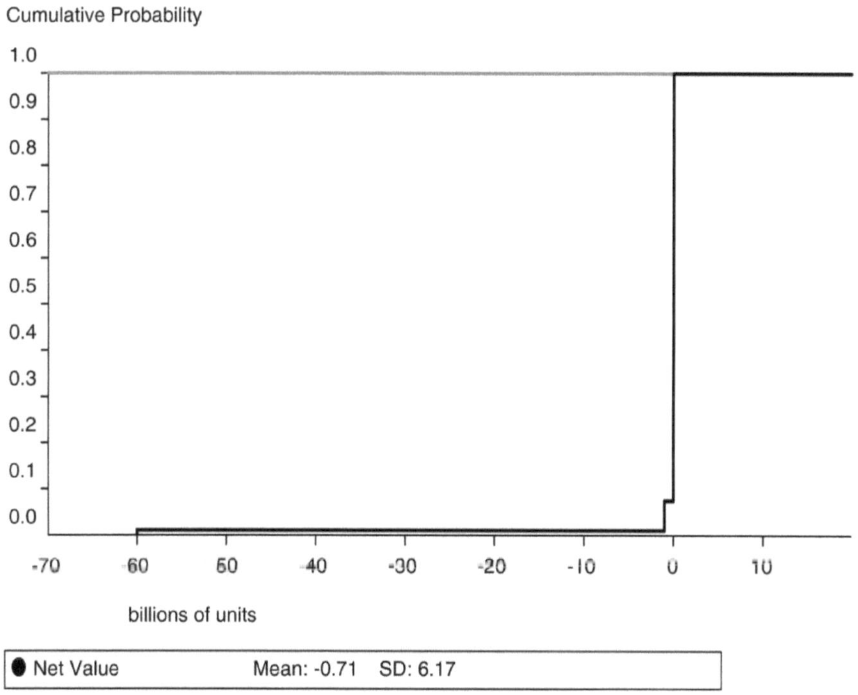

Fig. 3 Illustrative calculation of probability distribution of loss, with supply interruption scenarios valued in monetary units, such as Euros. SD = standard deviation. Distribution shown is output from Decision Advisor.

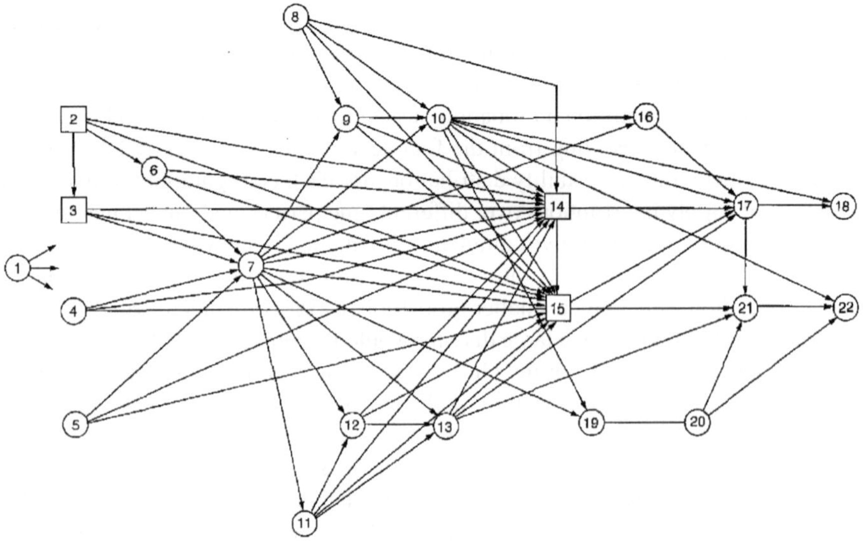

Fig. 4 Influence diagram for Persian Gulf political events (from [7, p. 145]; the original version was produced by Allen Miller and colleagues in 1973–1974).

equipment failure are widely carried out by gas and electric utilities. Application to the threat posed by terrorism and political decisions to interrupt supply are not widely practiced. It may be useful to carry out such an analysis as the basis for dialogue among the concerned parties. For example, some parties in Russia may assert that the threat of terrorist attack in Lithuania and other Baltic countries led to a preference for the Baltic undersea pipeline route over the cheaper overland route. Figure 4 and influence diagrams of similar complexity in [14] illustrate how complex sequences of events may be represented suing influence diagrams as a means of assessment expert judgement for probabilistic risk analysis.

Analytical Tools Area #2: Energy Models for Long-range Planning: Planning the future of a complex energy system on a time scale of many decades is a daunting challenge. Dating back at least to period in the 1970s when the United States was considering large investments in new energy systems, such computer planning models are increasingly being used by businesses and government agencies.

Stanford University has since the 1970s had an activity called the Energy Modeling Forum, in which leaders in the field of energy modeling bring their models (often proprietary computer software, developed at considerable investment expense) and come together to analyse case exercises, from regional planning to global response in the context of global climate change. Reference [15] is a report on dealing with natural gas supply in North America. One of the models used in the exercise described in [15], the North American Gas Model (NARG), is a direct descendent of the SRI-Gulf model used for the synthetic fuel commercialisation analysis described in [3] and [4]. In [15] this model was being used on behalf of the California

Table 1 Reproduced from [7, p. 146].

Table 1 Sequence of Decisions and Events in the Influence Diagram Example

I. Information received from the intelligence activity
   1. The activity reports that an event will occur.
II. Initial U.S. policy decisions
   2. The United States continues its strong diplomatic support for Israel and continues to supply significant amounts of military equipment to Israel.
   3. The United States moves the naval force it currently has in the Persian Gulf to a base in the Indian Ocean.
III. Soviet Influence
   4. The Soviets substantially increase the amount of military equipment they supply to revolutionary organizations in the Persian Gulf.
   5. The Soviets increase their own military presence in the Persian Gulf.
IV. Arab-Israeli conflict and revolutionary activity
   6. The Arabs and Israel reach a political settlement that is acceptable to most Arabs in the Persian Gulf states.
   7. There is a significant increase in the level of revolutionary organization and activity in the Persian Gulf states.
V. Instability and revolution in Saudi Arabia
   8. King Faisal dies.
   9. Considerable internal instability develops in Saudi Arabia
VI. Instability and "revolution" in Iran
   11. The Shah of Iran dies.
   12. Considerable internal instability develops in Iran.
   13. A revolutionary regime takes power in Iran.
VII. U.S. policy decisions to support Persian Gulf states
   14. The United States gives diplomatic support and large amounts of military equipment to Saudi Arabia.
   15. The United States gives diplomatic support and large amounts of military equipment to Iran.
VIII. Conflict between Saudi Arabia and Iran
   16. A revolutionary regime takes power in Qatar, Bahrain, or the U.A.E.
   17. Saudi Arabia and Iran engage in a serious military conflict.
   18. Iraq enters the conflict between Saudi Arabia and Iran.
IX. Conflict between Iraq and Kuwait or Iran
   19. A revolutionary regime takes power in Kuwait.
   20. Iraq engages in military conflict with Kuwait—possibly in an attempt to annex Kuwait.
   21. Iraq and Iran enter a serious military conflict.
   22. Saudi Arabia enters the conflict between Iran and Iraq.

Energy Commission; it has also been used for many leading US and international oil and gas companies. The lead modeler, Dale Nesbitt, worked with the author on the 1975 synthetic fuel analysis ([3] and [4]). The generalised equilibrium methodology involves an extension from mathematical programming to finding the fixed point solution x to an equation $f(x) = x$, where x is a vector of very high dimensionality describing flow quantities and prices of energy materials at different locations in a network connecting energy resources in the ground to energy end use demands, and at different time points from the beginning to the end of the planning horizon [16,17]. The NARG model and its European counterpart, the European Gas Model, are not unique to one vendor. Argonne National Laboratory (ANL) and Rice University have developed similar models based on the generalised equilibrium approach. Other modeling and simulation approaches can be used to forecast how the competition may evolve between natural gas from Russia, natural gas from other sources, and competing fuels and technologies for meeting Europe's needs.

Understanding how the European energy system may evolve over a period from now to the middle of the 21st century can be greatly aided by the use of advanced energy modeling tools. Energy models are particularly useful for projecting changes as energy prices and the availability of energy technologies change over time. The models can also be used to analyse upset conditions, such as those that occurred in the US during the last half of 2005 because of hurricane damage.

III. Combining Both Analytical Tool Areas: The 1975 Synthetic Fuels Analysis

The 1975 analysis [3, 4] of the US decision on investing billions of dollars to create a million-barrel-per-day industry to make gas and liquid fuels from coal and oil shale showed that this investment was a poor idea. The program proposed in the President's State of the Union was first scaled down to a smaller program of one commercial scale plant of each technology, a total capacity of 1/3 of a million barrels per day. This smaller program was proposed by President Gerald Ford but voted down by the US Congress. Under the next President, Jimmy Carter, a small synthetic fuels program was established. The program failed badly, for reasons that were foreseen in the analysis: The increased prices of oil and other energy materials that occurred in the mid-1970s led to reduction in demand, and also to expansion of conventional oil and gas supplies, with the result that there was no market for high-cost synthetic fuel made from shale and coal. The Great Plains Coal Gasification plant was built, but it was not able to sell its gas product, because the price of gas was below the operating cost to produce gas from this plant. There was no return of profit to offset the high capital cost of the plant. The plant was therefore abandoned, with the loss of the capital cost of its construction at the expense of the US government.

The analytic tools used for the analysis [3, 4] included the decision tree shown in Figure 5 describing important uncertainties and how future decisions to expand capacity would depend on the immediate decisions to be taken in the mid-1970s. Many thousands of scenarios were represented in the decision tree, and across this large number of scenarios (with some approximations via curve fitting) an energy model was used to assess how prices, technology choices, and energy quantities might evolve for that scenario.

A number of other analyses were carried out in the US in the late 1970s using multiple scenarios and energy models. A review of some of these analyses appears in [18]. The synthetic fuels analysis was one of the most elaborate, and it was tightly focused toward one specific government policy decision. It should be a useful example to Russian and European Union experts in analysis of specific alternatives, such as new pipelines and LNG transport facilities, for supplying reliable natural gas and other energy materials to Europe in the coming decades.

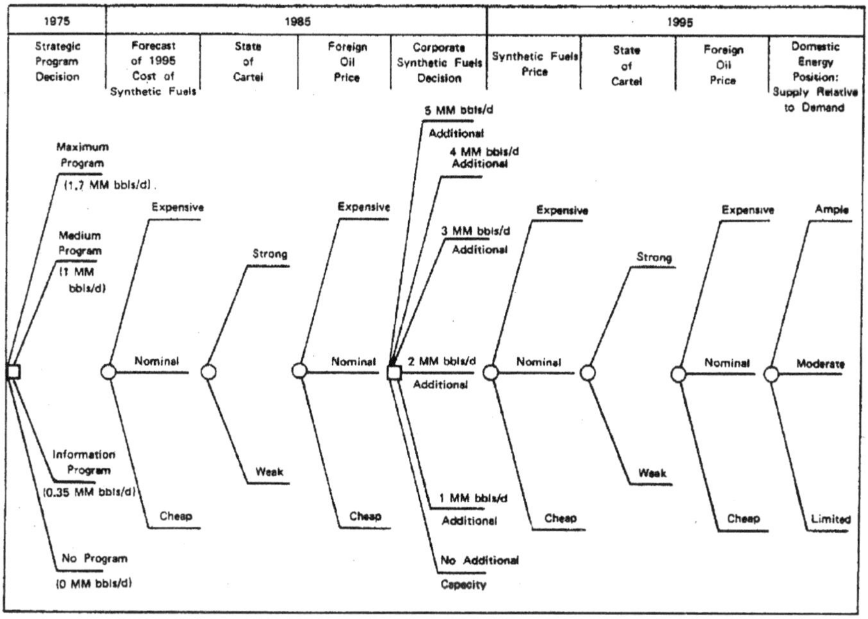

Fig. 5 Synthetic fuels decision tree.

IV. Needs for Future Planning and Decision Making

In planning Europe's energy future, the leadership in energy companies and governments need to work effectively together, and to overcome legacies of mistrust and misunderstanding that come from many historical events and from differences in institutions and cultures. Effective planning and decision making require not only the mastery of analytical methods for dealing with the complexity and uncertainty of energy markets and technological development, but also learning how effective governance can be achieved among a multiplicity of stakeholders – national governments, the European Union and the G-8, energy companies, and concerned citizens in many countries. It is not just a technical problem – it is also an extreme social and political challenge! Several recent reports have suggested how decision analysis methods can support improvements in risk governance in order to improve the transparency and effectiveness of national and international dialogue [19, 20, 21].

In the Preface to the National Energy Strategy for Lithuania [1, p. 5], Dr. Jurgis Vilemas[3] writes, 'Decision-making will hopefully, in the end, be based on economic reasoning, realistic demand forecasts and the latest achievements in development of energy generation technology'. This is an excellent place for planning to start,

[3] Dr. Jurgis Vilemas was Director for 23 years of the Lithuanian Energy Institute. Dr. Eugenijus Uspuras is now the LEI Director, and Dr. Vilemas remains active at LEI as Chairman of its Council. Dr. Vilemas, Dr. Uspuras, and Dr. Juozas Augutis from LEI attended the RIMS 2006 Conference, February 1–2, 2006.

and the author looks forward to working with Dr. Vilemas and his colleagues on a study of how to meet the energy needs of the Baltic Region. This analysis will be broadened from a focus on purely economic reasoning to include the four types of risk issues that have been listed in this paper and illustrated in the simple influence diagram: extreme weather, equipment failure, terrorism and sabotage, and political interference in energy trade. The author hopes that the RIMS-2006 participants, the management of VNIIGAZ and GAZPROM, and the leadership of energy planning for Europe share this expanded view on what needs to be done to support good decision making for the energy future of Russia and Europe.

References

1. J. Vilemas, V. Miškinis, and A. Galinis, *National Energy Strategy*, Prepared for, and approved by, the Seimas (Lithuanian Parliament) on 10 October 2002, Lithuanian Energy Institute, 2003.
2. Tristana Moore, "Gazprom's Global Ambitions", BBC News, Tuesday 20 December 2005. Available at http://news.bbc.co.uk/1/hi/business/4531578.stm
3. Synfuels Interagency Task Force, *Recommendations for a Synthetic Fuels Commercialization Program*, report prepared for the President's Energy Resources Council, November 1975, Washington, DC: US Government Printing Office.
4. S. N. Tani, "Decision Analysis of the Synthetic Fuels Commercialization Program", *The Principles and Applications of Decision Analysis*, R. A. Howard and J. E. Matheson, editors, Menlo Park, CA: Strategic Decisions Group, 1989.
5. Stephen M. Barrager, Bruce R. Judd, and D. Warner North, *The Economic and Social Costs of Coal and Nuclear Generation*, Report prepared for the National Science Foundation by Stanford Research Institute, Menlo Park, CA, March 1976.
6. Allen C. Miller, III, James. E. Matheson, Miley W. Merkhofer, Thomas R. Rice, and Ronald A. Howard, *Development of Automated Aids for Decision Analysis*, Final technical report prepared for the Defense Advanced Research Projects Agency (DARPA) by Stanford Research Institute, Menlo Park CA, May 1976.
7. Ronald A. Howard and James E. Matheson, "Influence Diagram Retrospective", *Decision Analysis* **2**:144–147, 2005.
8. Ronald A. Howard and James E. Matheson, "Influence Diagrams", *Decision Analysis* **2**:127–143, 2005.
9. D. Warner North, "EPA's Draft Reports to Congress on Global Warming: An Overview from 1990", available at http://www.northworks.net/w_epasab1990.htm.
10. D. Warner North and Stephen H. Schneider, "Global Climate Change: A Survey of the Science and Policy Implications", Conference Report Publication; available at http://www.northworks.net/w_pub_stuttgart.htm.
11. Robert T. Clemen, *Making Hard Decisions: An Introduction to Decision Analysis*, Belmont, CA: Duxbury Press, 1991.
12. Ross Shachter, "Evaluating Influence Diagrams", *Operations Research* **34**:871–882, 1986.
13. *Reliability and Decision Making*, Richard E. Barlow, Carlo A. Clarotti, and Fabio Spizzichino, editors, London: Chapman and Hall, 1993.
14. Elisabeth Paté-Cornell and Seth Guikema, "Probabilistic Modeling of Terrorist Threats: A Systems Analysis Approach to Setting Priorities Among Countermeasures", *Military Operations Research* **7**(4):5–23, 2002.
15. Stanford Energy Modeling Forum, *Natural Gas, Energy Diversity, and North American Energy Markets*, report on EMF 20, Stanford University, September 2003. Available at: http://www.stanford.edu/group/EMF/publications/doc/emf20summary.pdf. A list of other EMF reports is available at: http://www.stanford.edu/group/EMF/publications/order.htm.

16. Edward G. Cazalet, *Generalized Equilibrium Modeling: The Methodology of the SRI-Gulf Energy Model*, report prepared by Stanford Research Institute for the Federal Energy Administration, May 1977.
17. Horace W, Brock and Dale M. Nesbitt, *Large Scale Energy Models: A Methodological Analysis*, report prepared by Stanford Research Institute for the National Science Foundation, May 1977.
18. James Just and Lester Lave, "Review of Scenarios of Future US Energy Use", *Annual Review of Energy* **4**:501–536, 1979. http://arjournals.annualreviews.org/doi/pdf/10.1146/annurev.eg.04.110179.002441
19. National Research Council, *Understanding Risk: Informing Decisions in a Democratic Society*, Washington DC: National Academy Press, 1996. Available from: http://books.nap.edu/catalog/5138.html.
20. D. Warner North and Ortwin Renn, "Decision Analytic Tools and Participatory Decision Processes", "State of the Science" paper prepared for the National Research Council Panel on Public Participation in Environmental Assessment and Decision Making, March 2005. Available on the web through: http://www7.nationalacademies.org/hdgc/Workshop%20Materials.doc. A final report from this Panel of the National Research Council is expected in 2008.
21. Ortwin Renn, *Risk Governance: Towards an Integrative Approach*, White Paper #1, International Risk Governance Council, Geneva, Switzerland, September 2005. Available from: http://www.irgc.org/spip/IMG/pdf/IRGC_WP_No_1_Risk_Governance_(reprinted_version).pdf

Chapter 13
Nanotechnology Risk Governance

Mihail Roco[1], Ortwin Renn[2] and Alexander Jäger[3]
[1]*National Science Foundation, Washington, DC, USA*
[2]*University of Stuttgart, Stuttgart, Germany and DIALOGIK gGmbH, Stuttgart, Germany*
[3]*Interdisciplinary Research Unit on Risk Governance and Sustainable Technology Development (ZIRN), Stuttgart, Germany*

Introduction

Purpose and Background

This final case study summarises the major findings of the IRGC's White Paper on 'Nanotechnology Risk Governance' as prepared by the nanotechnology group lead by Mihail Roco and Ortwin Renn (IRGC 2006). Since its publication in June 2006, the White Paper provoked reactions from members of the academic community as well as from risk managers, regulators and representatives of non-governmental organisations (NGOs). Further comments were collected during two IRGC workshops on the risk management of nanotechnology in April and July 2006 and are presented here. The chapter will conclude with a discussion of the application and future development of the IRGC risk governance framework for nanotechnology.[1]

In contrast to most of the case studies described in this part of the volume, the risks from nanotechnology constitute a newly emerging field of research. Hence, the risk governance framework discussed in this article does not provide a lessons-learned perspective but a prospective and proactive one. Based on a careful assessment of nanotechnology's current status in the context of the regulatory environment, the level of science-policy interface and other aspects, IRGC's proposed framework presents decision-makers with a systematic and integrated approach to analysing and managing the anticipated risks, challenges and opportunities of nanotechnology.

[1] The 'White Paper on Nanotechnology Risk Governance' is the product of a collaborate effort for which input was provided on two initial workshops in May 2005 and January 2006 and which was advised by the IRGC's Nanotechnology Working Group and a number of external experts. In addition, the results of four stakeholder surveys undertaken as part of the project in the second half of 2005 are incorporated The surveys were concerned with the role of governments, industry, research organisations and NGOs and have been published as separate volumes on the IRGC website http://www.irgc.org/irgc/projects/nanotechnology/. On the same page, the full White Paper can be downloaded as well.

O. Renn and K. Walker (eds.), Global Risk Governance: Concept and Practice Using the IRGC Framework, 301–327.

The framework includes three major innovative concepts for the risk governance of nanotechnology:

- First, the risk governance strategies presented anticipate four generations of nanotechnology product development. Corresponding to the level of complexity of nanostructures and nanosystems, their behaviour dynamics and the level of knowledge about them, these four generations are divided into two levels of risk perception and represent two separate frames of reference: Frame 1: Passive Nanostructures (generation 1) – where complexity of a nanostructured component in a system is a typical characteristic, and Frame 2: Active Nanostructures and Nanosystems (generations 2–4) – where possible system uncertainty and a high degree of ambiguity are anticipated.
- Second, the framework integrates a scientific risk-benefit assessment (including environment, health, and safety and ethical, legal and other social issues) and concern assessment (an assess- ment of risk perception and the societal context of risk). The framework also includes the risk concerns about the educational gap issues, political and security issues and longer-term human development issues.
- Third, the authors elaborate risk management strategies that are based on a corrective and adaptive approach and take into account the level and extent of available knowledge and a societal balancing of the predicted risks and benefits. The proposed risk management escalator and stakeholder involvement are functions of the risk problem and quality of data and can be adapted to the level and nature of the risk situation.

Inherent in all three of these concepts and, indeed, throughout the whole risk handling chain is the need for all interested parties to be effectively engaged, for risk to be suitably and efficiently communicated by and to the different actors, for decision-makers to be open to public and expert concerns and, in cases of high ambiguity, for effective public engagement, and for anticipation of the need to build governance capacity early on in the process.

The final section of the chapter will conclude with high-level risk governance recommendations and suggestions for their implementation. The recommendations based on the IRGC framework are addressed to governmental, business, scientific, civil and communication actors who each share concerns about and responsibilities for the complex and interdependent field of nanotechnology governance. However, the focus of attention will be on governments, both individually and collectively, as they are responsible for developing and implementing the policies which will enable the maximum benefit to derive from nanotechnology with the minimum of risk. Before opening up the field of risk governance, a brief review of the promises of nanotechnology will follow.

Promises of Nanotechnology

Nanotechnology refers to the development and application of structures, materials, devices and systems with fundamentally new and valuable properties and functions which derive from the size of their structure in the range of about 1 to 100 nanometres (nm) (Siegel et al. 1999). It involves the manipulation and/or creation of material structures at the nanoscale in the atomic, molecular and supramolecular realm. At the nanoscale the physical, chemical, and biological properties of matter can be significantly changed as compared to properties of individual atoms and molecules or bulk matter, particularly under 10–20 nm, because of properties such as the dominance of quantum effects, confinement effects, molecular recognition and an increase in relative surface area. Nanotechnology is still in an early phase of development analogous to the state of information technology in the 1960s and of biotechnology in the 1980s.

However, because it allows fundamentally new characteristics and foresees almost unlimited applications, nanotechnology has the potential to become one of the defining technologies of the 21st century. It offers significant benefits to manufacturing, human health, energy conversion and to the environment and can act as a major driver of economic growth. In 2000, the US National Science Foundation (NSF) estimated that $1 trillion worth of products worldwide would incorporate nanotechnology in key functional components by the year 2015 (Figure 2; Roco and Bainbridge 2001). The corresponding industries would employ about 2 million workers directly in nanotechnology, and about three times as many in supporting activities. These estimates were based on a broad industry survey and analysis in the Americas, Europe, Asia and Australia, and continue to hold in 2006. The scientific discoveries at the nanoscale and this economic potential has encouraged a dramatic rise in research and development (R&D) expenditure in over 60 countries. Government R&D investments in each of the US, Japan, EU and the ‘Rest of the world’ (including Canada, China, Australia, Korea, Taiwan, and Singapore) totalled about or over $1 billion in 2005, with the fastest growth occurring in the ‘Rest of the world’. In 2006, industry R&D, with about $6 billion R&D investment, exceeded corresponding total government R&D expenditures of about $5 billion.

Nanotechnology has many characteristics which both increase its potential benefits while creating new issues for global risk governance. It:

- offers a broad technology platform for industry, biomedicine and environment as well as an almost infinite array of potential applications;
- holds promises for applications which have the potential to manage many technical, economic, ecological and social problems;
- allows manipulation at the basic level of organisation of atoms and molecules, where the fundamental properties and functions of all manmade and living systems are defined;
- has become one of the main drivers for technological and economic change and is already stimulating considerable industrial competition.

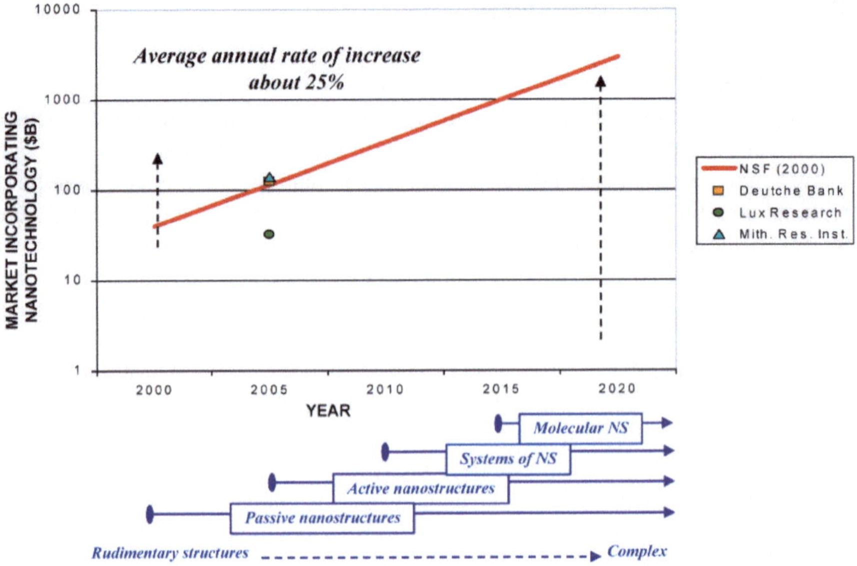

Fig. 1 Worldwide market affected by nanotechnology (NSF estimation made in 2000, the estimation holds in 2006).

The implications of nanotechnology are broad because its applications are at the confluence of modern biology, the digital revolution and cognitive sciences.

Reflecting the specific characteristics of nanotechnology, national R&D programmes established during the last five years have become highly integrative, involving multiple funding agencies. However, risk governance approaches specific to nanotechnology seem to be lagging behind as it has happen in other emerging technologies and there is a perception that the present speed and scope of R&D exceeds the capacity of regulators to assess potential human and environmental impacts.

Risk Governance of Nanotechnology: An Application of the IRGC Risk Governance Framework

Pre-Assessment: Categorisation of Nanotechnology into Two Frames of Reference

The first phase of the IRGC risk governance framework, pre-assessment, constitutes a preliminary assessment of what major societal actors (such as governments, companies, the scientific community, NGOs, communication organisations and the general public) define as risk problems, either because of their anticipated impacts or because they are areas of concern for other reasons. For nanotechnology, risks and opportunities are commonly associated with changes in the chemical reactivity, mechanical, optical, magnetic and electronic properties of downsized material structures as compared to a bulk structure with the same chemical elements. Additionally, the potential for confluence with modern biology, the digital revolution and cognitive sciences means that we can expect nanotechnology to penetrate and permeate through nearly all sectors and spheres of life (e.g. communication, health, labour, mobility, housing, relaxation, energy and food) and to have implications for socio-economical development and the environment on a global scale.

These emerging and integrated characteristics of nanotechnology lead to a situation whereby the risk perception of one application may drive apprehension about other applications with the label, nanotechnology, that, in reality, require quite different risk governance strategies. For that reason, the White Paper proposed that nanotechnology development not be viewed as a single consolidated concept but as comprising four overlapping generations of new nanotechnology products and processes, each generation having its own unique characteristics. We have defined these as (1) passive nanostructures, (2) active nanostructures, (3) complex nanosystems, and (4) molecular nanosystems (see Figure 2).

Furthermore, a second distinction can be made in terms of risk perception between the first generation and the following three generations. For the first generation of passive nanostructures, the ability to control nanostructure behaviour is easier to be done withing a system and more advanced than for the following three generations and it is therefore put in Frame 1. For generations two, three and four, potential social and ethical consequences are expected to be more transformative and they are put in Frame 2. Structuring nanotechnology risk governance into these two broad frames of reference allows for research and decision-making pathways to be adapted to the characteristics of each frame and, equally, for risks and concerns to be identified separately. Both, the categorisation in four generations and into two frames are depicted in Figure 2.

The distinction between Frame 1 and Frame 2 technologies is fundamental to the whole governance framework presented here and will further be elaborated in the following discussion of the current deficits of the nanotechnology risk governance, the risk appraisal and the risk management strategies.

On the basis of the Frame 1 and Frame 2 distinction it is possible to categorise the risk-related knowledge as part the IRGC risk governance framework, namely: simple risk, component complexity, system uncertainty, and ambiguity. Risk-related knowledge for Frame 1 can be best characterised as complex for those system components which contain nanostructures with new physical, chemical and/or biological properties and functions while impacts on the societal system are expected to be less substantial. Risk-related knowledge for Frame 2 can be best characterised as uncertain for active system components (second generation) and nanosystems as a whole (third and fourth generations). For the more ambiguous large nanostructured systems (third generation) and molecular nanosystems (fourth generation), an appropriate evaluation will require the use of conflict resolution methods in order to resolve problems of perception and interpretation between stakeholder groups.

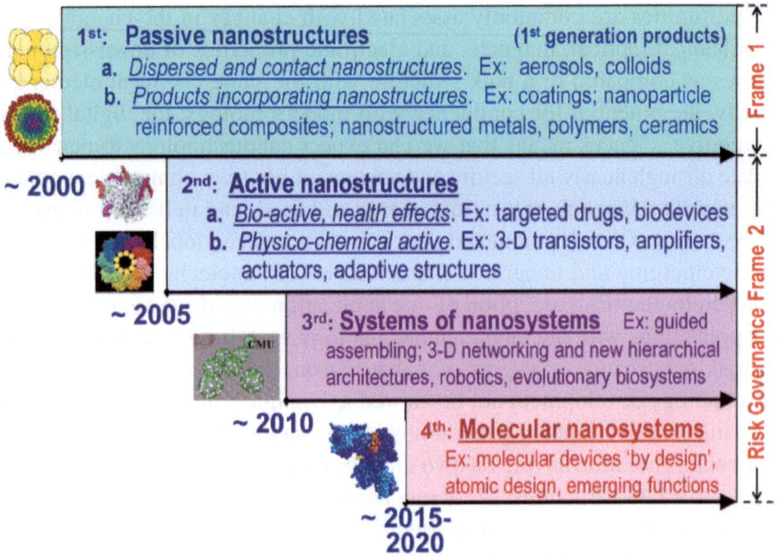

Fig. 2 Timeline for the beginning of industrial prototyping and nanotechnology commercialisation: Four generations of products and production processes.

After framing the governance processes in the pre-assessment section of the framework, current deficits of the nanotechnology risk governance processes can now be identified.

Deficits in Nanotechnology Risk Governance Today

The main risk governance deficits for the Frame 1 (first generation of passive nanostructures: nanoparticles, coatings, nanostructured materials) are a relatively low level of understanding of the new properties and functions of toxicity and bioaccumulation, limited knowledge of nanomaterials exposure rates and gaps in regulatory systems at national and global levels. The main risk governance deficits for the second to fourth generations of nanoproducts (including active nanodevices, nanobio applications and nanosystems) are the uncertain and/or unknown implications of the evolution of nanotechnology and its potential human effects (e.g. health, changes at birth, brain understanding and cognitive issues and human evolution), and the lack of a framework through which organisations and policies can address such uncertainties. In the following we list the main deficits anticipated for nanostructures. It is assumed that where deficits and recommendations are referred to as applying to first generation nanostructures these will also apply to later generations. However, there are specific deficits that are more unique to the second to fourth generations

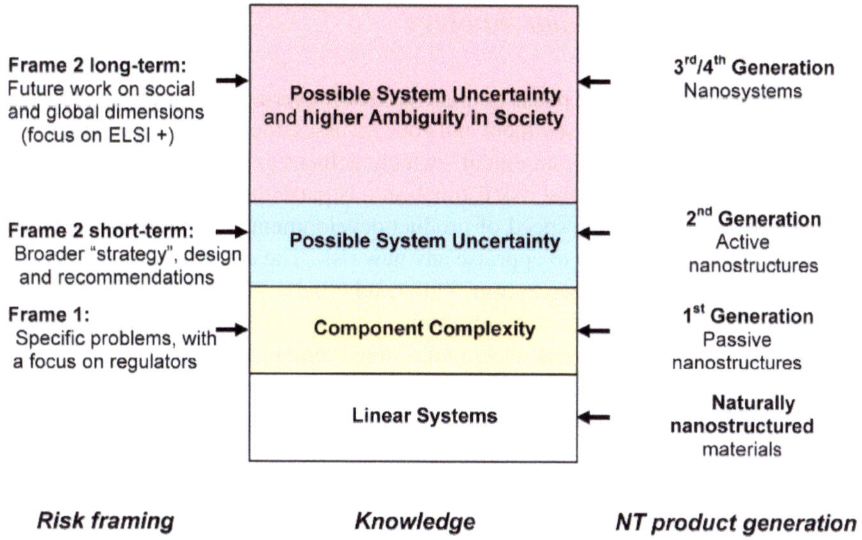

Fig. 3 Strategies as a function of the generation of nanoproducts: Application to Frame 1 and Frame 2.

due to their expected complex and/or evolving behaviour and, where this is the case, it is specifically mentioned.

General deficits more specific for Frame 2 are:

- uncertain or unknown implications mostly because the products are not yet fabricated;
- potential human effects of 2nd–4th generation nanoproducts;
- nanotechnology innovation proceeds ahead of the policy and regulatory environment: expected gaps in long-term for Frame 2;
- lack of a specific framework to address such issues.

Technical and organisational (environmental health and safety):

- limited knowledge on hazards and exposures, need for specific metrology, international transport, etc.

Institutional deficits (societal infrastructure, social and political):

- relatively fragmented government institutional structure, relatively simple cause and effect approach, weak coordination among key actors, etc.

Risk communication deficits:

- Gap between science communities, between science communities and manufacturers/regulators/public/NGOs/industry/media/public.

Risk Appraisal for Nanotechnology

Risk appraisal is the second phase of the IRGC risk governance framework and comprises two elements: risk assessment and concern assessment. During risk appraisal, the classic risk assessment component – which includes generally hazard identification, exposure assessment and risk estimation – is particularly important for Frame 1 nanostructures where the speed of product development and application exceeds the ability of risk assessors to appraise any new risk. The concern assessment component – focused on risk perception and stakeholders' concerns – is particularly important for Frame 2 where less substantive knowledge is available and actors are more concerned with the social desirability of the anticipated innovations. The following two sections summarise the current levels of knowledge available and the key risk appraisal requirements for each frame.

Risk Appraisal of Frame 1 Nanotechnology Applications: A Focus on Risk Assessment

Only a limited understanding exists of the potential environmental, health and safety risks of nanomaterials. Further studies are required for: (1) hazard identification, in areas such as toxicity, ecotoxicity, carcinogenicity, volatility, flammability, persistence and accumulation in cells; and (2) exposure, including the potential for oral, cutaneous and inhalation uptake of nanomaterials during production, transport (in air, water, soil and biosystems), decomposition and/or waste disposal. Some of these risks are:

- Human health risks. Several studies have shown that: (1) due to the high surface-area-to-volume ratio and frequently higher reactivity of nanostructures, large doses can cause cells and organs to demonstrate a toxic response (in particular inflammation) even when the material itself is non-toxic; (2) some nanosized particles are able to penetrate the liver and other organs and to pass along nerve axons into the brain; (3) nanomaterials may combine with iron or other metals, thereby increasing the level of toxicity and presenting unknown risks; (4) engineered nanomaterials raise particular concerns because of the unknown characteristics of their new properties and their potential use in concentrated amounts; and (5) some nanomaterials may have similar characteristics to known high-risk materials at the microscale.
- Explosion risks. The higher surface reactivity and surface-area-to-volume ratio of nanopowders increases the risk of dust explosion and the ease of ignition.
- Ecological risks. The impact of nanostructures on the environment may be significant because of the potential for: (1) bioaccumulation, particularly if they absorb smaller contaminants such as pesticides, cadmium and organics and transfer them along the food chain; and (2) persistence, in effect creating non-biodegradable pollutants which, due to the small size of the nanomaterials, will be hard to detect in situ in the environment.

In addition to the EHS risks the following societal impacts of nanotechnology development have been raised for Frame 1:

- Political and security risks. Decisions taken about the direction and level of nanotechnology R&D may result in: (1) insufficient investment in key areas to benefit future economic development; (2) an uneven distribution of nanotechnology risks and benefits among different countries and economic groups; (3) use in criminal or terrorist activity; and (4) a new military technology race.
- Educational gap risk. If the knowledge within professional communities is not appropriately shared with regulatory agencies, civil society and the public, and, consequently, risk perception is not based on the best available knowledge, innovative opportunities may be lost.

As we will see, the societal impacts described for Frame 1 are also, in part, valid for Frame 2.

Risk Appraisal of Frame 2 Nanotechnology Applications: A Focus on Concern Assessment

As already mentioned, the risk appraisal of Frame 2 is more concerned with the social desirability of the anticipated innovations. Several characteristics of the risk appraisal for Frame 2 are:

- Inclusion of the societal context from the risk pre-assessment to risk management is needed: risk-benefit analysis of what is a desirable investment; increased science–society interactions because societal implications are broad and political involvement is necessary.
- Capabilities to safely use converging technologies must be developed since nanotechnology is applied in conjunction with other technologies.
- Anticipatory and integrative measures based on both scenarios development and building capacity are needed to address them in an corrective, adaptive system; evaluation criteria coming form various communities.

The potential for environmental health and safety risks identified in Frame 1 are also relevant to Frame 2 where one expects a large impact and there is a lower level of knowledge and understanding of the nanostructures and nanosystems and their behaviours (see Figure 3). The risks requiring further study in Frame 2 are primarily related to the assessment of the more complex behaviours and prioritisation of stakeholder concerns which in part rest on different value-judgements. In the White Paper the following most significant potential risks were identified:

- Risks to human biological and societal development. Societal apprehension exists about the use of nanotechnology to change biological and socioeconomic dimensions. Examples include: (1) economic impact of mass application of nanotechnology; (2) changes to the environment, human safety and quality of life; (3) genetic modification to control factors such as sex or eye colour; (4) devices to control the human brain and body; and (5) new technological and cultural environment based on the ability to purchase new revolutionary products and access cognitive technologies and life extension promises.

- Society structural risks. Risks may be dampened but also induced and amplified by the effect of social and cultural norms, structures and processes, such as: (1) the inability of the regulatory envi- ronment to react rapidly to new technologies; (2) the unintended availability to the mass market of products based on applications developed by and for the military (e.g. tiny airborne surveillance devices); and (3) lack of propr educational and communication organisations to address development of nanotechnology and unexpected events.
- Public perception risks. Recent surveys have shown that public concern is currently less linked to any particular application or risk but rather to the capacity for human misuse, to unexpected technological breakouts, or to nanotechnology's potential to exacerbate existing social inequalities and conflicts. This situation may change if nanotechnology becomes associated with specific incidents and, meanwhile, there remains a deep suspicion of the motives of industry and doubts regarding government's ability (or desire) to act if required. The increasing impact of the mass media on risk perception (such as in movies and books) need to be fully considered.
- Transboundary risks. The risks faced by any individual, company, region or country depend not only on their own choices but also on those of others. Evidence that control mechanisms do not work in one place may fuel a fierce debate in other parts of the world about the acceptability of nanotechnology in general.

Risk Appraisal of Nanotechnology: A Summary

An overview of the various potential risks and other impacts in the development of nanotechnology is provided in Figure 4. The potential risks may be caused either by new processes and products (placed on the left column in Figure 4), by societal implications (central column), and/or global interactions (right column). The negative consequences may be directly related to harm by unintended effects (first row in Figure 4) or by the risk of missing the benefits (bottom row).

Characterisation and Evaluation of the Benefits and Risks

Arriving at a balanced judgement about the acceptability or tolerability of nanotechnology means that nanotechnology will deliver sustainable added value for society, economy and industry only if it is possible to achieve an acceptable balance between its projected benefits and the management of its unintended impacts and risks at a societal level. It is not sufficient to focus only on the physical, chemical or biological risks of nanotechnology, because they address only part of what is at stake within culturally plural, morally concerned and educated societies (AEBC 2001; Grove-White et al. 2000). Stakeholders play an important role in defining acceptable to

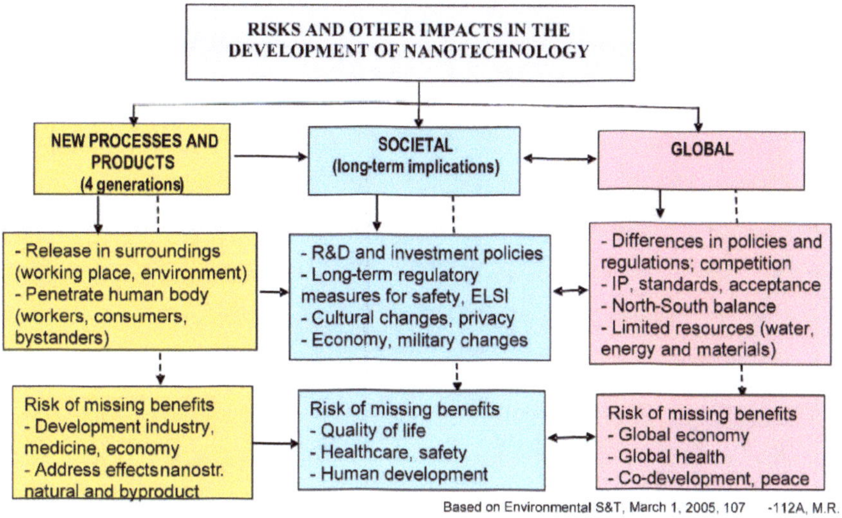

Fig. 4 Risks and other impacts in the development of nanotechnology (negative implications, including not taking advantage of the benefits).

intolerable by considering among other factors the balance between risk and benefits and the probability of extreme events. Figure 5 brings together the findings of the previous sections in the traffic light model of the IRGC risk governance framework (Chapter 1). As in the general IRGC framework on risk governance, 'acceptable' – attributed to most naturally nanostructured materials – refers to an activity where the remaining risks are so low that additional efforts are not necessary. 'Tolerable' – attributed to engineered nanostructures – refers to a technology worth pursuing yet requiring additional efforts for risk reduction. Intolerable risks would, for example, be constituted by explosive nanomaterials designed for other purposes.

Risk Management Strategies for Frame 1 and Frame 2

Risk management, the final phase of the risk governance framework, comprises the selection of a strategy or strategies designed to avoid, prevent, reduce, transfer or contain risks. For both frames there are factors particular to nanotechnology that will impact on the choice of measures. These include: (1) its multidisciplinary, cross-sectoral and multiple stakeholder nature; (2) its characterisation as more-or-less complex, uncertain or ambiguous (depending on the specific development or application under consideration); and (3) the need to ensure the consistent participation in the risk management process of key stakeholders including all countries concerned in nanotechnology research, development and application.

The White Paper proposes a risk management strategy based on adaptive and corrective measures rather than a simple cause-and-effect approach, as well as co-

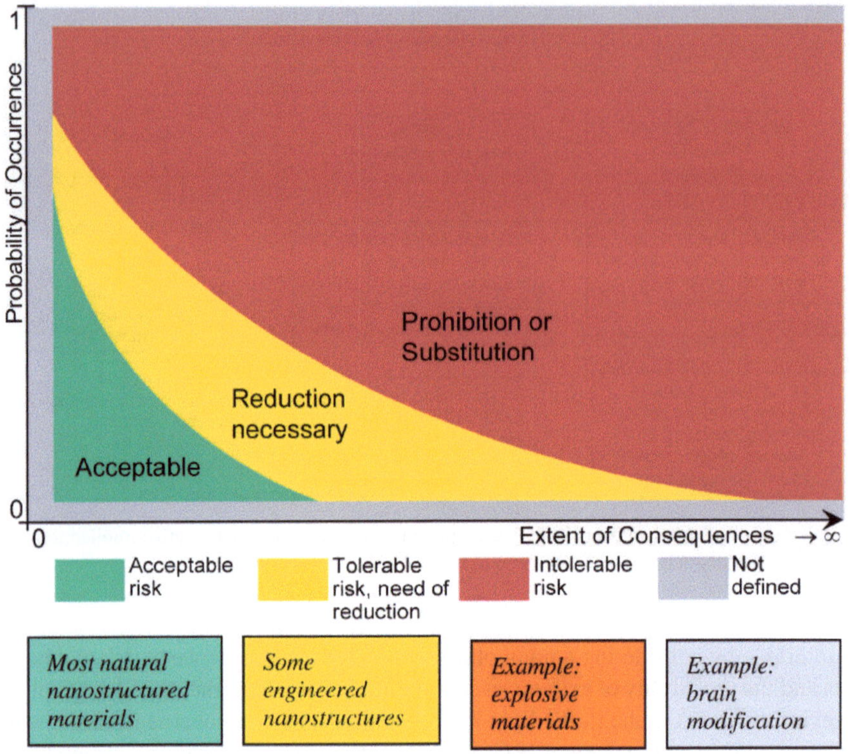

Fig. 5 Acceptable, tolerable, intolerable and undefined risks relative to benefits (Traffic Light Model, a stakeholder perspective).

ordination at the international level. Furthermore, it is recommended that risk management strategies include contingency plans for dealing with a wide variety of risk scenarios, so as to prepare for changes in the economy, the societal and political arena or in the available levels of knowledge. Decision-makers also need to distinguish between Frame 1 and Frame 2, designing risk management and communication programmes that provide adequate and effective strategies for the particular characteristics of each frame. Tables 1 and 2 provide detailed proposals of the risk management recommendations for Frame 1 and Frame 2. Both tables already contain information on stakeholder participation and risk communication that will be dealt with in the following paragraphs. The first part of recommendations addresses the classical risk assessment components – hazard, exposure and risk – the second part focuses on institutional, communication and transboundary issues.

Given the lack of nanotechnology-specific regulation, one of the most promising management strategies for Frame 1 is to establish internationally-applicable voluntary codes or rules for ensuring safety and risk control in the short term, allowing time for the necessary development and establishment of formal norms. The risk management strategies identified for Frame 1 will also be applicable for Frame 2.

Table 1 Risk government recommendations for Frame 1 passive nanostructures.

<table>
<tr><th colspan="4">Risk Management Recommendations for Frame 1</th></tr>
<tr><td rowspan="1">Classical Risk Assessment</td>
<td>Hazard Recommendations
– Testing strategies for assessing toxicity and eco-toxicity.
– Best metrics for assessing particle toxicity and eco-toxicity.
– A nomenclature which includes novel attributes, such as surface area.
– Pre-market testing, full lifecycle assessment and consideration of secondary risks.
– Disposal and dispersion methods for nano-engineered materials.
– Development of waste treatment strategies.</td>
<td>Exposure Recommendations
– Exposure monitoring methodologies.
– Methods for reducing exposure and protective equipment.</td>
<td>Risk Recommendations
– Risk assessment methodologies.
– Guidelines and best practices available internationally.
– Evaluation of the probability and severity of risks, including loss of benefits.
– Balanced knowledge-based communication and education of EHS and ELSI, including uncertainties and ambiguities.</td></tr>
<tr><td>Extended Risk Assessment</td>
<td colspan="3">Institutional Recommendations
– Systematic liaison between government and industry.
– Sufficient resources and capabilities for conducting concern assessments along with risk assessments.
– Information for consumers enabling them to make informed choices.
– Transparent decision-making processes for R&D and investment.
– Non-proprietary information on test results, impact assessments and their interpretations on the internet.
– Systematic feedback about the concerns and preferences of the various actor groups and the public at large.
– Incentives for promoting and sustaining international cooperation.
– Critical examination of intellectual property rights for basic natural processes and structures.
Risk Communication Recommendations
– Information about the benefits and non-intended side effects. Communication tools include: internet-based documentation of scientific research, product labelling, press releases and consumer hot lines.
– Public information on the principles and procedures used to test nanotechnology products, to assess potential health or ecological impacts and to monitor the effects.
– International disclosure of risk information by large transnational companies (not competitive information).
– Risk communication training courses and exercises for scientists.
– Integrated risk communication programmes for scientists, regulators, industrial developers, representatives of NGOs, the media and other interested parties.
Transboundary Recommendations
– Incentives for all countries to participate in risk governance. Possible tools include: policies by insurance companies, certification programmes, education programmes, R&D programmes, response to disruptive technological and economical developments, and international studies on cost and benefit/risk analysis.
– Explore the role of international organisations, international industry and academic organisations and NGOs.
– Public-private partnerships when participants are reluctant to adopt protective measures. Possible method include: government standards and regulations coupled with third party inspections and insurance.
– Global communication of international standards and best practices to both developing and developed countries in a reasonable timeframe.</td></tr>
</table>

In addition, given the ambiguity and lack of substantive knowledge associated with Frame 2 technologies, a more discursive and participatory approach is required in which all actors, including industry and NGOs should be involved from the beginning. Table 2 provides a more detailed proposal of risk management recommendations for Frame 2.

Table 2 Risk government recommendations for Frame 2 active nanostructures and nanosystems.

	Risk Management Recommendations for Frame 2 (in addition to those listed for Frame 1 in Table 2)		
Classical Risk Assessment	**Hazard Recommendations** – Identifying hazards using scenarios. – Matrix for assessing the identified hazards.	**Exposure Recommendations** – Estimation of exposure for events with great uncertainties using methods such as casual chain.	**Risk Recommendations** – Identifying, communicating and educating others on environmental health and safety, ELSI, Human Development Implications and Political and Security Issues. – Developing capacity to address uncertain/ unknown and ambiguous developments at national and global levels. – Identifying and analysing highly controversial developments.
Extended Risk Assessment	**Institutional Recommendations** – Communication platforms that help address the purposes for future technologies. – Common scenario development exercises for future applications of nanotechnology. – Common rules and standards for potentially high-impact, long-term projects for nanotechnology. – A process of periodic review of national and international institutional frameworks. **Risk Communication Recommendations** – Debate on the desirability of special applications of nanotechnology in the light of ethical and social issues. **Transboundary recommendations for Frame 2** – Use corrective, adaptive approach – Use integrative methods for converging technologies – Consider the co-evolution NT-ethics – Create incentives for all countries to participate in RG of emerging technologies in Frame 2 (certification programs, policies for insurance, research and education programs, and responses to disruptive technological and economical developments – Address ELSI + : ELSI, EGI, PSI, HDI – Increased role of risk communication between all stakeholders, from manufacturers to regulators and public; a special role of media and Internet – Responsible innovation: Include ELSI in training; Incentives in industry using certification; ELSI criteria in evaluating value of IPO		

Risk Management Strategies for Stakeholder Participation

A central aim of applying the IRGC model is to stimulate participatory innovation in anticipatory debates about emerging technologies, and to generate better and neutral platforms for stakeholder involvement. For this, it is again helpful to distinguish between simple, complex, high uncertainty and high ambiguity risk problems. In the case of nanotechnology, the four single levels of risk related knowledge and the respective technologies lead to the involvement of different types of actors and anticipate particular types of discourses (see Figure 6).

In Figure 6, the first two categories – naturally nanostructured materials/simple risk problems and engineered nanostructured materials/component complexity – are part of Frame 1, whereas the latter two categories – active nanostructures and systems/systems uncertainty induced and large and molecular nanosystems/ambiguity induced are part of Frame 2. In accordance to the risk related knowledge, specific types of conflicts can be identified that ask for appropriate ways to involve stakeholder groups. In the case of nanotechnology, the risk management escalator shows four different routes to which a particular technology may be allocated which then

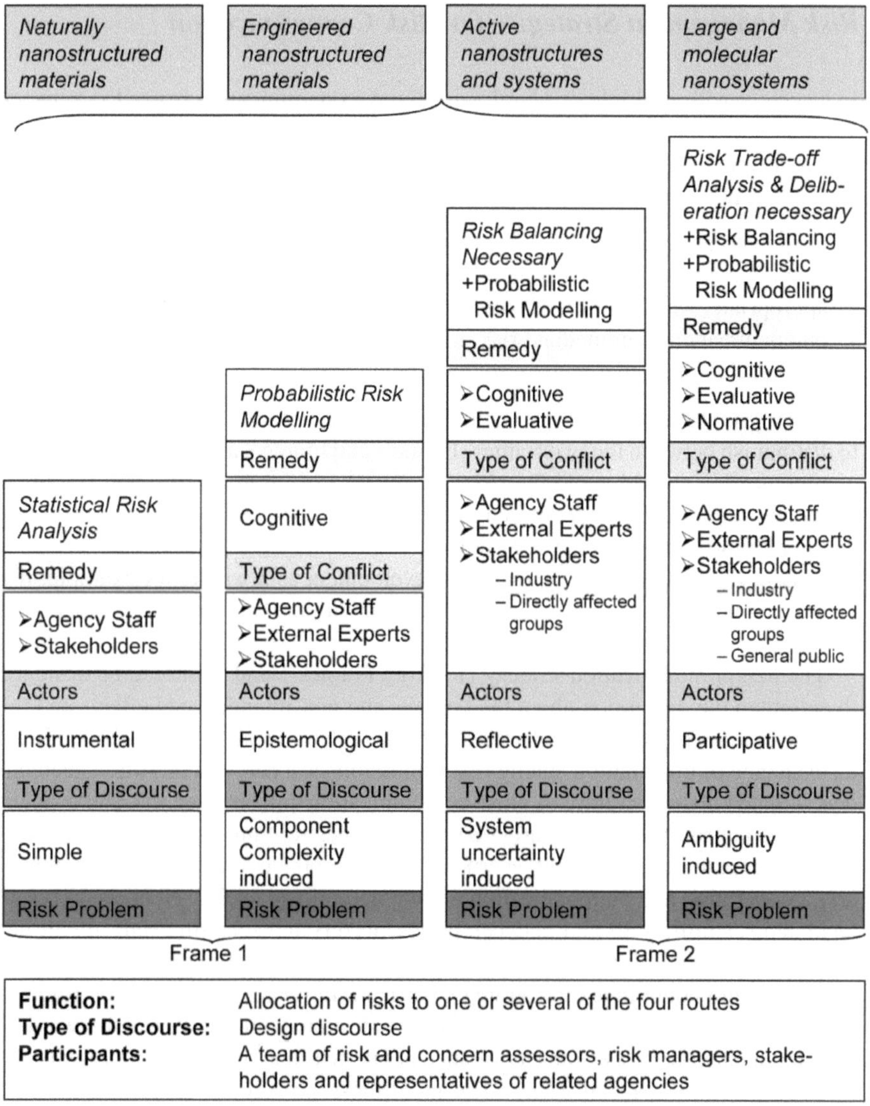

Fig. 6 The risk management escalator and stakeholder involvement (from simple systems via complex and uncertain to ambiguous phenomena) with reference to nanotechnology.

offer approaches to adequate stakeholder involvement. Each and one of the routes depict separate situations and remedies and do not build upon another.

Risk Management Strategies for Risk Communication

In a society with multiple or plural values, risk communication is needed between key staholders throughout the risk handling chain, from the framing of the issue to the monitoring of the potential impacts of risk management strategies themselves. Risk communication is a means to ensure that:

- those who are central to risk framing, risk and concern assessment or risk management understand what is happening, how they are to be involved, and, where appropriate, what their responsibilities are ('open' communication), and
- others outside the immediate risk appraisal or risk management process are informed and engaged ('transparent' communication).

In designing the risk communication strategies for nanotechnology, it is essential to distinguish between the two frames (Frame 1 and Frame 2 as defined earlier). Risk communication should avoid the strategic mistake of grouping all applications of nanoscale technologies under the single descriptor 'nanotechnology'. This approach would blur the distinction between the two frames and their subcategories and runs the risk of discrediting nanotechnology development as a whole, for example, if a serious incident related to a specific application within one frame or the other were to occur.

The first communication strategy (for both Frames 1 and 2) should be designed to enlighten the discussion about the benefits and non-intended side effects and the methods to identify and quantify those effects. The first task should be to facilitate an exchange of information among risk professionals, a task that has often been underestimated in the literature. A close communication link between risk and concern assessors and risk managers, particularly in the phases of pre-assessment and tolerability/acceptability judgement, is crucial for improving overall governance. Similarly, cooperation and communication among natural and social scientists, between legal and scientific staff and between policy-makers and scientific staff are all important prerequisites for enhancing the risk management performance.

The second communication strategy, particularly for Frame 2, requires communicating risk appropriately to the outside world, and should be directed towards a broader debate on the desirability of special applications of nanotechnology in the light of ethical and social issues. The main message here could be that it is not nanotechnology that creates the problem but rather the use of this technology in a controversial application. It may certainly be legitimate to reject special applications (such as using neurochips in the human brain for control of its functions without a medical justification) without having to oppose the technology that makes such an application technically feasible.

This second strategy is very challenging to implement. Many representatives of stakeholder groups, particularly members of the affected and non-affected public, are often unfamiliar with the approaches used to assess and manage risks. They may find it difficult to differentiate between the potentially harmful properties (i.e. hazards) of a nanotechnology product and the estimates of risk that depend not just on the hazards but on the scenarios that describe the ways in which the products might be used and on the potential for exposure to humans associated with those scenarios (Morgan et al. 2002). They may try to pursue their own specific agendas, trying to achieve extensive consideration of their own viewpoints.

Nonetheless, it is critical to provide public information on the principles and procedures used to test nanotechnology products, to assess their potential health or ecological impacts, and to monitor their effects. Candid discussions on the role of investment policies in research, development and production can also be helpful. If people have the reassurance that public authorities are taking special care to protect the population against unintended consequences of this new technology, they may be able to develop more trust than they demonstrate today in the capacity of society to control the risks and acknowledge and plan for remaining uncertainties. This is true for both the public and the private sector.

After diagnosing the needs of the two major audiences, the main form(s) of communication must be chosen. There are four basic forms of communication:

- *Documentation.* In a democratic society, it is absolutely essential that documentation in the field of nanotechnology products marketed to the consumers (such as sun creams) should more accurately reflect risks, show how and why decisions in risk management were made, which arguments were considered and what scientific bases were used. This serves transparency. Even if explanations are comprehensible only to a few, like with package inserts that describe the potential side effects of medicines that are understandable often only to the medically trained, they illustrate that nothing is being withheld.
- *Information.* Information should be prepared and compiled in such a way that it addresses key concerns of target groups and so that individuals within target groups can comprehend it and can integrate its message into their everyday lives.
- *Two-way communication or dialogue.* This form of communication aims at an exchange of arguments, experiences, impressions and judgements. There must be willingness on both sides to listen to and learn from the other.
- *Participation in risk analyses and management decisions.* In a pluralistic society people expect to be included adequately, directly or indirectly, in decisions which concern their lives. Not all affected people can participate in the risk governance, but it must be ensured that the concerns of the stakeholders will be represented in the decision-making process and that the interests and values of those who will later have to live with the risk effects will be taken up appropriately and integrated into the decision-making process. The previous section on stakeholder involvement provided some insights into this aspect of risk communication.

We acknowledge that public engagement does not necessarily solve all problems; it does, however, enlighten the public debate and provide incentives for mutual learning, gaining and sustaining trust. It also helps individuals to be more attentive to both benefits and risks. We believe that the present platform could act as a catalyst for achieving these goals.

IRGC recommends that the risk communication strategy be developed carefully within the context of the risk handling chain.

As part of the risk appraisal phase, in of risk assessment, risk management agencies should be encouraged to undertake an exercise to develop and characterise potential scenarios that might describe the diffusion of nanotechnology in their own countries and the likely social reactions to it. Academic researchers, developers, potential users and important other actors should be actively involved in the scenario workshops in order to get a adequate representation of societal forces that ultimately shape the future of nanotechnology in the world. The scenarios suggested in the IRGC White Paper on Nanotechnology (IRGC 2006) may serve as default options for designing more specific scenarios that relate to specific uses and contextual conditions in each participating country.

The second key component of the risk appraisal step is the concern assessment in which it is essential to investigate and explore the social and cultural frames and the individual risk and benefit perception patterns in the respective countries or cultures. The relevant actors in each country need to be informed about the structure and strength of the various frames that individuals and groups associate with nanotechnologies. For this purpose, interviews should be conducted with the leading individuals of civil society groups such as consumer unions, environmental groups, religious communities, and others. On the basis of these empirical results, one can compare insights from one country with similar studies in different other countries and conduct a systematic evaluation in terms of intensity of concerns, types of concerns and willingness to act.

Such an analysis is not only a means for identifying potential barriers and obstacles to the diffusion of the technology, but it is also an important input for refining potential scenarios and for the identification of potential opportunities based on revealed preferences. In addition, both risk management agencies and corporations would understand better the factors that govern the perception process for each nanotechnology frame and would be better equipped to design appropriate risk management and risk communication strategies.

Together, the scenario exercise and the concern study and can provide input to a targeted and effective communication programme designed to foster public understanding of technical issues and to address the perceptions and concerns of the key actors. The programme for Frame 1 could include internet presentations, brochures, press releases, consumer product labels and others. If the concern assessment concludes that the Frame 2 is also quite relevant, other communication means will be needed such as an open forum on the use and abuse of nanotechnology for medical, military or other controversial purposes. In addition, citizen panels or joint action committees (including consumer associations, unions, employers, etc.) could be convened to draft legislation that would inhibit the potential misuse of nanotechnologies. All these activities would be able to preserve or even restore trust in the risk managing agencies.

Risk Governance Strategies and the Potential Future Role for International Bodies

To summarise insights provided by the prior analysis, Table 3 gives an overview of the characteristics and the risk governance context of the four generations of nanotechnology developments put in Frame 1 and Frame 2.

On the base of the detailed analysis, the following general risk governance strategies can be recommended:

- Distinguish between Frame 1 and Frame 2 debates and design corresponding risk management and communication programmes.
- Ensure that the interests of all those potentially affected by nanotechnology are addressed and understood by decision-makers.
- Be cognisant of and, where appropriate, responsive to other global governance systems.
- Adaptive and corrective approaches need to be applied to societal system.
- Develop an inclusive risk governance framework addressing both short and long-term applications of nanotechnology.

However, it is clear that a reasonable risk governance framework needs a number of motivated and concerned stakeholders that put the recommendations into practice. The IRGC White Paper on Nanotechnology (IRGC 2006) has presented one of the first conceptual frameworks for nanotechnology governance that may serve as a basis for designing more specific programmes tailored to specific applications or countries. However, ongoing involvement of and debate amongst academic researchers, nanotechnology developers, potential users, regulatory or other decision-making authorities, and other important actors is essential to ensure the inclusion of an adequate representation of societal forces that ultimately shape the future of nanotechnology.

With the goal of fostering such debate and improvement to the risk governance process for nanotechnology, IRGC organised an important international conference on risk management of nanotechnology which was held at the Swiss Re Centre for Global Dialogue in Ršchlikon, Switzerland on the 6th and 7th of July 2006.[2] A broad diversity of stakeholders from may societal sectors, international organisations and various countries participated assuring perspectives from a large diversity of representative stakeholders for nanotechnology development worldwide. The section which follows summarises the key goals of that conference and the responses received regarding the IRGC's risk governance framework for nanotechnology.

[2] A conference programme and the presentations can be downloaded from http://www.irgc.org/irgc/events/conferences/ (30 September 2006).

Table 3 The different generations and frames of nanotechnology development.

	Four Generations	Generation Characteristics	Risk Governance Context
FRAME 1	**First Generation** – passive (steady function) nanostructures, e.g. nanostructured coatings and non-invasive; invasive diagnostics for rapid patient monitoring *From 2000 -*	***Behaviour:*** inert or reactive nanostructures which have stable behaviour and quasi-constant properties during their use. ***Potential risk:*** e.g. nanoparticles in cosmetics or food with large scale production and high exposure rates.	***Current context for Frame 1 products and processes:*** interested parties are seeking to develop knowledge about the properties of nanomaterials and their EHS implications so that risks can be characterised internationally. Debates are focused on the design and implementation of best practices and regulatory policies. ***Risk characterisation:*** the nanoscale components of the nanoscale products and processes result in increased system component complexity. ***Strategies:*** the establishment of an internationally reviewed body of evidence related to toxicological and ecotoxicological experiments, and simulation and monitoring of actual exposure. ***Potential conflict:*** the question of how much precaution is necessary when producing the nanomaterials (focusing on changes to best practices and regulation) and over their use in potential applications.
FRAME 2	**Second Generation** – active (evolving function nanostructures) e.g. reactive nanostructured materials and sensors; targeted cancer therapies *From 2005 -*	***Behaviour:*** the structure, state and/or properties of 'nanostructures' are designed to change during use so behaviour is variable and possible unstable. Successive changes in state may occur, either intended or as an unforeseen reaction to the external environment. ***Potential risk:*** e.g. nanobiodevices in the human body; pesticides engineered to react to different conditions.	***Current context for Frame 2 products and processes:*** interested parties are considering the social desirability of anticipated innovations. Debates are focused on the process and speed of technical modernisation, changes in the interface between humans, machines and products, and the ethical boundaries of intervention into the environment and living systems (such as possible changes in human development and the inability to predict transformations to the human environment). ***Risk characterisation:*** the nanoscale components and nanosystems of the Frame 2 products and processes result in knowledge uncertainty and ambiguity. ***Strategies:*** stakeholders must achieve understanding, engage in discussion about ethical and social responsibility for individuals and affected institutions and build institutional capacity to address unexpected risks. Projected scenarios need to be explored that show plausible (or implausible) links between the convergence of technologies and the possible social, ethical, cultural and perception threats. A major challenge is that decisions need to be undertaken before most of the processes and products are known.
	Third Generation – integrated nanosystems (systems of nanosystems) e.g. artificial organs built from the nanoscale; evolutionary nanobiosystems *From 2010 -*	***Behaviour:*** passive and/or active nanostructures are integrated into systems using nanoscale synthesis and assembling techniques. Emerging behaviour may be observed because of the complexity of systems with many components and types of interactions. New applications will develop based on the convergence of nanotechnology, biotechnology, information technology and the cognitive sciences (NBIC). ***Potential risk:*** e.g. modified viruses and bacteria; emerging behaviour of large nanoscale systems.	

Table 3 Continued

Four Generations	Generation Characteristics	Risk Governance Context
Fourth Generation – heterogeneous molecular nanosystems e.g. nanoscale genetic therapies; molecules designed to self-assemble *From 2015/2020 -*	***Behaviour***: engineered nanosystems and architectures are created from individual molecules or supramolecular components each of which have a specific structure and are designed to play a particular role. Fundamentally new functions and processes begin to emerge with the behaviour of applications being based on that of biological systems. ***Potential risk***: e.g. changes in biosystems; intrusive information systems.	are known. ***Potential conflicts***: the primary concern of Frame 2 is that the societal implications of any unexpected (or expected but unprepared for) consequences and the inequitable distribution of benefits may create tensions if not properly addressed. These concerns about technological development may not be exclusively linked to nanotechnology but are, at least partially, associated with it and will impact upon stakeholder perceptions and concerns.

Reception of the IRGC Risk Governance Framework for Nanotechnology: Feedback from an International Conference

The basis for the discussions at the international conference was IRGC's White Paper on Risk Governance for Nanotechnology, with a particular goal of the conference being to identify the strengths and weaknesses of the IRGC approach. To meet this objective, working groups on different aspects of the framework were organised and participants presenting at the conference were asked to give their opinions on the White Paper. The feedback from the conference provides a valuable insight into the reception of the framework by different stakeholder perspectives.

Framing the Debate on Potential Risks from Nanotechnology: Views on Frame 1 and Frame 2

Overall, the introduction of the Frame 1 and Frame 2 distinction and the four generations of nanotechnology development received considerable affirmation. It was felt that the frames provide a good basis for the debate, especially when highlighting the differences in levels of knowledge.

However, it was felt that there were a number of important similarities between Frame 1 and Frame 2 that the White Paper did not acknowledge or stress. In particular, it was felt that the two frames shared a number of ethical, legal and social issues. Accordingly, it was felt that the participation methods for both frames should not differ. Often it was also mentioned, that the close relation between Frame 1 and Frame 2 should be emphasised. This could be implemented in the form of a conceptual continuum of complexity of societal and technical questions and governance issues. In addition, a transition period between Frame 1 and Frame 2 technologies was asked for.

Whereas the Frame 1/Frame 2 distinction was generally seen as useful for analytic and academic purposes, it was perceived to be difficult for communication purposes to the public.

Only a few participants rejected the frame altogether. In these cases, the participants felt that the frames failed to cover broad socio-economic questions or were too dominated by the emphasis on knowledge problems.

Risk Management Recommendations

Frame 1

The issue raised the most regarding the Frame 1 risk management recommendations was that there should be a greater emphasis on ethical, legal and social issues. While the Nanotechnology White Paper properly highlights higher importance of these issues for Frame 2, these types of issues are seen as already critically important today, and especially so when combined with questions of environment, health and safety. Regarding issues of the Frame 1 knowledge level, it was further recommended to monitor possible uncertainties in future which would change the Frame 1 status. In this way the problem of surprise and changing boundaries of knowledge could be emphasised.

The most critical recommendations for Frame 1 issues were perceived to be (1) the development of concrete risk (hazard and exposure) assessment methodologies, and (2) the development of international standards. The participants expressed the hope that an internationally agreed set of standards would arise in the near future that could provide the basis for adaptable systems of regulation. Widely approved risk assessment procedures were perceived to be important inputs for such a set of standards.

Further recommendations were to revisit the transboundary recommendations, to make best practice guidelines available internationally and to pay careful attention to whole supply chain of nanotechnologies. IRGC's recommendations on self-regulation were perceived as unclear and industry commentators suggested that they should be exchanged for voluntary programmes.

Frame 2

The forward looking approach of Frame 2 was very much welcomed. For a number of commentators, however, Frame 2's foresight could be strengthened even further by emphasising the importance of detailed scenario development, and by addressing the methodological problems of uncertainty and ambiguity, future hazards and stakeholder problems. These steps could also include stakeholder and public debates on the potential goals of the technologies and the values attached.

As for the definition of Frame 2 technologies, it was suggested that second generation nanotechnologies can be complex and active without being evolving or unstable as mentioned in Table 3.

Especially for NGOs representatives, the most critical points of Frame 2 technologies were the application of nanotechnologies for military and surveillance purposes and their potential role in the economic development – in particular for the global south. These issues were perceived to be important political and security risks but were believed to be underemphasised in the document. It was therefore recommended that IRGC looks into existing work in this area and encourage development of international treaties.

Although participants welcomed the forward looking perspective offered Frame 2, some noted that Frame 2 developments will very much depend on responses to Frame 1 challenges and issues. Others were concerned that the Frame 2 perspective might be too non-specific for nanotechnology and actually could be applied to any technology.

Implementation of the Recommendations from the Framework

Comments received urged that the roles of the different stakeholders be clearly defined and openly communicated. Regulating institutions are seen as having a leading position in the field and should adhere to principles of transparency for building public trust. All parties involved, however, carry responsibilities and should openly share their perspectives and data. The development of an international data base or information clearing house with a regular update on existing products was seen as useful.

Participants felt that in order to be implementable, IRGC recommendations would need to be further specified and supported by research on technical standards as well as on ethical and societal implications. Commentators from industry expressed willingness to consider voluntary programmes though noted that their implementation could be difficult in practice. Generally, international cooperation and broadly agreed standards were seen as very crucial factors for the implementation of a governance process for nanotechnology.

Risk Communication

Transparent risk communication was perceived as a very important issue for building up public trust in nanotechnologies. A long-term strategy to achieve this goal would involve starting before the actual production of specific nanotechnologies and would raise public awareness at the research and development (R&D) stage. Multi-stakeholder dialogues and broad public involvement, as taking place for example in the field brain sciences, were seen as appropriate tools by most participants.

Industry representatives thought it especially important to communicate about nanomaterials that are currently used today. Challenges for risk communication were seen in the need to be responsive to cultural sensitivities and to the communication needs of specific stakeholders. It was also seen as crucial to involve scientists trained for communication purposes in communication efforts.

Non-First-World-Perspective

NGO representatives urged greater focus on the implications of nanotechnology for developing countries and emerging economies. They felt that IRGC would have to address questions of how nanotechnology could contribute to reducing poverty and how global innovation processes could be linked with local processes in developing countries. These discussions were part of the basic questions of how nanotechnologies could benefit people and how they could be directed towards that goal. In addition to the perspective of developing countries, it was noted that smaller emerging economies especially in the Asian region are struggling with nanotechnology issues. On the one hand, political institutions aim at attracting foreign companies. On the other, they need to develop legal frameworks that provide protection for the people but also ensure legal stability for companies. Major issues here are about environment, health and safety as well as about intellectual property rights. In general, actors in emerging economies are in a different position then those in highly developed countries. Academia, for example, has fewer capabilities for studies on environmental health and safety and/or for communication to the public. Emerging, economies often depend on corporate leadership from abroad and public awareness of developments in nanotechnology is very low.

Benefits of Nanotechnology

Another issue raised frequently by participants from different stakeholder groups was that the IRGC Risk Governance Framework should emphasis more on the benefits and opportunities of nanotech- nologies. This view was articulated as a general recommendation for the framework but also as a factor in understanding of public perception; people do perceive new technologies very much as positive and develop hopes and expectations towards them but these expectations are then balanced with concerns. The result is an overall evaluation process. Therefore, commentators thought it would be beneficial to include these positive perceptions in a risk governance approach. However, it was also noted that under these conditions it might be difficult to maintain a focus on risk governance.

Despite points of criticism raised over the two day conference, participants found the IRGC White Paper on Nanotechnology to be a valuable contribution to further development of the debate on risk governance of nanotechnology. The document's advocacy of an integrated approach to risk governance and recommendations that issues be dealt with at an early stage of developments were very much appreciated. The strong focus on communication issues and public participation was also warmly welcomed.

Concluding Remarks

Risk governance for nanotechnologies remains an important issue for all nations. Most countries are experiencing rapid changes and economic transformations that have been welcomed by many but have also increased people's concerns about the potential side-effects of technological change and its impacts on consumers and society at large. Given the prominence of both viewpoints, the danger exists that the public authorities as well as private risk management institutions may not be responsive enough to adequately address the needs those actors who hold them and hence will lose perceived competence and, hence, trust. It is important that all institutions dedicated to nanotechnology risk governance are well prepared to consider all the stages of the risk governance process and develop tools that address the challenges in each step of the process. This implies that resources should be invested in risk governance and that the persons dealing with this issue be adequately trained and prepared.

With respect to institutional responses to the nanotechnology debate, IRGC is promoting the following actions:

- Development of systematic liaisons between governmental agencies, academic, industrial, NGOs and other actors to share risk information and to promote socially responsible outcomes beyond the present cooperation. It is crucial that relevant information be shared and necessary actions coordinated. Public visibility and potential for stakeholder input into the present assessment proc- esses must also be assured.
- Provision of sufficient resources and capability for conducting concern assessments along with the risk assessments in order to identify concerns early in the risk governance process for nanotechnology. As far as we know, there has not been a systematic survey comparing risk perceptions, social concerns, and public attitudes on nanotechnology in a variety of countries. Such a study would be very helpful to design appropriate management and communication strategies on a global scale.
- Organisation of systematic feedback from the various actor groups and stakeholders, including the general public. Such rounds of feedback could provide valuable information about the concerns, hopes, worries, visions and preferences of the various actor groups and the public at large. Among the many instruments to perform such feedback rounds are stakeholder dialogues, round tables, citizen fora and citizen juries (OECD 2002). The present international activities, especially within OECD, in this direction are already a valuable attempt to collect feedback. It would be advisable, however, to ensure that other relevant actors

such as consumer groups, NGOs and other civil society groups have the opportunity to raise their concerns and to provide input to global governance.

- Provision of information to consumers so they are better able to make informed choices regarding the products that they purchase. This task could be delegated to internationally operating consumer groups. Package inserts or leaflets that are handed out to consumers together with their purchased goods, special articles in *Consumer Reports* and other popular journals are just some of the information strategies that could be used.
- Making decision-making processes on nanotechnology R&D and investment transparent so that stakeholders and the public are aware of how decisions are made and what evidence they are based on. Stakeholders can contribute to framing the issues related to the risks of nanotechnology by adopting a proactive approach. For example, collaboration should take place among various specialised organisations to create and maintain databases for knowledge on toxicity for nanomaterials, regulations, R&D needs and investment needs.
- Increasing transparency of decision-making by publishing all non-proprietary information on test results, impact assessments and their interpretations on the internet or in other forms.
- Establishing appropriate communication forums that help address the purposes for different actors in society want future technologies to be developed. Such discourse activities should be conducted prior to development of the new technologies or their applications. A targeted and effective communication programme is necessary and should include suggestions for a special educational initiative in the context of the worldwide activities to enhance public understanding of scientific, technological and humanitic implications of nanotechnology development at confluence with other emrging technologies.
- Involvement of different actors in the joint development of scenarios for future applications of nanotechnology, particularly referring to third and fourth generation products and processes. National or international exercises for constructing scenarios that appear relevant to the context of the diffusion of nanotechnology and the likely social reactions to it should be encouraged. The scenarios suggested in the White Paper may serve as starting points for designing more specific scenarios that relate to the specific situation and the contextual conditions of the countries selected for the analysis. These scenarios could act as catalysts for public debate and consensus-seeking exercises.
- Promotion of international cooperation for establishing common rules and standards for potentially high-impact, long-term projects in nanotechnology. Incentives should be provided for promoting and sustaining international cooperation.

The IRGC's White Paper on Nanotechnology Governance presents for the first time a conceptual framework for nanotechnology risk governance at an international level for short and long-term issues which are upstream of specific implementation policies. By considering the particularities of nanotechnology as an emerging technology, the proposed conceptual framework and guidelines on risk governance provide a step forward in assisting risk management agencies as well as private companies to integrate scientific assessments and concern assessments into one appraisal process and to select the appropriate risk management and stakeholder involvement strategies.

To further contribute to the discussion on nanotechnology governance, IRGC has published a report on the nanotechnology conference at Rüschlikon in July 2006 (SwissRe 2007) and a policy briefing based on its White Paper on Nanotechnology Risk Governance (reference needed). In 2007, IRGC is also undertaking an in depth analysis of risk governance of the use of nanotechnology in food, food packaging and cosmetics. The application of IRGC's proposed framework to concrete nanotechnology applications will constitute a further effort of IRGC to contribute to the beneficial governance of nanotechnology.

References

Agriculture and Environment Biotechnology Commission (AEBC), 2001, *Crops on Trial*, AEBC, London.

Grove-White, R., Macnaghten, P. and Wynne, B., 2000, *Wising up: The Public and New Technology*, CSEC, Lancaster.

International Risk Governance Council (IRGC), 2005, *White Paper on Risk Governance*, IRGC, Geneva.

International Risk Governance Council (IRGC), 2006, *White Paper on Nanotechnology Risk Governance*, IRGC, Geneva (available on the IRGC website: http://www.irgc.org/irgc/projects/nanotechnology/).

IRGC Policy Brief on Risk Governance for Nanotechnology, to be completed.

Morgan, M.G., Fischhoff, B., Bostrom, A. and Atman, C.J., 2002, *Risk Communication: A Mental Models Approach*, Cambridge University Press, Boston, MA.

OECD, 2002, *Guidance Document on Risk Communication for Chemical Risk Management*, OECD, Paris.

Roco, M.C. and Bainbridge, W.S. (eds.), 2001, *Societal Implications of Nanoscience and Nanotechnology*, NSET Workshop Report, March 2001, Virginia; based on National Science Foundation (NSF) Workshop in September 2000, Kluwer Academic Publishers, Dordrecht, the Netherlands.

Siegel, R.W., Hu, E. and Roco, M.C. (eds.), 1999, *Nanostructure Science and Technology*, Kluwer Academic Publishers, Dordrecht, the Netherlands (also available at http://www.wtec.org/loyola/nano/).

SwissRe, 2007, *The Risk Governance of Nanotechnology: Recommendations for Managing a Global Issue*, 6–7 July 2006, Conference Report, SwissRe Centre for Global Dialogue, Rüschlikon.

Part 4. A Framework for Risk Governance: Lessons Learned

Chapter 14
Lessons Learned: A Re-Assessment of the IRGC Framework on Risk Governance

Ortwin Renn[1] and Katherine Walker[2]
[1] *University of Stuttgart, Stuttgart, Germany and DIALOGIK gGmbH, Stuttgart, Germany*
[2] *IRGC, Geneva, Switzerland*

Introduction

The IRGC risk governance framework presented in Chapter 1 is a work in progress, a new model in Löfstedt's and van Asselt's words. Like any new model, its intellectual rigor and ultimately, its progression from the theoretical to the practical are indebted to open debate and constructive criticism. The many contributors to this volume provided a large variety of critical yet constructive comments and suggestions for improving the IRGC framework (Part 2). The diverse case studies in which the framework was applied retrospectively provided valuable insight to the practical utility of the framework (Part 3).

In the following sections, we would like to address first the conceptual issues that were raised in the formal comments because they are fundamental to the purpose of the framework. We then address comments about specific crucial components and phases of the risk governance framework, drawing on both the formal comments and observations from the case studies. In our response to comments, our goal at this point is time is to acknowledge where we may have fallen short and to provide clarification using language that will be more accessible to our readers. We know that there are some substantive issues that will clearly require additional theoretical and practical work but these are beyond the scope of this chapter.

Conceptual Issues

Underlying Concept of Risk in the IRGC Framework

A number of commentators have raised questions about IRGC's conceptual definition of risk:

> Risks are *mental 'constructions'* (OECD 2003). They are not real phenomena but originate in the human mind. Actors, however, creatively arrange and reassemble signals that they

O. Renn and K. Walker (eds.), Global Risk Governance: Concept and Practice Using the IRGC Framework, 331–367.

> get from the 'real world' providing structure and guidance to an ongoing process of reality enactment. So risks represent what people observe in reality and what they experience. The link between risk as a mental concept and reality is forged through the experience of actual harm (the consequence of risk) in the sense that human lives are lost, health impacts can be observed, the environment is damaged or buildings collapse. The invention of risk as a mental construct is contingent on the belief that human action can prevent harm in advance. Humans have the ability to design different futures, i.e. construct scenarios that serve as tools for the human mind to anticipate consequences in advance and change, within constraints of nature and culture, the course of actions accordingly. (IRGC 2005)

Their views are captured best in the detailed critique given by Eugene Rosa in Chapter 5 which is therefore the focus of our comments. Rosa expressed the concern that IRGC's definition of risk was inherently inconsistent and incoherent – that it defines risk as a 'mental construct' yet relies on 'signals ... from the "real world"'. He asks, 'If risk originates in the human mind and is not a real phenomenon, how can "real world" signals be a part of that origination?' In asking this question, he is touching on an ongoing debate about the nature of knowledge, and in particular, on the philosophical question of constructivism versus realism.[1] In its simplest form, the issue is whether scientific risk estimates represent 'objective' probabilities of harm or reflect only conventions of an elite group of professional risk assessors that may claim no more degree of validity or universality than competing estimates of stakeholder groups or the lay public.

Throughout the development of this framework, Rosa has provided important guidance on the need for a healthy balance between a relativist and realist view of risk (see Rosa 1998). However, as important as this debate is, we wish to clarify that it is not the intention of IRGC's risk governance framework to take a decisive stand on the controversial issue of constructivism versus realism of evidence and values, although it continues to be an active topic of discussion. Those interested in going into greater depth on this issue should also explore reviews of the implications of a constructivist versus a realist concept of risk in Bradbury (1989) and Renn (1992). A pronounced constructivist approach can be found in Hillgartner (1992), Luhmann (1993), Adams (1995) and Hannigan (1995). Realist perspectives in the social sciences on risk and environment can be found in Catton and Dunlap (1978), Dunlap (1980), Dunlap et al. (1994), Rosa (1998), Campbell and Currie (2006) Hacking (1999), Mayo and Hollander (1991), Bradbury (1989), Douglas (1990), Shrader-Frechette (1991b, 1995), Wynne (1992), Laudan (1996) and Jasanoff (2004).

Ultimately, whether the evidence collected represents human ideas about reality or depicts representations of reality is of no importance for the distinction between evidence and values that is suggested throughout the risk governance framework.

The framework tries to avoid the naïve realism of risk as a purely objective category as well as the relativistic perspective of making all risk judgements subjective reflections of power and interests. What IRGC's framework emphasises is that risk governance must deal with both the 'physical' and 'social' dimensions of risk. It is important to expand the set of criteria for assessing, characterising, evaluating, and

---

[1] For a philosophical review of the two 'risk camps', see Shrader-Frechette (1991a); see also Bradbury (1989), Clarke and Short (1993: 379–382), Burningham and Cooper (1999), Horlick-Jones and Sime (2004), Horlick-Jones (2007).

managing risks beyond the largely technologic or scientific factors that have dominated earlier models of risk governance. Public values, concerns, perceptions of risk are often equally important for identifying, understanding, and managing risks and must be included. If specific perceptions are clearly inconsistent with the best scientific knowledge about the likely effects of events, technologies or human actions, it is the task of risk managers to provide evidence-based information that help people to understand the causal relationships that they may have misjudged. A vast majority of studies on risk perception and concerns tend to show, however, that most of the worries are not related to blatant errors or poor judgement but to divergent views about how much uncertainty they are willing to tolerate, short-term versus long term impacts, the trustworthiness of risk regulating or managing agencies, and the experience of inequity or injustice with respect to the distribution of benefits and risks. These are valid inputs to determining what is relevant for consideration in the initial *pre-assessment* phase, in particular in the framing of an issue. They are also why the framework emphasises the need for both risk assessment and concern assessment in the *risk appraisal* phase.

Risk assessments are therefore 'mental constructs' but may be based on observations/perceptions or models of the world that can be justified by logical reasoning (e.g. that reflect varying degrees of knowledge and are consistent with fundamental axioms of mathematics and probability) or can be verified by comparisons with what actually happens. Public values, perceptions, and social concerns can act as the driving agents for identifying those topics for which risk assessments are judged necessary or desirable and for evaluating ultimately the acceptability or tolerability of those risks. Whether based on scientific predictions or public perceptions, estimates for the magnitude of risks, however, should reflect sound knowledge and technical expertise as much as possible, since the implications of taking action – for health, the environment, or the economy – may be very real.

It follows that managing risks will inevitably be directed by both evidence claims (e.g., what are the causes and what are the effects?) and normative claims (e.g., what is good, acceptable and tolerable?). We make this distinction despite the common belief that providing evidence is always contingent on existing normative axioms and social conventions. Likewise, normative positions are always enlightened by assumptions about reality (Ravetz 1999). The fact, however, that evidence is never value-free and that values are never void of assumptions about evidence does not compromise the need for a functional distinction between the two. In managing risks one is forced to distinguish between what is likely when selecting option x rather than option y, on one hand, and what is more desirable or tolerable, the consequences of option x or option y, on the other hand. It is hence highly advisable to maintain the classic distinction between evidence and values. We maintain this distinction in the framework by having both *risk characterisation* and *risk evaluation* as inputs to judgements about tolerability and acceptability.

Risk Governance: Defining Different Concepts and Levels

IRGC argues that the *risk governance* process includes, but also extends beyond, the three conventionally recognised elements of *risk analysis* (risk assessment, risk management and risk communication).[2] It requires consideration of the legal, institutional, social, and economic contexts in which a risk is evaluated and involvement of the actors and stakeholders who represent them. Thus, the framework includes several other dimensions including concern assessment and explicit discussion of stakeholder participation.

Several critical remarks sought clarification on how multiple levels of governance were incorporated into the framework. Robin Cantor (Chapter 3), among others thought that the framework is not quite clear about the key actors and their interactions on the horizontal and vertical levels of risk governance. The vertical governance axis defines the political arena which ranges from the local to the global level. The horizontal axis comprises various actors including national and international agencies, economic entities, academic and other experts, and civil society representatives. On each vertical level, different actors from the horizontal axis can join the governance process and contribute either knowledge or values to the process.

The degree of involvement or intersection of these vertical and horizontal elements can depend strongly on the governance model. Millstone et al. (2004) suggests three broad categories of models, outlined in Figure 1, each progressively more inclusive of the actors from the horizontal axis of governance:[3]

- *'Technocratic' model* (Figure 1a). In this model, objective science is seen to directly inform policy making; scientists are the best judges for the tolerability of risks and inform policy makers directly about what they should do. Risk managers should possess technical or scientific expertise because it is assumed that this knowledge will guide them to make not only valid statements about the risk itself but also prudent value decisions.
- *'Decisionistic' model* (Figure 1b). This model corresponds closely, although not exactly, to that illustrated by the NRC's Red Book (National Research Council 1983). In this model, policy making requires inputs other than science, such as socio-political, economic, and other legitimate factors, to inform decisions. The Red Book in 1983 established the division between the scientific aspects ('risk assessment') and political and value aspects ('risk management') within the overall process of risk analysis. This division, and several other aspects of the 'Red Book' model, have been adopted across a wide variety of risk management fields (Omenn 2003).

[2] See National Research Council (1996), Codex Alimentarius Commission (2005); Regulation (EC) No. 178/2002 (OJ 2002, L31/1) as amended by Regulation (EC) No. 1642/2003 (OJ 2003, L 245/4).

[3] The following insights were partially taken from the result of a large European Project called Safe Food in which the basic elements of the IRGC framework were adapted and transferred to the area of food safety. For more information, see Renn et al. (2006).

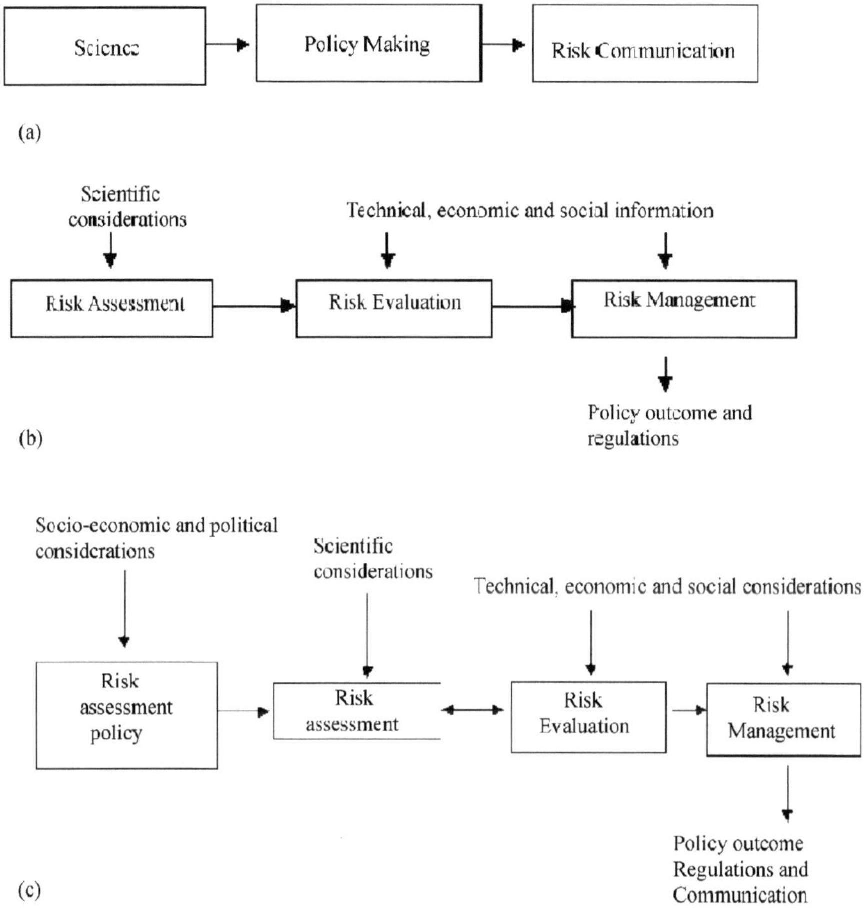

Fig. 1 Models of governance (from Millstone et al. 2004). (a) The 'technocratic' model. (b) The 'decisionistic' model. (c) The inclusive governance model.

- *'Inclusive governance' model* (Figure 1c). This model is inspired by the 1996 NRC report on risk characterisation in which the interface between assessment and management is stressed and in which science, politics, economic actors, and representatives of civil society are invited to play a role in both assessment and management (Stern and Fineberg 1996).

It is this third model of inclusive governance that IRGC has embodied in its framework on risk governance.

Examining the Purpose and Scope of the IRGC Risk Governance Framework

Purpose

Many comments focused on the purpose of the IRGC framework. Although most contributors appreciated the idea of a consistent terminology across different risk areas and of guidance on the necessary steps for conducting risk assessments and undertaking risk management decision, their comments also addressed discrepancies between their expectations and what they thought IRGC had delivered. Some critics would have liked a more concrete manual of how to accomplish good governance. Others felt that IRGC had designed an ideal process that was too detached from reality to be useful for risk management purposes. Still others felt that the terminology was too directive and rigid and left little room for flexibility.

To understand the purpose of the IRGC framework, it is important to recall the original motivation for this work. The founders of IRGC recognised that, historically, failures in risk governance have diverse origins:

- Incomplete or inappropriate framing of the issue from the outset so important risks or concerns are not addressed.
- Incomplete or inadequate analysis of the risks and of risk management actions (including their intended as well as unintended consequences).
- Failure to anticipate the power of public perception in debates about risk.
- Lack of clear goals and plans for the role of stakeholders.
- Poor communication in numerous forms.
- Absence of the necessary infrastructure and capabilities to analyse risks, implement management solutions, and to monitor their effectiveness.
- Inability or reluctance to incorporate new information about risks, concerns, the effectiveness of management solutions or other new knowledge into the risk governance process.

The case studies in Part 3 of the book provide numerous examples of these and other deficits in risk governance. As several of these cases are current or relatively recent, they demonstrate that similar mistakes continue to be made. The growing complexity created by globalisation, the scale of the problems faced, and the entry of new and greater numbers of players into any given risk governance arena make it likely that the same mistakes will be repeated.

The existence of these kinds of pitfalls for risk governance has in many cases been known for years. A glimpse into the published literature from the social sciences, economics, engineering, operations research, medicine, environmental science, law, etc. reveals that investigators from each of these fields have recognised one or more of these issues – or in Warner North's words, have hold of one part of the elephant.[4] But, as long as these key reservoirs of knowledge and experience remain separated, the opportunity for a more comprehensive understanding of a risk, its broader context, and its solutions, remains elusive.

The IRGC framework, then, is first and foremost a broad conceptual framework incorporating a set of key principles for sound risk governance. It provides a structure within or around which particular risks may be investigated, discussed by stakeholders, communicated, and managed. By laying a clear rationale for taking into account not only scientific evidence, economic considerations but also risk perceptions, social concerns and societal values, the IRGC framework attempts to provide a more comprehensive and integrated view of risk governance than alternative approaches. At the same time, it builds on the foundations of many important previous frameworks that we reviewed in preparation of the original White Paper (IRGC 2005).

Some commentators would have preferred to see more analytical detail. We recognise that by referring to 'an integrated analytic framework' at the outset, we may have misled our readers. However, a checklist or a primer on analysis was never our intent. No one framework could begin to dictate the particular analytic methods or approaches to appraising the risks and concerns, designing communication strategies, involving stakeholders, and finding the risk management solutions for a broad range of problems. We have provided illustrative examples of analytical methods, management strategies, etc. for various phases of the risk governance process but a close reading of the text will show that we did not intend these to be restrictive.

The appropriate tools for understanding and developing solutions for particular risk issues will be dictated by the nature and context of the risk. Excellent guidebooks and manuals are already available on individual elements of the framework.[5] But ultimately, experts with specialised training and skills relevant to those elements will be needed. IRGC does not want to usurp the responsibility of all involved to think and work creatively in response to the needs of a particular issue. We however do hope that this more comprehensive framework will contribute to the development of more balanced, more inclusive, and more effective risk governance strategies – ones that avoid the pitfalls of the past.

---

[4] North is referring to the old Indian parable of the blind men and the elephant. Each of the blind men grasps hold of a different part of the elephant and describes what he sees. Each has a different perspective, but none can perceive the whole.

[5] For example, a sample of guidance is included in IAEA (1995), IEC (1993), Kolluru and Brooks (1995) and National Research Council (1983). See also resources from WHO, UNEP, the European Commission, and national government agencies.

Scope

What Risks Is the IRGC Framework Designed for?

Several commentators (see, for example, Löftstedt and van Asselt, North) specifically raised the questions about the nature and scope of risks the IRGC framework is intended to address. Does the framework cover all types of risks or is it meant to apply to chemicals, technologies, and food only? Some people felt it was not suited for natural disasters, others felt the same for malicious acts and still others for critical infrastructures.

As discussed earlier, we recognise that no framework can give justice to the particularities of all applications. Yet we would still argue that our framework could be used for the purpose we described – as broad conceptual guidance on the critical elements of the risk governance process that need to be considered. This guidance is relevant for the various types of risks we identified in the original document: physical agents (e.g. noise), chemical agents (e.g. environmental pollutants), biological agents (e.g. viruses), natural forces (e.g. earthquakes), social-communicative hazards (e.g. terrorism and sabotage) and multiple hazards (i.e. when several hazards can simultaneously impact the same receptor). We did not claim to cover financial, economic and cultural, psychological, or other non-physical risks but believe that many of our guidelines would also apply to them.

Several commentators expressed concern that we had not given adequate attention to the definition and assessment of vulnerabilities. It is true that the terminology used in the framework derives somewhat more from risk assessment for public health rather than from engineering applications. But in addition to the more philosophical definition of risk discussed at the beginning of the chapter, we have defined risk for practical purposes as:

> [T]he uncertain consequence of an event or an activity with respect to something that humans value. Risks always refer to a combination of two components: the likelihood or chance of potential consequences and the severity of [those] consequences ... (Kates et al., 1985)

In our view, understanding vulnerability of a target – whether it is a system, an individual, a community or a nation – is an important part of estimating risk. The US Nuclear Regulatory Commission defines vulnerability as:

> The condition determined by physical, social, economic and environmental factors or processes, which increase the susceptibility of a community to the impact of hazards.

Thus, vulnerabilities can increase risk, either by influencing the likelihood of some event or the severity of the consequences should it occur, or both.

The issue of vulnerability has been specifically addressed in the IRGC paper by making a clear distinction between an 'agent' such as an earthquake or a chemical and the 'risk absorbing system' such as a building or an organism. Vulnerability refers to the ability of the risk absorbing system to withstand the effect(s) of the agent to which it may be exposed, for example, the ability of an organism to mount an immunological defence to a viral infection, of a building to withstand a seismic shock, or, as the definition above suggests, of an entire community to withstand the various impacts of a hazardous agent. Thus, a risk absorbing system may include a complex chain of interacting elements starting with physical entities such as buildings and ending with the availability of effective disaster relief organisations.

As with other elements of risk analysis, the IRGC framework does not specify how to assess or measure vulnerabilities but it provides the conceptual elements that are needed to integrate the results into the larger risk governance frame.

Simple, Complex, Uncertain, or Ambiguous Risks?

Another facet of the question, 'What risks?' pertains to the categories of risks – that is, simple, complex, uncertain, or ambiguous – that IRGC intended to be the focus of the framework. Löfstedt and van Asselt suggested that IRGC limit the risk governance framework to risks that are complex, multi-sectoral, and transboundary. North also suggested a focus on global risks. Indeed, IRGC is mostly interested just these types of risks that require input from governments, the corporate sector, scientists or other experts, and civil society actors. These will tend to be risks that are more complex, uncertain and/or ambiguous where implementation of good governance practices is particularly critical. Yet our taxonomy is built to include the simple as well as the complicated cases. However we note that the framework will require clearer discussion of these characteristics and their implications for analysis and governance. We have initiated this discussion later in the chapter.

Institutional Focus

Several commentators were unclear about what institutions the IRGC framework was primarily directed to. Is the framework directed towards governments, the corporate sector or NGOs? Many commentators believed that the corporate sector was not given due respect in what they contribute to risk governance. Robin Cantor diagnosed a lack of attention to the complex regulatory and legal climate which affects the risk management decisions faced by the financial services and insurance industries. She suggested that the framework might extend its applicability from *ex ante* evaluation and decision making, processes that precede or accompany a new risk, to *ex post* analysis, focusing on the governance lessons to be learned from the decisions or actions that were taken. Since the time Cantor made these suggestions, IRGC commissioned the retrospective analyses embodied in the case studies. However, we agree with Cantor that further studies would be valuable to assess the utility of the framework for risk management programmes being implemented in the corporate and insurance sectors, for example, Enterprise Risk Management programmes.[6]

[6] See, for example, http://www.coso.org/Publications/ERM/COSO_ERM_ExecutiveSummary.pdf.

We do agree that the present framework focuses more on risk governance processes driven by major government agencies or international oversight organisations. The role the framework should play for decision makers in the private sector is not well-articulated. The reason behind this focus has been our emphasis on the new structure of governance, that is, the need for cooperation between the four major players: governments, corporate sector, academic institutions, and civil society. These categories of players of course include major government agencies, individual corporations, communities and NGOs.

Target Audience for the Framework

In addition to remarks on the substantive scope of the IRGC framework, many commentators had questions about whether the style of the document was suitable for senior risk managers and other decision makers the White Paper includes in its target audience. Many critics felt that that the language was sometimes overly academic and yet, at others, almost simplistic. We agree. We offer the explanation that the original White Paper was written predominantly in an academic style because the authors and initiators believed that this new framework must first be exposed to the test of academic discussion and rigorous intellectual scrutiny. We anticipate creating alternative versions more suited to senior risk managers and decision makers in the future.

The Role of Benefits Assessment in the IRGC Framework

Many comments raised the question: 'Why focus on risk and not on risk and benefits?' Several individuals even suggested renaming the IRGC the International Risk and Benefit Governance Council, the IRBGC!

The language of the IRGC risk governance framework deliberately emphasises risks. In part, this choice reflects our focus on the adverse nature of many of the global risks that we hope the framework will help address: global warming, terrorism, natural disasters, pandemics and the like. The benefits associated with many of these events are not obvious, if they exist at all. But many of the risks we talk about in the IRGC framework are associated with technologies or activities that also have benefits. However, historically, much greater emphasis has often been given to the benefits of proposed technologies or activities (e.g. jobs development) as opposed to risks (e.g. health or environmental impact). The neglect of these risks is often what has given rise to government initiatives or agencies focused on protection of the environment, on safety of the food supply, and on the safety and efficacy of pharmaceuticals. So, in part, the language of the IRGC framework reflects an emphasis on having a clear discussion about the risks, which can sometimes be less obvious, downplayed or delayed.

Although benefits are not the focus of the IRGC framework, the reality is that the framework anticipates the need to discuss benefits in virtually every phase of a project: in the framing of a risk problem, in the assessment of risks and concerns, in tolerability assessment, and ultimately in risk management. The emphasis that the IRGC framework places on the involvement of stakeholders representing all sides of an issue also indicates that attention is being given to both risks and to benefits, real and perceived.

Indeed, each of the case studies illustrates some of the tradeoffs decision makers face in balancing the potential risks and benefits in setting policies:

- *Genetically modified (GM) crops*: ecological, cultural and land use impacts versus more cost-effective food production and other benefits of the reduced use of pesticides.
- *Nature-based tourism*: safety risks to humans, risks to local cultures and to ecological systems versus the enjoyment experienced by nature tourists, the economic benefits for local communities, and, ironically, the protection of environmental and ecological resources.
- *Listeria in raw milk soft cheese*: public health protection versus preservation of food quality and cultural traditions.
- *Nagara River Estuary conflict*: environmental and fisheries impacts versus secure water supply and flood control.
- *Acrylamide in food*: carcinogenic risks versus availability of food desirable to the public and economic impacts on food producers.
- *Energy security for the Baltic Region*: benefits to countries of tying in to the Gazprom gas supply systems versus security issues of doing so; the benefits of improved energy security and reduced climate impacts of relying on domestic nuclear power versus risks of nuclear power facilities and their wastes.
- *Nanotechnology*: potential health and environmental risks versus economic, medical, other potential benefits.

On an individual level, it is plausible that decisions should be made on the basis of a balanced comparison of potential costs and benefits, of risks and opportunities. Market economies assume that each individual is the best judge for striking the correct balance. Whether someone purchases a 20th pair of shoes or 7 pyjamas remains his or her choice.

Balancing risks and benefits is more problematic when decision makers are faced with weighing collective or private benefits that impose risks on others or on the environment (Jaeger et al. 2001). The question then becomes: What risk can a society impose on those who do not share the benefits in full or in part? (see Rayner and Cantor 1987; MacLean 1986). This question is in principle independent of the benefits that the one individual enjoys by obtaining the desired good. Yet all societies are willing to impose some risks on its members if some broader groups in society will benefit. This relationship, however, is sometimes far from being symmetrical (Schrader-Frechette 1984) so the distribution of benefits and risks must be scrutinised very carefully. Regulatory action may be necessary to prevent major inequities.

Of course, when making regulatory decisions a society should also include positive external effects, for example, on the labour market, the competitiveness of an economy, the effects on social justice, etc. The IRGC risk governance framework anticipates the need for this deliberation during the phase of *risk evaluation*. This is the point at which risk acceptability and tolerability is addressed and the likely benefits to society, whether in whole or in part, must be included in the balancing procedure.

Distinctions between Complexity, Uncertainty, and Ambiguity

IRGC's discussion of the distinctions between complexity, uncertainty, and ambiguity and their implications for designing risk appraisal approaches, informing risk evaluations and developing risk management options was perhaps the least understood and most controversial component of the framework. Several commentators, including North, Löfstedt and van Asselt, and case study authors, Kuenzi and McNeely and Okada felt that these distinctions were too rigid and insufficiently specific to be of practical use. Kuenzi and McNeely echoed others by noting that 'clear-cut compartmentalisation hardly reflects reality' and that 'these risk classes may simply describe aspects of the same risk'. In his conclusions from the Nagara River Estuary Barrage case study, Okada worried whether the limited knowledge decision makers and stakeholders often have at the outset of an analysis would result in mischaracterisation of the risks and lead to management decisions built on a faulty premise. Wouldn't it be a safer strategy, he asks, just to assume that the risk was uncertain and thus involve a broader involvement of potential stakeholders?

These critiques prompted us to give greater thought to these terms and to their significance for the framework of risk governance.

Why Make These Distinctions?

The basic motivation for drawing these distinctions was to create more precision in the language we use to characterise the various challenges to our knowledge or understanding of a problem. The term 'complexity' or 'complicated' is used in common parlance to describe the entire portfolio of factors that make something difficult to understand – lots of different actors, stakeholders, numerous potential impacts, uncertainty about individual variables or about the system as a whole, a plurality of views on how and what outcomes should be valued. Such an overarching concept is too broad to have any real value in risk governance. However, we do believe that it can be useful to be as clear as possible about what aspect of this 'portfolio' we are talking about as we work through a risk problem.

- *Simple*. A simple case could be illustrated by drug has been developed that is effective for the treatment of a serious disease. The metabolic pathway is well understood, the dose response well defined, and there are no side effects. The decision about whether and how to treat the disease is clear.
- *Complex*. Complexity is introduced when this simple causal relationship is actually better described as a 'causal web' where multiple factors may interact to

affect the efficacy, and even safety, of the drug. For example, age, health status, dietary factors, genetic variation in key metabolic enzymes might all play a role in mediating the efficacy of the drug. Change in one factor or more factors might reduce the efficacy of the drug in treating the disease in some individuals or in other cases might cause some individuals to experience serious side effects. So long as this network of relationships is known and the inter-individual variation in the population is well characterised, the decision on how to treat a particular patient can still be made. It may require more tests and information to ascertain the status of the patient with respect to the additional mediating factors.

- *Uncertainty*. The less well defined and understood this causal web is, the more uncertainty is introduced to the system. What happens, for example, if the clinical trials for the drug were done predominantly in one population with a particular age, health, dietary and genetic profile and then use of the drug is extended to another population, perhaps in a developing country, that differs in critical ways from the first population? Greater uncertainty may exist about the amount of disease reduction that will be achieved or about the incidence and severity of side effects that might be observed. Decisions now must consider more carefully the benefits and costs of the treatment, the need perhaps for more clinical trials in the new population, the availability and efficacy of alternative drugs, and/or the need for a careful monitoring system for potential side effects, for example.
- *Ambiguity*. Ambiguity arises when differences exist in how individual actors or stakeholders value some input or outcome of the system. For instance, what if the drug in our example has side effects but it is the most affordable alternative for a serious disease in the new population seeking treatment? Health authorities or drug manufacturers might be reluctant to authorise use of the drug while individuals might be willing to risk the side effects. What the 'right' decision is and who gets to make it is not so clear.

This illustrative example suggests that the boundaries between these characteristics should be considered more flexible. Not all simple systems or risks are necessarily well understood; nor are they necessarily unambiguous. We know, for example, that bike helmets save lives, but the policy decision to require them has been surprisingly controversial. The more complex a system, the more uncertainty, and in some cases ambiguity, may be introduced. At the same time there may be simple, or even complex risks where the relationships between cause and effect, action and outcome, are well understood (Graham and Rhomberg 1996).

Why Do These Distinctions Matter to Risk Governance?

Ultimately, the details of how a risk is analysed, appraised, evaluated and managed must be dictated by the problem and its context. We recognise that the two dimensional tables we have used to discuss simple, complex, uncertain and ambiguous risks and their implications for various phases of risk governance (e.g. Table 6 in Chapter 1) imply a kind of fixed, deterministic relationship between a given characteristic and particular actions. As a closer reading of our text will show, it was not our intent. Nor did we intend to limit the types of assessment tools to be used or actions that should be taken. However, we did want to make the point that these characteristics might signal to a decision maker that different types of strategies might be called for at different phases of the risk governance process.

At the pre-assessment phase, these characteristics may be used in the initial framing and screening of a risk. In a sense, they can assist with a form of triage:

- Is this a simple problem for which a decision can be made right away?[7]
- If not, what do the complexities of the problem suggest about how the risk appraisal should proceed – the methods, types of knowledge that needs to be brought to bear, and thus the types of participants who need to contribute to the process?
- How much uncertainty is there likely to be? Is the risk totally new and unknown? Are the consequences such that a decision must be made before uncertainty can be reduced? Or, must uncertainty simply be dealt with in the usual course of trying to understand the issue through the risk assessment process?
- Are there likely to be serious ambiguities that the framing and later stages of the analysis and management are going to have to deal with? What do these imply about who must be included in discussions early on? Are debates about management so likely to be dominated by differences in values that no amount of analysis will resolve them? Or will early discussions about possible mitigation or management strategies help?

How influential one or more of these characteristics may be on the course of risk governance for a problem may depend on a number of contextual factors: the urgency with which a decision about a risk must be made; the institutional capacity and other resources available both to appraise and evaluate the risk; and ultimately, the resources available to implement risk management solutions. One can imagine that in cases where the urgency is high, and/or the resources available to assess a risk are more limited, this triage step may help guide a decision maker to types of risk management solutions that have been effective for risks with similar characteristics.

Re-assessment of a risk relative to these characteristics is also an element of what we call the tolerability and acceptability judgement phase of the framework (discussed in greater detail in a later section of this chapter), in particular that component the framework refers to as risk characterisation. The risk characterisation is the 'evidence based' component of this phase and involves 'collecting and summarising all relevant evidence' from both the risk assessment and the concerns assessment. It is appropriate at this stage to take stock of what the appraisal phase has achieved:

- How well has complexity, as we have defined it above, has been defined? Have we identified all the key variables and relationships necessary to predict a risk?

[7] For example, US Office of Management and Budget developed a policy whereby it would prompt agencies to take action when it thought regulation could lead to highly beneficial outcomes (Stimulating SmarterRegulation 2002). See also presentation by John D. Graham available on the AEI-Brookings website: http://www.aei.brookings.org/admin/authorpdfs/page.php?id=1210.

What are the key benefits or other social and economic impacts that must be considered? Do we understand what the most influential intervention points might be for risk management?
- How much uncertainty remains? About the risk as a whole or about individual variables and relationships that are used to predict it?
- Are there significant ambiguities – for example, differences in values, in distribution of risks and benefits, or other equity issues – that must be considered in the evaluation and ultimately the management of risks?

Again, how well we understand the dimensions of the problem, how certain we are, can influence whether we proceed with confidence to particular risk management solutions, or consider more precautionary approaches, or rely more strongly on negotiation.

A Note on Aleatory versus Epistemic Uncertainty

A few commentators took issue with our discussion of different types of uncertainty. Some thought it was too much detail. Others like, Eugene Rosa, in particular (Chapter 5), just thought we got it wrong. While we think that this issue does approach the kind of analytic detail that the framework is not suited to address, we think is important to clarify these terms more carefully and why this type of distinction might ultimately be important in risk management. Readers interested in exploring uncertainty analysis and its role in policy analysis in greater detail are advised to consult the abundant literature on this topic (see, for example, Morgan and Henrion 1990; Hoffman and Hammond 1994; Paté-Cornell 1996; Frey and Cullen 1999).

Aleatory Uncertainty – Decision and risk analysts also refer to aleatory uncertainties as, 'those that stem from variability in known (or observable) populations and therefore, represent randomness in samples' (Paté-Cornell 1996). A simple example of this type of variability is represented by the distribution of possible values from the roll of a fair set of dice. A random process gives rise to any particular value at any point in time, but in the long run, with a large enough sample or rolls of the dice in our example, the distribution of possible values can be well characterised. Collecting additional information or increasing the sample sizes can help to characterise this distribution more exactly, but they cannot reduce its fundamental parameters.

Epistemic Uncertainty – Epistemic uncertainty arises from 'from basic lack of knowledge about fundamental phenomena' (Paté-Cornell 1996).[8] The impacts of global warming, for example, have been considered to be very uncertain for just these reasons. Epistemic uncertainty can, in principle, be reduced by the generation

[8] Paté-Cornell notes that some analysts also refer to epistemic uncertainty as ambiguity but she has argued for a cleaner separation. We prefer to use ambiguity to refer to differences in both what a risk means for those affected and the values to be applied when judging whether or not something needs to be done about it. See also Hoffman and Hammond (1994).

of knowledge, the collection of samples, or other forms of research appropriate to the particular issue.

In reality, as Rosa correctly points out, what is often loosely referred to as 'uncertainty' is some combination of aleatory and epistemic uncertainty.

Why are these distinctions important for risk assessment and risk management? For the risk assessor, these distinctions between types of uncertainty can be helpful in developing an approach to characterising uncertainty. If the assessor knows that some well-defined random process gives rise to an outcome, as in the case of the role of a pair of dice or in the combination process that gives rise to genetic variability in some trait, this information may be all that is needed to characterise uncertainty (that is, he or she is certain that this *aleatory uncertainty* or variability tells the whole story). If the assessor does not know if such a random process actually is responsible or the probabilities are not well understood, epistemic uncertainty exists and there may be need to collect more data. Bayesian analysis, in which probabilities are defined as degrees of belief, can also be used for this purpose. Degrees of belief, also referred to as subjective probabilities, are grounded in evidence as well as judgement about how good or relevant that evidence is. Thus they can combine both aleatory and epistemic uncertainties (Paté-Cornell 1996).

Decision makers care about uncertainty because it one of the factors that make decisions difficult to make. If one is certain that a decision to take a particular action will reduce a risk by the desired amount for a predictable cost, the decision is easy. If the effect of the action on the risk is uncertain and/or the cost of taking the action is not well known, the decision to act requires more of a gamble.

Although they may not always know it, decision makers implicitly care about the distinction between aleatory and epistemic uncertainty. Gamblers exploit knowledge of aleatory uncertainty in games of chance; they place bets on the basis of estimating what the odds are in their favour. But they also have epistemic uncertainty about, say, what the true likelihood is that the next card will be what they need. They adjust their bets accordingly. But they may also try to reduce epistemic uncertainty by collecting information that allows them to refine their odds, say by counting cards in the game of blackjack, and thereby increase their winnings. This is why gambling houses typically prohibit counting cards!

Decision makers make similar judgements. If epistemic uncertainty about a risk is high, and the costs of taking action are great, a decision maker may choose to wait and do additional research that might reduce uncertainty about the likelihood or magnitude of the risk. The field of decision analysis, in particular value of information analysis, has developed to help decision makers examine the tradeoffs between taking immediate action in the face of uncertainty or waiting to collect information that they hope will help them make a better, less risky decision. The gamble is that collecting the information might not actually change the original decision that would have been made, or even if it does, may contribute to delays that increase risks in the meantime.

In the end, epistemic uncertainty may not be reducible for many risks. In these cases, decision makers have to face the tradeoffs implicit in taking actions that have uncertain outcomes.

The Structure and Content of the Overall Risk Governance Framework

The formal comments and the experience of the case studies provided broad support for the basic IRGC framework which builds upon the logical structure of four consecutive phases called: *pre-assessment*; *appraisal*; *characterisation/evaluation*; and *management*. A fifth phase, *risk communication*, accompanies all four. Each phase specifies activities that IRGC believes are important elements for good governance. This simple framework is in line with almost all other alternative concepts which assures the compatibility of the framework with professional codices and risk governance legislation. Moreover, it has transformed the linear structure more commonly found in other contemporary conceptions of risk governance (Prime Minister's Strategy Unit/UK Cabinet Office 2002; National Research Council 1983, 1996; RCEP 1998) – into an open, cyclical, iterative and interlinked process that we show again in Figure 2.

The five phases correspond to the two major challenges of risk governance: generating and collecting knowledge about the risk and making decisions about how to mitigate, control or otherwise manage it. These two challenges are illustrated by the two activities portrayed on the horizontal axis: assessment and management. However, there are two additional phases in which knowledge and values are closely intertwined: pre-assessment and characterisation/evaluation. These two phases are located on the vertical axis and constitute interfaces between knowledge and values. This design avoids the naïve separation of facts here and values there but also escapes the quandary created by post-modern relativity by honouring the analytical distinctions between the factual world and the world of values even if they clearly interact. Each of the phases discussed in greater detail below.

Pre-Assessment

Both the formal commentaries and the case studies strongly supported the role of the pre-assessment phase of the framework, in particular the critical role of framing. Many of the conflicts around risks, and the risk management decisions made about them, can be traced to the sometimes fundamentally different frames through which the issues are viewed.

The genetically modified (GM) crops (Tait) and nanotechnology (Roco, Renn and Jäger) case studies highlight the importance of framing for new technologies and the impact they can have on product development. Tait compares the very different fate of GM crops in the US and Europe based on a difference in regulatory framing she summarises as 'product v. process'. The US, in line with OECD, framed GM crops as analogous to existing *products* developed through conventional plant breeding programmes and subject to existing regulations. The EU, in part reflecting some inconsistencies in how the industry itself framed the issue, framed GM crops

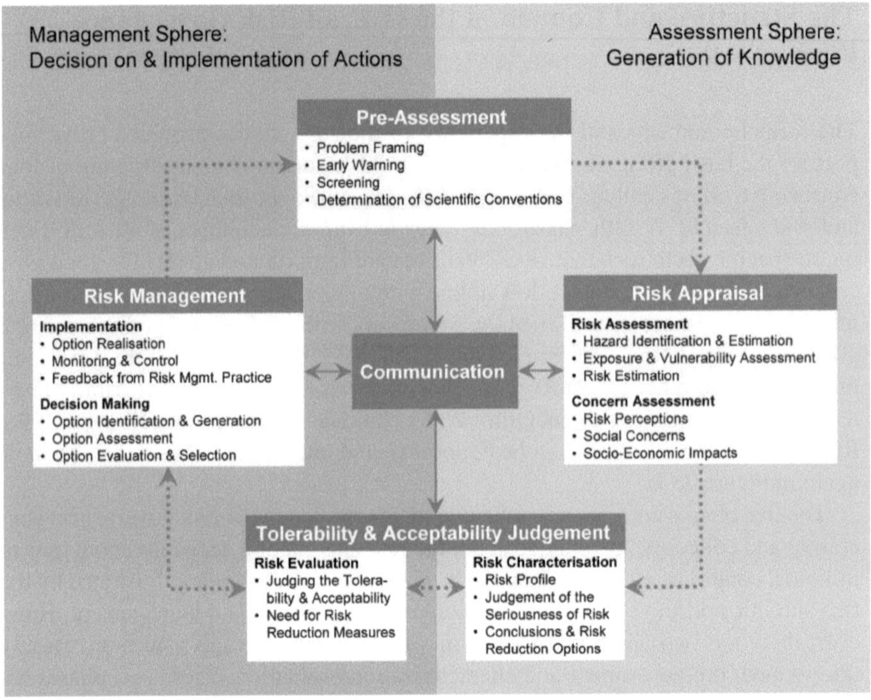

Fig. 2 The IRGC risk governance framework.

as a new *process* (involving genetic engineering) that required a new regulatory approach. While GM crops have achieved widescale agricultural application in the US and worldwide, their development and use have been extremely limited in Europe.

For similar reasons, the nanotechnology case study argues the importance of separating discussions on first generation uses of nanotechnology (Frame 1) from the future generations (Frame 2). (These two frames correspond to the relative levels of complexity of the generations, their behavioural dynamics, and the available level of knowledge.) Whereas products from Frame 1 are already in such diverse goods as paints, sunscreens and cosmetics, Frame 2 includes evolving function nanostructures and nanosystems where possible system uncertainty and ambiguity have been identified. Although subject to considerable research and development effort, few Frame 2 products have yet to reach the market. The advantage of framing the different generations of nanotechnology separately, Roco et al. argue, is that there is less potential for 'painting' all nanotechnology products with the same risk 'brush'. The increased conflict potentially associated with treating Frame 1 risks as if they belonged to Frame 2 could place an excessive burden onto the governance process without enhancing safety or reliability. In this way, it might echo the example of GM crops.

The case study on Listeria in raw milk illustrates a classic division between a 'consumer sovereignty' and an 'illness prevention' approach to viewing risk. The *consumer sovereignty* frame is based on autonomy and freedom of choice. In the Listeria case, it represents the views of two groups of people; well-informed consumers who place a premium on more 'natural' or 'quality' products derived from unpasteurised milk and perhaps less well-informed consumers whose choice of products is dominated by cultural heritage. The *illness prevention* frame is based on the use of standardised regulation to implement an across-the-board legal framework whose purpose is to ensure consumer safety.

The US adoption of the *illness prevention* frame, and with it a view of unpasteurised milk as unsafe *per se*, has limited the ability of policy makers to consider regulations of cheese making *processes*. Such an approach might still allow raw milk cheeses to be available and could furthermore prevent illness in those subpopulations that ignore the blanket prohibition on the use of raw milk. Understanding these different viewpoints and the implications they have for a more comprehensive *appraisal* of the risks and for the selection of effective *risk management* solutions is critical.

Other framing issues may be drawn from each of the other case studies:

- The gas pipeline issue looks very different from the perspective of securing gas imports to Western Europe or the perspective of geopolitical power distribution if countries like the Baltics and Poland are left out.
- Nature tourism offers quite different perspectives depending on the viewpoint of the stakeholder. Seen from the perspective of a service economy, tourism adds wealth to a nation. Seen from the perspective of social justice, tourism creates more social inequalities than it is able to bridge. Seen from the perspective of biodiversity, tourism can help place a value on biodiversity, since tourists are willing to pay for wilderness, yet it can also contribute to destroying natural habitats.
- Bonneck's acrylamide case was a good example of the effect of framing a food contaminant as naturally occurring, rather than introduced by human activity. Once acrylamide was characterised as a naturally occurring by-product of high temperature treatment of starch-containing foods, much of the 'heat' was taken out of the debate. Public responses to the subsequent detection of high acrylamide levels in bread and other bakery products were quite calm.

IRGC received fewer direct comments on the role of the other activities identified as part of pre-assessment – early warning, the determination of applicable conventions, and the initial screening of the risk issues. However, Kuenzi and McNeely, in their chapter on Nature Tourism, took us to task for a lack of clarity about what these activities specifically involved which they felt made it difficult for these ideas to be incorporated in their case study. Cantor's concern that the framework's scope did not include the '[litigation-driven] high-profile risk experience of the US' faced by corporate decision makers (Chapter 3), may also reflect this lack of clarity.

Each of the other activities can contribute to the way in which a problem is initially perceived, framed, appraised, evaluated (with respect to their tolerability or acceptability) and ultimately managed. However, as indicated in Chapter 1, none of them is necessarily restricted to the pre-assessment phase.

- *Early warning* – if early warning systems exist, by which we mean systems that exist within government, business, or civil society to identify unusual events or phenomena (e.g. disease registries, biodiversity indices, climate indices, environmental quality monitoring), then these may provide some initial insight into the extent or severity of an emerging risk.
- *Identification of applicable conventions* – in response to several comments, we would like to expand our definition of conventions from 'scientific conventions'[9] in the original framework to include existing legal, policy, social, or economic conventions. First, any one of these conventions can affect the way in which a new risk is framed, evaluated and managed at all phases of the risk governance process. They are all part of the context within which the risk governance process occurs (see further discussion of context later in this chapter). Also, by acknowledging the need to take into account the broader array of conventions that often govern risks, we hope to make it clearer that the framework can also be suitable for approaching the types of corporate problems raised by Robin Cantor.
- *Screening* – Consistent with the expansion of the definition of 'convention', screening must also be expanded from dealing solely with the applicability and implications of scientific conventions for risk assessment, concern assessment, risk evaluation and risk management to those of legal, policy, social and economic conventions as well.

We do need to acknowledge that we, along with most of our commentators, have assumed that pre-assessment will help reduce overall risk by preventing decision makers from neglecting key risks or concerns and facing unpleasant surprises later on. Corporate risk managers may view the choices differently; for example, they may need weigh the benefits of being proactive about identifying potential risks of a new product against the risk of inadvertently generating information that increases future liability (the 'what did you know when?' dilemma alluded to by Robin Cantor). Again, the context of the problem is critical.

Risk Appraisal

The Role of Concern Assessment

Virtually all of the commentators praised IRGC's efforts to include concern assessment as part of the appraisal of risks. However, comments from Knight et al. (Chapter 9) suggest the need for greater clarification of how it is defined. They observe:

> It seems intuitive that a framework incorporating social science would be an improvement to existing risk governance procedures and decision making. However, examining concern

[9] Scientific conventions can include a variety of characteristics, by which the hazard of a risk agent might be judged (Zinn and Taylor-Gooby 2006).

assessment with anecdotal evidence may lead to a position that a particular frame has more public support than it actually does.

IRGC wishes to make clear that *concern assessment* is a social science activity aimed at providing sound insights and a comprehensive diagnosis of concerns, expectations and perceptions that individuals, groups or different cultures may link to the hazard (Hyman and Stiftel 1988). This social scientific analysis should be submitted to the same kind of methodological scrutiny and peer review as any other scientific activity. It should not be confused with eliciting stakeholder feedback or providing platforms for participatory processes.

Understanding these different concerns, expectations, and perceptions can be important in the design of successful communication and management strategies. In some cases, failure to take them into account can lead a potential increase in risk. Surveys of the public regarding the safety of irradiated food, for example, indicated that substantial numbers of people thought that irradiated food could be safely left unrefrigerated, a practice that could actually increase the risk of food-borne illness. Public communication programmes, instead of just addressing concerns that irradiated food might be radioactive, had also to advise the public to continue safe food handling practices.

Some of the case studies provided examples of the problems encountered by failing to anticipate, assess, and address the concerns and perceptions of the public:

- The Listeria case study points out that the *illness prevention* framing of the risks posed by soft raw milk cheeses set the US on a particular regulatory course to ban the sale of soft raw milk cheeses. Because there are large substantial subgroups in the US for whom preparation of cheeses from raw milk is part of a cultural tradition, and because research indicates that the preparation process, rather than the cheese itself is key to preventing Listeria contamination, the outright ban is not wholly effect at protecting these subgroups. The case study raises the question whether or not a better understanding of the concerns and cultural traditions of particular subpopulations might have allowed for a different framing, educational process, and health outcome.
- The Nagara River Estuary Barrage case offers a compelling example of how incorporating concern assessment into the process both in the beginning and over time, might have allowed a broader set of stakeholders and their concerns to be addressed more effectively. The barrage was designed with control of one set of risks in mind, the provision of an adequate water supply for projected industrial development and protection against flooding of downstream communities. If they were voiced early on, concerns about impacts on the ecology of the estuary and related fisheries appear not to have been heard. Of course, the barrage was first proposed at a time when public officials were more apt to make decisions in what they believed to be the best interest of the public. However, the lack of a process to recognise and address these concerns either in the beginning or as the original need for the barrage came into doubt, contributed to lawsuits and to substantial delays in the project.

Both these examples illustrate the general point that any risk governance strategy may need to revisit the original basis for the decisions that were made as new information arises that raises serious questions about the validity or effectiveness of those decisions.

Characterising and Evaluating Risks: The Need for a Simpler Risk Evaluation

Comments on this stage of the framework targeted both its conceptual underpinnings and its representation in the 'traffic light model'. Some commentators thought that the effort to separate out 'evidence' and 'values' as distinct inputs to this decision was artificial. Several commentators thought it was already covered in the management or appraisal phases. Since there was so much confusion, we think it is necessary to explain this step of the framework more carefully.

The step, which follows the careful assessment of risks and concerns, is undertaken for two main purposes:

- First, to reach a balanced judgement on the tolerability/ acceptability of a given risk.
- Second, to initiate (if deemed necessary) a management process and make preliminary suggestions for the most suitable management approach.

Reaching a balanced judgement on tolerability or acceptability of a given risk requires consideration of several inputs, beginning with the evidence or information provided from the appraisal phase. The framework describes development of what it calls 'a multi-criteria profile' of the risk. This profile is essentially a summary of the essential findings and characteristics of risks from the appraisal processes in the previous step:

- From the risk assessment, the profile may include point estimates or distributions of risks, descriptions of remaining uncertainties, results of any scenario analyses for the health, ecological, technical or other outcomes of interest. It may include the characterisation of the risks of preliminary management alternatives that might have been considered as part of the risk assessment (for example, the risks of substituting one chemical or process for another).
- From the concern assessment, the profile should focus on the critical concerns and perceptions about the risk itself, the evidence regarding inequitable distribution of risks and benefits, or other factors that may need to be taken into account in decisions about management of the risk.

This part of the evaluation is not just a repetition of the basic findings of the appraisal, but should lead to a deeper understanding of the findings and their implications. It touches back to those characteristics (complexity, uncertainty, ambiguity) discussed earlier that influence how the risks might need to be handled. How serious are these risks or concerns? How do they compare to any scientific conventions, legal standards, international guidelines or other 'benchmarks' of risk that may have been identified in the pre-assessment phase?[10] What are the particular complexities involved in understanding how to manage this risk – what factors, technical or human, are most influential on the outcomes people care about? How uncertain are we about any predictions that have been made? Is the degree of uncertainty likely to affect the decision about the tolerability or acceptability of a risk? How feasible is it to generate data that would reduce uncertainty sufficiently to change a decision that might otherwise be made now? Are there likely to be significant ambiguities, fundamental differences in values that affect discussions of what is tolerable or acceptable?

Ultimately, judgements about tolerability and acceptability also require balancing wider social and economic factors. These include, for example, consideration of the benefits or opportunities associated with the risk being evaluated, quality of life factors, and sustainability issues, among others. Trade-offs between all these factors are almost inevitably involved.

Representing this more involved set of deliberations in a two-dimensional figure is challenging and IRGC's simple analogy of a traffic light (HSE 2001; Bandle 2007) is an inevitable oversimplification. In the 'traffic light model' depicted in Figure 2 of the IRGC framework (Chapter 1), red signals intolerable, yellow tolerable, and green acceptable.[11] *Intolerable* signals a situation in which the risks exceed the benefits. The term '*tolerable*' refers to a situation where the benefits may be worth the risk if the risk can be sufficiently reduced or mitigated. The term 'acceptable' refers to the situation where any residual threat is so low that additional measures for reducing or mitigating the threat are not seen as necessary. The relative areas of the intolerable, tolerable, and acceptable regions are not meant to represent the proportion of risks that fall into these categories.

Drawing the line between 'intolerable' and 'tolerable' or between 'tolerable' and 'acceptable' is one of the most difficult tasks in risk governance (Fairman 2007). Nonetheless, it is a decision that ultimately has to be made at some point. Although not explicit in the framework's discussion of the traffic light model, tolerable risks could also include situations in which a decision maker is willing to live with the risk in the short term while further data is collected that he or she hopes will better inform the decision. The IRGC framework emphasises that the tolerability/acceptability judgement should be made as transparent as possible to all interested parties and that the organisations responsible for this judgement should have the skills, assets, knowledge, and sensitivity to arrive at an informed, balanced and fair judgement.

---

[10] IRGC acknowledges that the evaluation of the seriousness of the risks may rely on conventions, for example, safety factors, acceptable cancer risk levels, that themselves can embed particular degrees of risk tolerance. In this case, it is true that the separation of evidence and values is not so clear.

[11] Note that the original figure was presented in colour.

Risk Management

IRGC received varied comments on the risk management phase. Some felt that the management section followed a very traditional process (e.g. identification and generation, assessment, implementation and monitoring of risk management options) that was not well integrated into the concept of risk governance presented by the framework. Others would have liked to have seen more explicit discussion of particular management approaches, for example, mitigation both as it is understood in engineering but also in insurance as providing ways of reducing or compensating for potential risks. The more controversial aspect of IRGC's risk management phase has been how the degree of complexity, uncertainty and ambiguity of the risk is used in defining approaches to risk management.

First, IRGC incorporated a fairly traditional process of selecting and evaluating risk management options into the framework because it is one that is theoretically sound and empirically proven. The basic elements of the process do not necessarily change. What is different is how they are integrated into the decision making process. Historically, and under other risk management frameworks, these steps might be carried out independently of the risk appraisal or evaluation phases for example. In the IRGC framework the other phases of the risk governance cycle help inform the choice and evaluation of risk management options.

In response to those commentators who were concerned that we did not go into sufficient detail on particular risk management strategies, we want to reiterate a point that we made early on. The IRGC framework was never intended to be a comprehensive manual that would provide solutions for all types of problems. Many of our tables did try to provide specific examples of the types of management strategies that might be relevant to risks with particular characteristics (for example, Table 6 in Chapter 1). Though these strategies were intended as illustrative, the comments we have received make us realise that they were interpreted sometimes as comprehensive and exclusive.

Finally, the critiques both from commentators and from case study authors on this subject directed most of their criticism toward the risk management escalator (Figure 4, Chapter 1). The escalator in particular embodied their concerns about the potential rigidity of the risk characteristics (simple, complex, uncertain, and ambiguous) discussed in a previous section of this chapter and their implications for risk management. The Nature Tourism case study offered advice that echoed the comments of others:

> Risk classes may simply describe different aspects of the same risk. Risk management would therefore have to consist of a mix of the offered strategies and instruments as well as possibly others.

At the same time, the case study on nature-based tourism provides an example of the challenges to risk governance posed by a dynamic and fragmented and thus, complex policy making arena. Numerous actors were involved — national and local governments and authorities, local communities, the businesses who constitute the tourism-industry, myriad international organisations including those concerned with environmental protection, the tourists themselves and society as a whole. Each viewed the risks and benefits of nature tourism from very different perspectives and often acted independently from each other. This case study was therefore helpful in showing that finding solutions to some of the problems created by nature tourism will require a multi-faceted approach that recognises, but develops different ways to work with these perspectives.

Commentators also observed that the availability and quality of knowledge often changes over the course of time making initial characterisations obsolete. This observation is certainly consistent with Norio Okada's lesson from the Nagara River Estuary Barrage project. Decision makers may not know enough about a risk early on, leading them to mischaracterise it and thus, potentially to run into some of the same governance problems the framework is trying to help them avoid.

We agree. In our earlier discussions about the concepts of complexity, uncertainty, and ambiguity, we have tried to clarify how these characteristics might be helpful indicators of the dimensions of a problem. They are not a substitute for a detailed understanding of a problem.[12] We certainly do not wish to imply that they obligate the decision maker to a particular course of action.

Risk Communication

All of the case studies demonstrated the crucial function of *risk communication* (Atman et al. 1994). In the nature-based tourism case, the authors argued for a mutual exchange of information among all relevant actors; they posited that if the tourists and the local actors in particular were more aware of the impacts of their own behaviours, a shift towards more sustainable tourism would be possible. Likewise, the study on nanotechnologies emphasised the need for better communication. For most of society, knowledge of and about nanotechnology is very limited. Best practice risk communication should be inclusive of all relevant stakeholders and should involve objectively stating information about social benefits and risks of nanotechnology. The international disclosure of risk information by large multi-national companies and an integrated risk communication programme for scientists, regulators and industrial developers, could both facilitate the development of new products. They could also foster the emergence of open forums to discuss the most suitable regulatory options for enjoying the benefits while reducing the risks of nanotechnologies. The global nature of technology development would require involvement of all nations, encouraging public-private partnership, and sharing of standards and best practices.

[12] As noted earlier, it is possible that decision makers whose resources for undertaking a comprehensive risk governance process are limited may find it useful to rely on these characteristics as indicators of approaches to consider.

Stakeholder Involvement and Public Participation

IRGC's openness to stakeholder input has been one of the most criticised parts of the framework. Löfstedt and van Asselt argued that stakeholder involvement should not be considered a panacea but, in some cases, a hindrance to the governance process. At the same time, they questioned whether input from stakeholders during the appraisal stage was too limited (and therefore would not benefit from contribution to knowledge and uncertainty stakeholders might offer). Other commentators questioned the need for any stakeholder participation in the appraisal phase (should be left to 'sound science'). For the phases of evaluation and management, Löfstedt and van Asselt thought stakeholder engagement was too broad and inclusive.[13]

The lessons from the case studies echo these concerns about the benefits and challenges of involving stakeholders in the risk governance process. Knight et al.'s Listeria case study raises the question of whether earlier stakeholder involvement might have led to a regulatory process and outcome that reflected more of a balance between the consumer safety goals of the government and the food quality and cultural values of the consumers of raw milk cheeses. A more focused solution than an outright ban might have more effectively limited the health risks to the small sub-populations for whom the making of raw milk soft cheeses is a cultural tradition. The ban in these communities has often been ignored with occasionally serious health consequences.

Joyce Tait's case study on GM crops suggests that stakeholder involvement can sometimes contribute more to the problem than the solution. Tait expressed concern about the framing power of NGOs in the GMO debate and their influence on public opinion. Her cautious warning about the evidence and tactics used by some NGOs and their resistance to compromise echoes the recommendations of Löfstedt and van Asselt to be more sensitive to the problems of inclusive governance. Tait argues strongly for the need to establish a set of standards or guidelines for effective engagement of stakeholders that would cover responsible and unbiased use of evidence and the need for compromise.

What are IRGC's reasons for and concepts of stakeholder engagement and public participation for each of the phases of the framework?

As discussed earlier, IRGC's framework is founded on an inclusive governance model – that not just political bodies, but experts, the corporate sector, and civil society have roles to play. This choice reflects IRGC's assessment that many deficits in the governance of risks have had their roots, in one way or another, in how these various actors, stakeholders and the broader public have been dealt with. Overly narrow framing of issues, risk assessments or management plans that neglect potentially important sources of expertise and knowledge, inadequate understanding of and planning for differing stakeholder concerns, perceptions, and values and poor communication can lead to lack of trust in decision makers, lack of confidence in the decisions made, and difficulties in implementation of management measures, not to mention outright opposition and legal interventions.

[13] Compare also Gethmann (2001).

In keeping with this view, we have briefly summarised the purpose of stakeholder and public engagement at various phases of the risk governance process:

- *During pre-assessment*: In keeping with the purpose of the pre-assessment phase of the IRGC risk governance framework, the goal of stakeholder engagement is to assist with the initial framing of the problem – defining boundary conditions, applicable scientific, political, social and other conventions, and making a preliminary assessment of the nature of the complexities, uncertainties, and normative ambiguities that assessors and decision makers are likely to face. To avoid the kinds of problems we have seen from the case studies in this volume, care should be taken to involve the range of expertise and stakeholders necessary to define the important dimensions of the risk problem that will need to be considered.
- *During appraisal*: We agree with Löftsted and van Asselt that not only technical experts (e.g., senior risk assessors, scientists, engineers, economists, other specialists) but also stakeholders (e.g., affected communities, industries, governments) can offer knowledge that is valuable for assessing risks, their related uncertainties, and possible approaches to managing them. The experience of local communities potentially affected by a risk, of engineers involved in the design of manufacturing processes, etc. can provide important realistic input. It is important to note that it is *not* the task of stakeholders at the appraisal stage to deal with normative questions pertaining to the tolerability of the risk or risk management options.
- *During risk characterisation and evaluation*: The purpose of stakeholder engagement here is to assure that, in addition to the factual evidence from the appraisal process, the values and preferences of those who will be affected are made clear to those ultimately responsible for deciding the tolerability or acceptability of a risk and how it might be managed.
- *In risk management*: The main purpose of participation here is again to assure that relevant expertise and knowledge, as well as values and preferences, are considered in the evaluation and selection of management measures.

The degree and design of stakeholder involvement at any particular stage needs to take into account the fundamental issues raised and characteristics of the risk problem that need to be addressed:

- The degree of *complexity* of the problem, for example the range of interacting variables, agents, processes, etc., will guide risk managers to decide what kinds of expertise and knowledge are needed to understand the complex causal connections. Delphi methods, meta-analysis conferences, scientific consensus conferences, expert judgement elicitation, and other knowledge-oriented procedures are examples of methods that may be considered here.
- The nature and magnitude of *uncertainty* anticipated may again dictate the expertise and processes used to characterise uncertainties and to discuss how decisions should be made in the face of irresolvable uncertainties. Some of the

same methods used to characterise the complexities of a problem are also relevant here. However, when uncertainties remain it is also important to discuss the level of protection, precaution or compensation that should be taken into account in a decision. Round tables, open forums, negotiated rule-making exercises, mediation or mixed advisory committees including scientists and stakeholders are examples of some of the methods that can be used (Susskind et al. 1978; Amy 1983; Moore 1996; Owen 2001; Gregory et al. 2001).

- When high *ambiguity* exists, that is, fundamental differences in values or preferences about risks and how they should be managed, entirely different stakeholder engagement processes may be required. Deliberative processes for these kinds of problems include citizen forums, citizen panels, citizen juries, consensus conferences, ombudspersons, citizen advisory commissions, and similar participatory instruments in addition to classic stakeholder engagement processes (Armour 1995; Crosby et al. 1986; Dienel 1989; Joss 1999; see reviews in Hagendijk and Irwin 2006; Lynn 1990; Renn 2004).

The design of participatory procedures at any phase should display these basic features (Laird 1993; Webler 1995, 1999; Adler et al. 2000; US-EPA-SAB 2001: 3; Renn 2004b; Rowe et al. 2004: 93; Blackstock et al. 2007; Goldschmidt and Renn 2006):

- *transparency* from the point of view of third parties, in documenting how stakeholders were selected to participate, and how their views were taken into account. Participants should also have a clear understanding of the process, the communicative procedures, the methods for reaching agreements and the future use of the results;
- *competence* in terms of ensuring that the state of the art in knowledge about the risk issue is considered during the deliberations and that all participants are made literate in the issue itself and in the use of deliberative reasoning;
- *fairness* in terms of both an adequate representation of the key constituents in the process and in equal speaking and deliberating opportunities among the participants;
- *efficiency* in terms of a balance between resources invested into the participatory activities and the envisioned outcome as well as the cost-effective use of deliberative techniques and methods;
- *clear mandate* of what is being expected from the participatory exercise including the time table involved, the scope and range of options being considered, and the nature and the future use of the outcomes of the deliberations;
- *diversity* in terms of multiple perspectives and disciplines bearing on the risk in question; and
- *professionalism* in terms of structuring, moderating and facilitating the process, and summarising and disseminating the results.

Exactly how the roles for various stakeholders or other actors should be defined and incorporated into the governance of a specific problem remains a topic for debate and depends on the particular context. And we acknowledge in the framework (Chapter 1) that meaningful public engagement is not an easy goal to accomplish:

> Reaching consensus and building trust on highly complex and [contentious] subjects such as global climate change is ... much more difficult. Being inclusive and open ... does not guarantee, therefore, constructive cooperation by those who are invited to participate.

At the same time, however, ample evidence exists to show that well-designed stakeholder participation can improve risk governance. See for example, several empirical studies on the effectiveness of pubic participation in the environmental and other fields (US-EPA/SAB 2001; Beierle 2000; Beierle and Cayford 2002; Rowe et al. 2004).

IRGC recognises that the responsibility for the final decision ultimately lies with the agency or organisation responsible for management of the risk. The intent of our inclusive governance model is to help make sure that decision makers have asked all the right questions and thus have the most complete information available with which to make their decisions.

The Importance of Context

Throughout this chapter we have noted the importance of context – the legal, political, scientific, social, and economic milieu – for the risk being evaluated. We have done so in acknowledgement of the many comments we received on this issue. Risk governance is not something that takes place in isolation. Nor is it something that can be applied in a standard way in all locations, political cultures, organisations and for all kinds of risks.

To reflect this point more directly in the framework, we have created Figure 3, which displays the dimensions of context affecting the risk governance process.

Organisational Capacity is specific to the organisation – or group of organisations – responsible for dealing with risks. This may be at the level of the individual, the company, the local or national government or at an international level (e.g. WHO) or a combination of any or all of them. The framework identifies three components of organisational capacity:

- the assets needed – rules, resources, competencies and knowledge, and organisational integration;
- the skills needed to use the assets in all circumstances – flexibility, vision and the ability to influence the external environment; and
- the institutional norms and capabilities for deploying the assets and skills – relationships, networks, and the legal statutes, regulations, and policies (i.e. the rules of the game).

The *Actor Network* will be different for every risk and every situation. It comprises both those involved in dealing with the risk (the organisational capacity) and those who have a potential stake in their management or outcomes. The actor network may include international bodies, national, regional and local governments, different sectors of industry, regulators, NGOs, the media, members of the general public. An understanding of all actors at each governance level is very helpful in

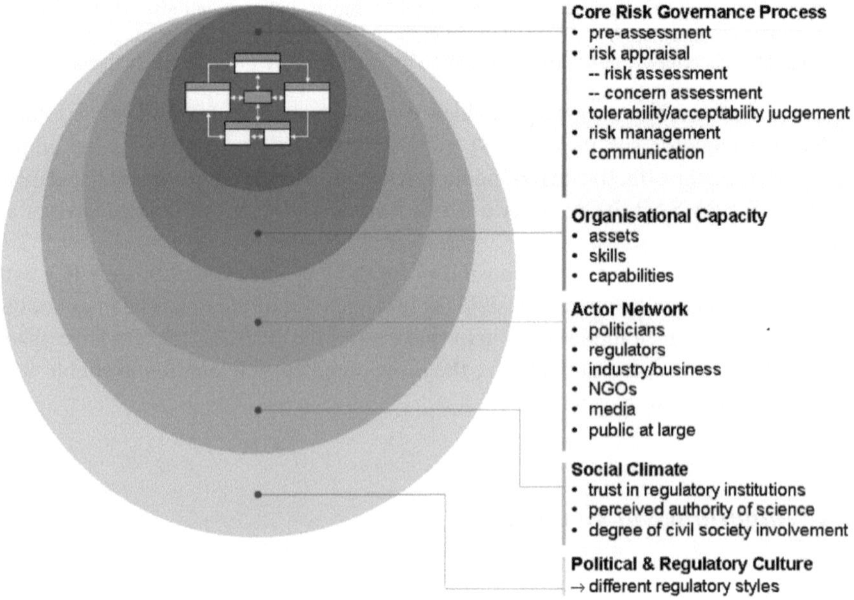

Fig. 3 The IRGC framework and contextual factors.

framing the risk, undertaking concern assessment, judging acceptability and tolerability, communicating throughout the governance cycle and taking appropriate risk management decisions.

Understanding *Social Climate* can help to judge the level of preparedness for the change that a risk might trigger or, alternatively, the potential acceptability of risk management decisions. Often, the factors that comprise a social climate will impact on perceptions of the fairness, appropriateness and feasibility of decisions. In turn, these perceptions can influence compliance with the decisions.

Giving consideration to *Political and Regulatory Culture* allows reference to how different countries or organisations within countries handle and regulate risks. There is no common, global regulatory system or set of methodologies for managing even relatively well-known risks, never mind newly emerging global ones. The same risk may be processed differently, and subject to a different management decision, depending on such factors as national culture, political tradition, regulatory systems, and social norms. In some environments, a top-down 'vertical governance' approach will dominate; in others, an inclusive 'horizontal governance' approach will be the norm.

As a final note, one commentator remarked that the framework misses the point. He argues that the real problem is distribution of power – who gets to decide and who benefits and who loses. This may be true; however, there is not much that professionals interested in improving the governance of risks can do to change the current power systems. Perhaps through a more open and inclusive process, we can help bring the disparities in power and equity to light, while working to improve the system from within.

Conclusions

The objective of this chapter was to review the lessons learned from the critical reviews and applications of the IRGC framework on risk governance. The comments we received ranged from fundamental questions about the concept of risk and the underlying vision of governance, to the distinctions between complexity, uncertainty, and ambiguity in risk knowledge, the process of risk evaluation, the role of stakeholders and other topics.

Many commentators found the features of the risk governance framework to be innovative and even provocative. Most alternative risk governance models still lack a pre-assessment stage and a separate stage for characterisation and evaluation. The idea of a concern assessment running parallel to the risk assessment is, as far as we know, unique to the IRGC model. The distinction of three knowledge-related characteristics of risk has been proposed before but has certainly not entered the policy arena until now (WBGU 2000). The inclusion of risk-based and precautionary risk management strategies in one consistent framework also sets a milestone in the ongoing debate about the roles of sound science and precaution in regulatory decision making.[14]

The case studies demonstrated that the IRGC risk governance framework facilitates a broad understanding of the risk governance issues facing a variety of different problems. It provides a viable and productive tool for identifying problems and deficiencies in the course of risk governance and suggesting alternative actions. The studies also highlight the need for carrying out the early phases of the IRGC framework (pre-assessment, risk appraisal) as comprehensively as possible in order to ensure that the remaining phases (tolerability/acceptability judgement, risk management and communication) are efficient and constructive. Thus, the framework supports policy makers' efforts to identify governance gaps as well as to design and implement structures and processes to overcome them in the future.

In this chapter we have also acknowledged that we have work to do. We have begun this work in this chapter by trying to clarify and respond to some of the major questions raised by our reviewers and case study authors. Some of the more substantive work that needs to be done in the future includes:

- *Development of a simpler, more straightforward version of the framework.* We recognise that the language of the framework is difficult for many and plan a simpler version. Some have also suggested publications targeted for different audiences, specifically risk managers in the private sector, public organisations, and organisations representing civil society (NGOs, opinion leaders, media). IRGC recently published a shorter article on the framework which in addition

[14] See the edited volume by Fisher et al. (2006) and Löfstedt (2004).

to summarising the framework, provides a practical checklist of questions that practitioners and decision makers can use to think through a risk problem (Bunting et al. 2007).

- *More discussion of risk governance from the standpoint of corporate decision makers.* We acknowledge that the IRGC framework, with its inclusive governance model, is oriented more toward large public institutions or international organisations.
- *Further clarification and development of the conceptual basis for the risk management component of the framework.* There are several dimensions to this including more discussion of the implications of complexity, uncertainty, and ambiguity for the practical development of risk management strategies. We also recognise that further elaboration of the function of liability, insurance and financial arrangements in risk management is also very important.
- *Prospective use of the framework in developing risk governance approaches to emerging risks.* We recognise that such retrospective analysis, or testing of the framework for cases that occurred in the past, should not be taken as a proof for its superiority over alternative frameworks. Any type of framework can appear to work well when tested retrospectively because one can always stress the parts of the framework that were either neglected or misjudged during the actual case and then offer a better solution. Ultimately, a true test of a model is how it performs when used prospectively.
- *Ongoing retrospective analysis of the framework.* While prospective use of the framework is critical, we also recognise that the framework can be improved by analysing actual risk governance regimes in areas we have not covered extensively. For example, some commentators have suggested more concrete application of the framework to the areas of vulnerability assessment, the assessment of malicious acts (e.g., terrorism), and multiple hazard analysis.
- *Development of practical advice on stakeholder involvement at various phases of the framework.* The IRGC framework emphasises the need for multidimensional and multi-actor decision making but does not provide yet a clear methodology for doing so. There is a particular need for the development of guidelines to encourage responsible stakeholder involvement and to govern the introduction of evidence to a debate. The insights from the commentaries and case studies support the conclusion that inclusive governance is neither a panacea for better risk governance nor a guarantee for more democratic processes.[15] Yet, given the right conditions and processes in place, inclusive governance has the potential to improve decision making and substantially add to the rationality and democratic quality of risk governance (Welp and Stoll-Kleemann 2006).
- *Development of practical advice on how to conduct effective and efficient risk communication programmes.* Communication is at the centrepiece of our frame-

[15] See the empirical anaylsis of successes and failures of public involvement proceses in Beierle and Cayford (2002).

work, but we need to demonstrate how it can be incorporated into individual phases of the framework using concrete examples.

Advances in technology, in the globalisation of economies, in the growth of our communities, and in our ability to transport ourselves to the far reaches of the planet all bring with them tremendous opportunities. But they can bring with them risks to human health, to the natural and built environment, and to human welfare. We know from our case studies that existing risk governance frameworks have not always been up to the task of assessing and managing even more conventional risks very well. Deficits in governance – in understanding the dimensions of a problem from the outset, the need for communication, the implications of neglecting key stakeholder concerns, or the unintended consequences or risks of making particular management choices – have often increased risks to many aspects of life that humans care about. The increased complexity and interdependence of an increasingly large number of risks promises to make the development and implementation of adequate risk governance strategies ever more difficult.

The IRGC was established as a result of widespread concern within the public sector, the corporate world, academia, the media and society at large that our current approaches to risk governance were not sufficient. Our first goal was to develop a more comprehensive and systematic approach to risk governance.

The framework for risk governance that we have presented in this book is a first step toward meeting that lofty goal we set for ourselves. What we have created is a broad conceptual framework that we believe is an advance relative to frameworks that that have existed for many years. It is not a manual that prescribes concrete steps for all parties involved in risk governance but rather is an overview of a dynamic and interative governance process. The framework cannot replace thinking or, for that matter, creativity. It is however hoped that, by building into conventional 'risk analysis' soft issues such as societal values, concerns as well as perceptions of risk and by looking into the interactions required between the various actors involved in the process, it can contribute to the development of better balanced, more effective and more inclusive strategies in risk governance for the kinds of difficult risks that we will face in the years ahead.

References

Adams, J., 1995, *Risk*, UCL Press, London.

Adler, P.S., Barrett, R.C., Bean, M.C., Birkhoff, J.E., Ozawa, C.P. and Rudin, E.B., 2000, *Managing Scientific and Technical Information in Environmental Cases. Principles and Practices for Mediators and Facilitators*, Contribution to the EPA Science Advisory Board. RESOLVE, US; Institute for Environmental Conflict Resolution and Western Justice Center Foundation, Washington, DC.

Amy, D.J., 1983, Environmental mediation: An alternative approach to policy stalemates, *Policy Sciences* **15**, 345–365;

Armour, A., 1995, The citizen's jury model of public participation, in: O. Renn, Th. Webler and P. Wiedemann (eds.), *Fairness and Competence in Citizen Participation. Evaluating New Models for Environmental Discourse*, Kluwer Academic Publishers, Dordrecht, pp. 175–188.

Atman, C.J., Bostrom, A., Fischhoff, B. and Morgan, M.G., 1994, Designing risk communication: Completing and correcting mental models of hazardous processes. Part 1, *Risk Analysis* **14**(5), 779–788.

Bandle, T., 2007, Tolerability of risk: The regulator's story, in: F. Boulder, D. Slavin and R. Löfstedt (eds.), *The Tolerability of Risk. A New Framework for Risk Management*, Earthscan, London, pp. 93–104.

Beierle, T., 2000, The quality of stakeholder-based decisions: Lessons from case study record, Discussion Paper 00-56, Resources for the Future, Washington, DC (November).

Beierle, T. and Cayford, J.C., 2002, *Democracy in Practice. Public Participation in Environmental Decisions*, Resources for the Future, Washington, DC.

Blackstock, K.L., Kelly, G.J. and Horsey, B.L., 2007, Developing and applying a framework to evaluate participatory research for sustainability, *Ecological Economics* **60**.

Bradbury, J.A., 1989, The policy implications of differing concepts of risk, *Science, Technology, and Human Values* **14**(4), 380–399.

Bunting, C., Renn, O., Florin, M.V. and Cantor, R., The IRGC risk governance framework, *John Liner Review* **21**(2), Summer 2007.

Burningham, K. and Cooper, G., 1999, Being constructive: Social constructivism and the environment, *Sociology* **32**(2), 297–316.

Campbell, S. and Currie, G., 2006, Against Beck: In defence of risk analysis, *Philosophy of the Social Sciences* **36**(2), 149–172.

Catton, W.R. and Dunlap, R.E., 1978, Environmental sociology: A new paradigm, *The American Sociologist* **13**, 41–49.

Clarke, L. and Short, J.F., 1993, Social organization and risk: Some current controversies, *Annual Review of Sociology* **19**, 375–399.

Codex Alimentarius Commission, 2005, *Procedural Manual* (fifteenth edition), Joint FAO/WHO Food Standards Programme, World Health Organization/Food and Agriculture Organization of the United Nations, Rome.

Crosby, N., Kelly, J.M. and Schaefer, P., 1986, Citizen panels: A new approach to citizen participation, *Public Administration Review* **46**, 170–178.

Cullen, A.C. and Frey, H.C., 1999, *Probabilistic Techniques in Exposure Assessment: A Handbook for Dealing with Variability and Uncertainty in Models and Inputs*, Plenum Press, New York.

Dienel, P.C., 1989, Contributing to social decision methodology: Citizen reports on technological projects, in: C. Vlek and G. Cvetkovich (eds.), *Social Decision Methodology for Technological Projects*, Kluwer Academic Publishers, Dordrecht, pp. 133–151.

Douglas, M., 1990, Risk as a forensic resource, *DEADALUS* **119**(4), Fall, 1–16.

Dunlap, R.E., 1980, Paradigmatic change in social science: From human exemptionalism to an ecological paradigm, *American Behavioral Scientist* **24**, 5–14.

Dunlap, R.E., Lutzenhiser, L.A. and Rosa, E.A., 1994, Understanding environmental problems: A sociological perspective, in: B. Bürgenmeier (ed.), *Economy, Environment, and Technology. A Socio-Economic Approach*, Sharpe, New York, pp. 27–49.

Fairman, R., 2007, What makes tolerability of risk work? Exploring the limitations of its applicabilitity to other risk fields, in: F. Boulder, D. Slavin and R. Löfstedt (eds.), *The Tolerability of Risk. A New Framework for Risk Management*, Earthscan, London, pp. 119–136.

Fisher, E., Jones, J. and von Schomberg, R., 2006, *Implementing the Precautionary Principle: Perspectives and Prospects*, Edward Elgar, Cheltenham.

Gethmann, C.F., 2001, Participatory technology assessment. Some critical questions, in: M. Decker (ed.), *Interdisciplinarity in Technology Assessment. Implementation and Its Chances and Limits*, Springer, Heidelberg, pp. 3–14.

Goldschmidt, R. and Renn, O., 2006, *Meeting of Minds – European Citizens' Deliberation on Brain Sciences. Final Report of the External Evaluation*, Stuttgarter Beitrge zur Risiko- und Nachhaltigkeitsforschung, Vol. 5. Social Science Department, University of Stuttgart, Stuttgart.

Graham, J.D. and Rhomberg, L., 1996, How risks are identified and assessed, in: H. Kunreuther and P. Slovic (eds.), *Challenges in Risk Assessment and Risk Management. The Annals of the American Academy of Political and Social Science*, Sage, Thousand Oaks, CA, pp. 15–24.

Gregory, R., McDaniels, T. and Fields, D., 2001, Decision aiding, not dispute resolution: A new perspective for environmental negotiation, *Journal of Policy Analysis and Management* **20**(3), 415–432.

Health and Safety Executive (HSE), 2001, *Reducing Risks, Protecting People: HSE's Decision-Making Process*, HSE, London.

Hacking, I., 1999, *The Social Construction of What?*, Harvard University Press, Cambridge, MA.

Hagendijk, R. and Irwin, A., 2006, Public deliberation and governance: Engaging with science and technology in contemporary Europe, *Minerva* **44**, 167–184.

Hannigan, J.A., 1995, *Environmental Sociology. A Social Constructivist Perspective*, Routledge, Camden.

Hillgartner, S., 1992, The social construction of risk objects: Or, how to pry open networks of risk, in: J.F. Short and L. Clarke (eds.), *Organizations, Uncertainties, and Risk*, Westview, Boulder, CO, pp. 39–53.

Hoffman, F.O. and Hammonds, J.S., 1994, Propagation of uncertainty in risk assessments: The need to distinguish between uncertainty due to lack of knowledge and uncertainty due to variability, *Risk Analysis* **14**(5), 707–712.

Horlick-Jones, T., 2007, On the signature of new technologoes: Materiality, sociality, and practical reasoning, in: R. Flynn and P. Bellaby (eds.), *Risk and the Public Acceptability of New Technologies*, Palgrave, Basingstoke.

Horlick-Jones, T. and Sime, J., 2004, Living on the border: Knowledge, risk, and transdisciplinarity, *Futures* **36**, 441–456.

Hyman, E.L. and Stiftel, B., 1988, *Combining Facts and Values in Environmental Impact Assessment*, Westview Press, Boulder, CO.

IAEA, 1995, *Guidelines for Integrated Risk Assessment and Management in Large Industrial Areas*, Technical Document IAEA-TECDOC PGVI-CIJV, International Atomic Energy Agency, Vienna.

IEC, 1993, *Guidelines for Risk Analysis of Technological Systems*, Report IEC-CD (Sec) 381 issued by the Technical Committee QMS/23, European Community, Brussels.

IRGC, 2005, *White Paper on Risk Governance: Towards an Integrative Approach*, IRGC, Geneva.

Jaeger, C.C., Renn, O., Rosa, E.A. and Webler, Th., 2001, *Risk, Uncertainty and Rational Action*, Earthscan, London, 112 ff.

Jasanoff, S. 2004, Ordering knowledge, ordering society, in: S. Jasanoff (ed.), *States of Knowledge: The Co-Production of Science and Social Order*, Routledge, London, pp. 31–54.

Joss, S., 1999, Public participation in science and technology policy- and decision-making: Ephemeral phenomenon or lasting change?, *Science and Public Policy* **26**, 290–373.

Kolluru, R.V and Brooks, D.G., 1995, Integrated risk assessment and strategic management, in: R. Kolluru, S. Bartell, R. Pitblade and S. Stricoff (eds.), *Risk Assessment and Management Handbook. For Environmental, Health, and Safety Professionals*, McGraw-Hill, New York, pp. 2.1–2.23.

Laird, F., 1993, Participatory analysis: Democracy and technological decision making, *Science, Technology, and Human Values* **18**(3), 341–361.

Laudan, L., 1996, The pseudo-science of science? The demise of the demarcation problem, in L. Laudan (ed.), *Beyond Positivism and Relativism. Theory, Method and Evidence*, Westview Press, Boulder, CO, pp. 166–192.

Löfstedt, R.E., 2004, *The Swing of the Pendulum in Europe: From Precautionary Principle to (Regulatory) Impact Assessment*, AEI-Brookings Joint Center for Regulatory Studies Working Paper 04-07, Kings College, London.

Luhmann, N., 1993, *Risk: A Sociological Theory*, Aldine de Gruyter, New York.

Lynn, F.M., 1990, Public participation in risk management decisions: The right to define, the right to know, and the right to act, *Risk-Issues in Health and Safety* **1**, 95–101.

Mayo, D.G. and Hollander, R.D. (eds.), 1991, *Acceptable Evidence: Science and Values in Risk Management*, Oxford University Press, Oxford.

MacLean, D., 1986, Social values and the distribution of risk, in: D. MacLean (ed.), *Values at Risk*, Rowman and Allanheld, Totowa, pp. 75–93.

Millstone, E., Van Zwanenberg, P., Marris, C., Levidow, L. and Torgersen, H., 2004, *Science in Trade Disputes Related to Potential Risks: Comparative Case Studies*, Institute for Prospective Technological Studies, Seville.

Moore, C., 1996, *The Mediation Process. Practical Strategies for Resolving Conflict*, Jossey-Bass, San Francisco, CA.

Morgan, G. and Henrion, M., 1990, *Uncertainty*, Cambridge University Press, Cambridge.

National Research Council, Committee on the Institutional Means for Assessment of Risks to Public Health, 1983, *Risk Assessment in the Federal Government: Managing the Process*, National Academy of Sciences, National Academy Press, Washington, DC.

National Research Council, 1996, *Understanding Risk: Informing Decisions in a Democratic Society*, National Academy Press, Washington DC.

Omenn, G.S., 2003, On the significance of 'The Red Book' in the evolution of risk assessment and risk management, *Human and Ecological Risk Assessment* **9**, 1155–1167.

Owen, H., 2001, *Open Space Technology*, Klett-Cotta, Stuttgart.

Paté-Cornell, M.E., 1996, Uncertainties in risk analysis: Six levels of treatment, *Reliability Engineering and System Safety*, **54**, 95–111.

Prime Minister's Strategy Unit/UK Cabinet Office, 2002, *Risk: Improving Government's Capability to Handle Risk and Uncertainty*, UK Cabinet Office, London.

Rayner S. and Cantor R., 1987, How fair is safe enough? The cultural approach to societal technology choice, *Risk Analysis* **7**, 3–13.

RCEP, Royal Commission on Environmental Pollution, 1998, *Twenty-First Report: Setting Environmental Standards*, Royal Commission on Environmental Pollution, London.

Renn, O., 1992, Concepts of risk: A classification, in: S. Krimsky and D. Golding (eds.), *Social Theories of Risk*, Praeger, Westport, CT, pp. 53–79.

Renn, O., 2004a, Perception of risks, *The Geneva Papers on Risk and Insurance* **29**(1), 102–114.

Renn, O., 2004b, The challenge of integrating deliberation and expertise: Participation and discourse in risk management, in: T.L. MacDaniels and M.J. Small (eds.), *Risk Analysis and Society: An Interdisciplinary Characterization of the Field*, Cambridge University Press, Cambridge, pp. 289–366.

Renn, O., Dreyer, M., Stirling, A., Ely, A., Vos, E. and Wendler, M., 2006, *A General Framework for the Precautionary and Inclusive Governance of Food Safety: Accounting for Risks, Uncertainties and Ambiguities in the Appraisal and Management of Food Safety Threats*, Interim Report of the Safe Foods Project, University of Stuttgart, Stuttgart.

Rosa, E.A., 1998, Metatheoretical foundations for post-normal risk, *Journal of Risk Research* **1**(1), 15–44.

Rowe, G., Marsh, R. and Frewer, L.J., 2004, Evaluation of a deliberative conference, *Science, Technology, and Human Values* **29**(1), 88–121.

Schrader-Frechette, K., 1984, Risk-cost-benefit methodology and equal protection, in: V.T. Covello, J. Menkes and J. Mumpower (eds.), *Risk Evaluation and Management*, Plenum, New York, pp. 275–296.

Shrader-Frechette, K.S., 1991a, *Risk and Rationality. Philosophical Foundations for Populist Reforms*, University of California Press, Berkeley, CA, 53 ff.

Shrader-Frechette, K.S., 1991b, Reductionist approaches to risk, in: D.G. Mayo and R.D. Hollander (eds.), *Acceptable Evidence: Science and Values in Risk Management*, Oxford University Press, Oxford/New York, pp. 218–248.

Shrader-Frechette, K.S., 1995, Evaluating the expertise of experts, *Risk: Health, Safety & Environment* **6**, 115–126.

Stern, P.C. and Fineberg, H.V., 1996, *Understanding Risk: Informing Decisions in a Democratic Society*, National Academic Press, Washington, DC.

Stimulating SmarterRegulation, 2002, *Report to Congress on the Costs and Benefits of Regulation*, Washington, DC, pp. 21–23.

Susskind, L.E., Richardson, J.R. and Hildebrand, K.J., 1978, *Resolving Environmental Disputes. Approaches to Intervention, Negotiation, and Conflict Resolution*, Environmental Impact Assessment Project, MIT Press, Cambridge, MA.

US-EPA/SAB Environmental Protection Agency, 2001, *Improved Science-Based Environmental Stakeholder Porcesses*, EPA-SAB-EC-COM-01-006, EPA Science Advsiroy Board. Washington, DC.

WBGU, Wissenschaftlicher Beirat der Bundesregierung Globale Umweltveränderungen, 2000, *World in Transition: Strategies for Managing Global Environmental Risks*, Annual Report 1998, Springer, Heidelberg.

Webler, Th., 1995, 'Right' discourse in citizen participation. An evaluative yardstick, in: O. Renn, Th. Webler and P. Wiedemann (eds.), *Fairness and Competence in Citizen Participation. Evaluating New Models for Environmental Discourse*, Kluwer Academic Publishers, Dordrecht, pp. 35–86.

Webler, Th., 1999, The craft and theory of public participation: A dialectical process, *Risk Research* **2**(1), 55–71.

Welp, M. and Stoll-Kleemann, S., 2006, Integrative theory of reflexive dialogues, in: S. Stoll-Kleemann and M. Welp (eds.), *Stakeholder Dialogues in Natural Resources Management. Theory and Practice*, Springer, Heidelberg, pp. 43–78.

Wynne, B., 1992, Risk and social learning: Reification to engagement, in: S. Krimsky and D. Golding (eds.), *Social Theories of Risk*, Praeger, Westport, CT, pp. 275–297.

Zinn, J.O. and Taylor-Gooby, P., 2006, The challenge of (managing) new risks, in: P. Taylor-Gooby and J. Zinn (eds.), *Risk in Social Science*, Oxford University Press, Oxford, pp. 54–75.

MIX
Papier aus verantwortungsvollen Quellen
Paper from responsible sources
FSC® C105338

If you have any concerns about our products,
you can contact us on
ProductSafety@springernature.com

In case Publisher is established outside the EU,
the EU authorized representative is:
Springer Nature Customer Service Center GmbH
Europaplatz 3, 69115 Heidelberg, Germany

Printed by Libri Plureos GmbH
in Hamburg, Germany